Gregory Levitin (Ed.)

Computational Intelligence in Reliability Engineering

Studies in Computational Intelligence, Volume 39

Editor-in-chief
Prof. Janusz Kacprzyk
Systems Research Institute
Polish Academy of Sciences
ul. Newelska 6
01-447 Warsaw
Poland
E-mail: kacprzyk@ibspan.waw.pl

Further volumes of this series can be found on our homepage: springer.com

Vol. 22. N. Nedjah, E. Alba, L. de Macedo Mourelle (Eds.)
Parallel Evolutionary Computations, 2006
ISBN 3-540-32837-8

Vol. 23. M. Last, Z. Volkovich, A. Kandel (Eds.)
Algorithmic Techniques for Data Mining, 2006
ISBN 3-540-33880-2

Vol. 24. Alakananda Bhattacharya, Amit Konar, Ajit K. Mandal
Parallel and Distributed Logic Programming, 2006
ISBN 3-540-33458-0

Vol. 25. Zoltán Ésik, Carlos Martín-Vide, Victor Mitrana (Eds.)
Recent Advances in Formal Languages and Applications, 2006
ISBN 3-540-33460-2

Vol. 26. Nadia Nedjah, Luiza de Macedo Mourelle (Eds.)
Swarm Intelligent Systems, 2006
ISBN 3-540-33868-3

Vol. 27. Vassilis G. Kaburlasos
Towards a Unified Modeling and Knowledge-Representation based on Lattice Theory, 2006
ISBN 3-540-34169-2

Vol. 28. Brahim Chaib-draa, Jörg P. Müller (Eds.)
Multiagent based Supply Chain Management, 2006
ISBN 3-540-33875-6

Vol. 29. Sai Sumathi, S.N. Sivanandam
Introduction to Data Mining and its Applications, 2006
ISBN 3-540-34689-9

Vol. 30. Yukio Ohsawa, Shusaku Tsumoto (Eds.)
Chance Discoveries in Real World Decision Making, 2006
ISBN 3-540-34352-0

Vol. 31. Ajith Abraham, Crina Grosan, Vitorino Ramos (Eds.)
Stigmergic Optimization, 2006
ISBN 3-540-34689-9

Vol. 32. Akira Hirose
Complex-Valued Neural Networks, 2006
ISBN 3-540-33456-4

Vol. 33. Martin Pelikan, Kumara Sastry, Erick Cantú-Paz (Eds.)
Scalable Optimization via Probabilistic Modeling, 2006
ISBN 3-540-34953-7

Vol. 34. Ajith Abraham, Crina Grosan, Vitorino Ramos (Eds.)
Swarm Intelligence in Data Mining, 2006
ISBN 3-540-34955-3

Vol. 35. Ke Chen, Lipo Wang (Eds.)
Trends in Neural Computation, 2007
ISBN 3-540-36121-9

Vol. 36. Ildar Batyrshin, Janusz Kacprzyk, Leonid Sheremetor, Lotfi A. Zadeh (Eds.)
Perception-based Data Mining and Decision Making in Economics and Finance, 2006
ISBN 3-540-36244-4

Vol. 37. Jie Lu, Da Ruan, Guangquan Zhang (Eds.)
E-Service Intelligence, 2007
ISBN 3-540-37015-3

Vol. 38. Art Lew, Holger Mauch
Dynamic Programming, 2007
ISBN 3-540-37013-7

Vol. 39. Gregory Levitin (Ed.)
Computational Intelligence in Reliability Engineering, 2007
ISBN 3-540-37367-5

Gregory Levitin (Ed.)

Computational Intelligence in Reliability Engineering

Evolutionary Techniques in Reliability Analysis and Optimization

With 99 Figures and 64 Tables

Dr. Gregory Levitin
Research & Development Division
The Israel Electric Corporation Ltd.
PO Box 10
31000 Haifa
Israel
E-mail: levitin@iec.co.il

ISSN print edition: 1860-949X
ISSN electronic edition: 1860-9503
ISBN 978-3-642-07218-5 e-ISBN 978-3-540-37368-1

Springer is a part of Springer Science+Business Media
springer.com

Softcover reprint of the hardcover 1st edition 2007

Cover design: deblik, Berlin

Preface

This two-volume book covers the recent applications of computational intelligence techniques in reliability engineering. Research in the area of computational intelligence is growing rapidly due to the many successful applications of these new techniques in very diverse problems. "Computational Intelligence" covers many fields such as neural networks, fuzzy logic, evolutionary computing, and their hybrids and derivatives. Many industries have benefited from adopting this technology. The increased number of patents and diverse range of products developed using computational intelligence methods is evidence of this fact.

These techniques have attracted increasing attention in recent years for solving many complex problems. They are inspired by nature, biology, statistical techniques, physics and neuroscience. They have been successfully applied in solving many complex problems where traditional problem-solving methods have failed.

The book aims to be a repository for the current and cutting-edge applications of computational intelligent techniques in reliability analysis and optimization.

In recent years, many studies on reliability optimization use a universal optimization approach based on metaheuristics. These metaheuristics hardly depend on the specific nature of the problem that is solved and, therefore, can be easily applied to solve a wide range of optimization problems. The metaheuristics are based on artificial reasoning rather than on classical mathematical programming. Their important advantage is that they do not require any information about the objective function besides its values corresponding to the points visited in the solution space. All metaheuristics use the idea of randomness when performing a search, but they also use past knowledge in order to direct the search. Such algorithms are known as randomized search techniques.

Genetic algorithms are one of the most widely used metaheuristics. They were inspired by the optimization procedure that exists in nature, the biological phenomenon of evolution. The first volume of this book starts with a survey of the contributions made to the optimal reliability design literature in the resent years. The next chapter is devoted to using the metaheuristics in multiobjective reliability optimization. The volume also contains chapters devoted to different applications of the genetic algorithms in reliability engineering and to combinations of this algorithm with other computational intelligence techniques.

The second volume contains chapters presenting applications of other metaheuristics such as ant colony optimization, great deluge algorithm, cross-entropy method and particle swarm optimization. It also includes chapters devoted to such novel methods as cellular automata and support vector machines. Several chapters present different applications of artificial neural networks, a powerful adaptive technique that can be used for learning, prediction and optimization. The volume also contains several chapters describing different aspects of imprecise reliability and applications of fuzzy and vague set theory.

All of the chapters are written by leading researchers applying the computational intelligence methods in reliability engineering.

This two-volume book will be useful to postgraduate students, researchers, doctoral students, instructors, reliability practitioners and engineers, computer scientists and mathematicians with interest in reliability.

I would like to express my sincere appreciation to Professor Janusz Kacprzyk from the Systems Research Institute, Polish Academy of Sciences, Editor-in-Chief of the Springer series "Studies in Computational Intelligence", for providing me with the chance to include this book in the series.

I wish to thank all the authors for their insights and excellent contributions to this book. I would like to acknowledge the assistance of all involved in the review process of the book, without whose support this book could not have been successfully completed. I want to thank the authors of the book who participated in the reviewing process and also Prof. F. Belli, University of Paderborn, Germany, Prof. Kai-Yuan Cai, Beijing University of Aeronautics and Astronautics, Dr. M. Cepin, Jozef Stefan Institute, Ljubljana , Slovenia, Prof. M. Finkelstein, University of the Free State, South Africa, Prof. A. M. Leite da Silva, Federal University of Itajuba, Brazil, Prof. Baoding Liu, Tsinghua University, Beijing, China, Dr. M. Muselli, Institute of Electronics, Computer and Telecommunication Engineering, Genoa, Italy, Prof. M. Nourelfath, Université Laval, Quebec, Canada, Prof. W. Pedrycz, University of Alberta, Edmonton, Canada, Dr. S. Porotsky, FavoWeb, Israel, Prof. D. Torres, Universidad Central de Venezuela, Dr. Xuemei Zhang, Lucent Technologies, USA for their insightful comments on the book chapters.

I would like to thank the Springer editor Dr. Thomas Ditzinger for his professional and technical assistance during the preparation of this book.

Haifa, Israel, 2006 Gregory Levitin

Contents

Recent Advances in Optimal Reliability Allocation[1]

Way Kuo and Rui Wan

Department of Industrial and Information Engineering, University of Tennessee, Knoxville, TN 37919, USA

Acronyms

GA	genetic algorithm
HGA	hybrid genetic algorithm
SA	simulated annealing algorithm
ACO	ant colony optimization
TS	tabu search
IA	immune algorithm
GDA	great deluge algorithm
CEA	cellular evolutionary approach
NN	neural network
NFT	near feasible threshold
UGF	universal generating function
MSS	multi-state system
RB/i/j	recovery block architecture that can tolerate *i* hardware and *j* software faults
NVP/i/j	N-version programming architecture that can tolerate *i* hardware and *j* software faults
LMSSWS	linear multi-state sliding-window system
LMCCS	linear multi-state consecutively connected system
ATN	acyclic transmission network
WVS	weighted voting system
ACCN	acyclic consecutively connected network
UMOSA	Ulungu multi-objective simulated annealing
SMOSA	Suppapitnarm multi-objective simulated annealing

W. Kuo and R. Wan: *Recent Advances in Optimal Reliability Allocation*, Computational Intelligence in Reliability Engineering (SCI) **39**, 1–36 (2007)
www.springerlink.com

PSA	Pareto simulated annealing
PDMOSA	Pareto domination based multi-objective simulated annealing
WMOSA	weight based multi-objective simulated annealing

Notations

$\mathbf{x}_j$	number of components at subsystem j
$\mathbf{r}_j$	component reliability at subsystem j
n	number of subsystems in the system
m	number of resources
x	$(\mathbf{x}_1,\ldots,\mathbf{x}_n,\mathbf{r}_1,\ldots\mathbf{r}_n)$
$g_i(\mathbf{x})$	total amount of resource i required for $\mathbf{x}$
R_S	system reliability
C_S	total system cost
R_0	a specified minimum R_S
C_0	a specified minimum C_S
α	system user's risk level, $0<\alpha<1$
$t_{\alpha,\mathbf{x}}$	system percentile life, $t_{\alpha,\mathbf{x}} = \inf\{t \geq 0 : R_S \leq 1-\alpha\}$
E_S	generalized MSS availability index
E_0	a specified minimum E_S
S_1	the set of optimal solutions of P_1
S_2	the set of optimal solutions of P_2
$U_i(z)$	U-function of component i
g_{1i}	output performance level of component 1, $i = 1,\ldots, I$
g_{2j}	output performance level of component 2, $j = 1,\ldots, J$
p_{1i}	$\lim_{t\to\infty}[\Pr\{g_1(t) = g_{1i}\}]$
p_{2j}	$\lim_{t\to\infty}[\Pr\{g_2(t) = g_{2j}\}]$
ω_s	$\omega(\cdot)$ of series components
$\omega_{s\gamma}$	ω_s for Type-γ MSS
ω_s^τ	ω_s in mode τ
ω_p	$\omega(\cdot)$ of parallel components
$\omega_{p\gamma}$	ω_p for Type-γ MSS

ω_p^τ	ω_p in mode τ
G_τ	output performance level of the entire MSS in mode τ
W_τ	system demand in mode τ
$F_\tau(G_\tau, W_\tau)$	function representing the desired relation between MSS performance level and demand in mode τ
$\mu_{\bar{D}}(x)$	membership function of the fuzzy decision
$\mu_{\tilde{f}_i}(x)$	membership function of ith fuzzy goal
r_{ij}	reliability of component *j* in subsystem *i*
c_{ij}	cost of component *j* in subsystem *i*
w_{ij}	weight of component *j* in subsystem *i*
τ_{ij}	pheromone trail intensity of (i, j)
η_{ij}	problem-specific heuristic of (i, j), $\eta_{ij} = r_{ij}/(c_{ij} + w_{ij})$
P_{ij}	transition probability of (i, j), $P_{ij} = \begin{cases} \dfrac{\tau_{ij}(\eta_{ij})^{0.5}}{\sum_{l=1}^{a_i} \tau_{il}(\eta_{il})^{0.5}} & j \in \text{the set of available component choices} \\ 0 & \text{otherwise} \end{cases}$
T	MSS operation period
$\mathbf{W}^*$	required MSS performance level
Pry	probability of failure from related fault between two software versions
Pall	probability of failure from related fault among all software versions due to faults in specification

1.1 Introduction

Reliability has become an even greater concern in recent years because high-tech industrial processes with increasing levels of sophistication comprise most engineering systems today. Based on enhancing component reliability and providing redundancy while considering the trade-off between system performance and resources, optimal reliability design that aims to determine an optimal system-level configuration has long been an important topic in reliability engineering. Since 1960, many publications have addressed this problem using different system structures, performance measures, optimization techniques and options for reliability improvement.

Refs [45], [93] and [123] provide good literature surveys of the early work in system reliability optimization. Tillman, *et al.* [123] were the first to classify papers by system structure, problem type, and solution methods. Also described and analyzed in [123] are the advantages and shortcomings of various optimization techniques. It was during the 1970s that various heuristics were developed to solve complex system reliability problems in cases where the traditional parametric optimization techniques were insufficient. In their 2000 report, Kuo and Prasad [45] summarize the developments in optimization techniques, along with recent optimization methods such as meta-heuristics, up until that time. This chapter discusses the contributions made to the literature since the publication of [45]. The majority of recent work in this area is devoted to

- multi-state system optimization;
- percentile life employed as a system performance measure;
- multi-objective optimization;
- active and cold-standby redundancy;
- fault-tolerance mechanism;
- optimization techniques, especially ant colony algorithms and hybrid optimization methods.

Based on their system performance, reliability systems can be classified as binary-state systems or multi-state systems. A binary-state system and its components may exist in only two possible states–either working or failed. Binary system reliability models have played very important roles in practical reliability engineering. To satisfactorily describe the performance of a complex system, however, we may need more than two levels of satisfaction–for example, excellent, average, and poor [46]. For this reason, multi-state system reliability models were proposed in the 1970s, although a large portion of the work devoted to MSS optimal design has emerged since 1998. The primary task of multi-state system optimization is to define the relationship between component states and system states.

Measures of system performance are basically of four kinds: reliability, availability, mean time-to-failure and percentile life. Reliability has been widely used and thoroughly studied as the primary performance measure for non-maintained systems. For a maintained system, however, availability, which describes the percentage of time the system really functions, should be considered instead of reliability. Availability is most commonly employed as the performance measure for renewable MSS. Meanwhile, percentile life is preferred to reliability and mean time-to-failure when the system mission time is indeterminate, as in most practical cases.

Some important design principles for improving system performance are summarized in [46]. This chapter primarily reviews articles that address either the provision of redundant components in parallel or the combination of structural redundancy with the enhancement of component reliability. These are called redundancy allocation problems and reliability-redundancy allocation problems, respectively. Redundancy allocation problems are well documented in [45] and [46], which employ a special case of reliability-redundancy allocation problems without exploring the alternatives of component combined improvement. Recently, much of the effort in optimal reliability design has been placed on general reliability-redundancy allocation problems, rather than redundancy allocation problems.

In practice, two redundancy schemes are available: active and cold-standby. Cold-standby redundancy provides higher reliability, but it is hard to implement because of the difficulty of failure detection. Reliability design problems have generally been formulated considering active redundancy; however, an actual optimal design may include active redundancy or cold-standby redundancy or both.

However, any effort for improvement usually requires resources. Quite often it is hard for a single objective to adequately describe a real problem for which an optimal design is required. For this reason, multi-objective system design problem always deserves a lot of attention.

Optimal reliability design problems are known to be NP-hard [8]. Finding efficient optimization algorithms is always a hot spot in this field. Classification of the literature by reliability optimization techniques is summarized in Table 1. Meta-heuristic methods, especially GAs, have been widely and successfully applied in optimal system design because of their robustness and efficiency, even though they are time consuming, especially for large problems. To improve computation efficiency, hybrid optimization algorithms have been increasingly used to achieve an effective combination of GAs with heuristic algorithms, simulation annealing methods, neural network techniques and other local search methods.

This chapter describes the state-of-art of optimal reliability design. Emphasizing the foci mentioned above, we classify the existing literature based on problem formulations and optimization techniques. The remainder of the chapter is organized as follows: Section 2 includes four main problem formulations in optimal reliability allocation; Section 3 describes advances related to those four types of optimization problems; Section 4 summarizes developments in optimization techniques; and Section 5 provides conclusions and a discussion of future challenges related to reliability optimization problems.

Table 1. Reference classification by reliability optimization methods

Meta-heuristic Algorithm	
ACO	[82], [92], [95], [116]
GA	[1], [7], [11], [14], [28], [35], [52], [53], [54], [56], [57], [58], [59], [60], [62], [63], [64], [65], [67], [68], [72], [73], [74], [78], [79], [84], [91], [128]
HGA	[33], [34], [48], [49], [50], [114], [115], [121], [122], [132], [133]
TS	[41]
SA	[1], [118], [124], [131]
IA	[9]
GDA	[107]
CEA	[109]
Exact Method	
[20], [27], [30], [83], [102], [103], [119], [120]	
Max-Min Approach	
[51], [104]	
Heuristc Method	
[105], [129]	
Dynamic Programming	
[97], [127]	

1.2 Problem Formulations

Among the diversified problems in optimal reliability design, the following four basic formulations are widely covered.

Problem 1 (P_1):

$$\begin{aligned}&\max R_S = f(\mathbf{x})\\ &\text{s.t.}\\ &g_i(\mathbf{x}) \le b_i, \text{ for } i = 1,\dots,m\\ &\mathbf{x} \in \mathbf{X}\end{aligned}$$

or

$$
\begin{aligned}
&\min C_S = f(\mathbf{x}) \\
&\text{s.t.} \\
&Rs \geq R_0 \\
&g_i(\mathbf{x}) \leq b_i, \qquad \text{for } i = 1, \ldots, m \\
&\mathbf{x} \in \mathbf{X}
\end{aligned}
$$

Problem 1 formulates the traditional reliability- redundancy allocation problem with either reliability or cost as the objective function. Its solution includes two parts: the component choices and their corresponding optimal redundancy levels.

Problem 2 (P_2):

$$
\begin{aligned}
&\max t_{\alpha,\mathbf{x}} \\
&\text{s.t.} \\
&g_i(t_{\alpha,\mathbf{x}}; \mathbf{x}) \leq b_i, \qquad \text{for } i = 1, \ldots, m \\
&\mathbf{x} \in \mathbf{X}
\end{aligned}
$$

Problem 2 uses percentile life as the system performance measure instead of reliability. Percentile life is preferred especially when the system mission time is indeterminate. However, it is hard to find a closed analytical form of percentile life in decision variables.

Problem 3 (P_3):

$$
\begin{aligned}
&\max E(\mathbf{x}, \mathbf{T}, \mathbf{W}^*) \\
&\text{s.t.} \\
&g_i(\mathbf{x}) \leq b_i, \qquad \text{for } i = 1, \ldots, m \\
&\mathbf{x} \in \mathbf{X}
\end{aligned}
$$

or

$$
\begin{aligned}
&\min C_s(\mathbf{x}) \\
&\text{s.t.} \\
&E(\mathbf{x}, \mathbf{T}, \mathbf{W}^*) \geq E_0 \\
&g_i(\mathbf{x}) \leq b_i, \qquad \text{for } i = 1, \ldots, m \\
&\mathbf{x} \in \mathbf{X}
\end{aligned}
$$

Problem 3 represents MSS optimization problems. Here, E is used as a measure of the entire system availability to satisfy the custom demand represented by a cumulative demand curve with a known **T** and **W***.

Problem 4 (P_4):

$$\begin{aligned} &\max\ z = [f_1(\mathrm{x}), f_2(\mathrm{x}), \ldots, f_S(\mathrm{x})] \\ &\text{s.t.} \\ &g_i(\mathrm{x}) \le b_i, \qquad \text{for } i = 1, \ldots, m \\ &\mathrm{x} \in \mathrm{X} \end{aligned}$$

For multi-objective optimization, as formulated by *Problem 4*, a Pareto optimal set, which includes all of the best possible trade-offs between given objectives, rather than a single optimal solution, is usually identified,.

In all of the above formulations, the resource constraints may be linear or nonlinear or both.

The literature, classified by problem formulations, is summarized in Table 2.

Table 2. Reference classification by problem formulation

P_1	[1], [2], [3], [9], [12], [13], [15], [20], [27], [30], [35], [36], [41], [48], [49], [50], [79], [81], [83], [95], [97], [100], [101], [102], [109], [119], [120], [121], [122], [124], [127], [129], [132], [133]
P_2	[11], [14], [39], [103], [132], [133]
P_3	[52], [53], [54], [56], [57], [58], [59], [60], [62], [63], [64], [65], [67], [68], [72], [73], [74], [78], [79], [84], [92], [99], [105]
P_4	[7], [16], [28], [91], [106], [114], [115], [116], [118], [131], [132]

1.3 Brief Reviews of Advances in P_1-P_4

1.3.1 Traditional Reliability-Redundancy Allocation Problem (P_1)

System reliability can be improved either by incremental improvements in component reliability or by provision of redundancy components in parallel; both methods result in an increase in system cost. It may be advantageous to increase the component reliability to some level and provide redundancy at that level [46], i.e. the tradeoff between these two options must be considered. According to the requirements of the designers, traditional reliability-redundancy allocation problems can be formulated either to maximize system reliability under resource constraints or to minimize the total cost that satisfies the demand on system reliability. These kinds of problems have been well-developed for many different system structures, objective functions, redundancy strategies and time-to-failure distributions. Two important recent developments related to this problem are addressed below.

1.3.1.1 Active and Cold-Standby Redundancy

P_1 is generally limited to active redundancy. A new optimal system configuration is obtained when active and cold-standby redundancies are both involved in the design. A cold-standby redundant component does not fail before it is put into operation by the action of switching, whereas the failure pattern of an active redundant component does not depend on whether the component is idle or in operation. Cold-standby redundancy can provide higher reliability, but it is hard to implement due to the difficulties involved in failure detection and switching.

In Ref [12], optimal solutions to reliability-redundancy allocation problems are determined for non-repairable systems designed with multiple *k*-out-of-*n* subsystems in series. The individual subsystems may use either active or cold-standby redundancy, or they may require no redundancy. Assuming an exponentially distributed component time-to-failure with rate λ_{ij}, the failure process of subsystem *i* with cold-standby redundancy can be described by a Poisson process with rate $\lambda_{ij}k_i$, while the subsystem reliability with active redundancy is computed by standard binominal techniques.

For series-parallel systems with only cold-standby redundancy, Ref [13] employs the more flexible and realistic Erlang distributed component time-to-failure. Subsystem reliability can still be evaluated through a Poisson process, though $\rho_i(t)$ must be introduced to describe the reliability of the imperfect detection/switching mechanism for each subsystem.

Ref [15] directly extends this earlier work by introducing the choice of redundancy strategies as an additional decision variable. With imperfect switching, it illustrates that there is a maximum redundancy level where cold-standby reliability is greater than, or equal to, active reliability, i.e. cold-standby redundancy is preferable before this maximum level while active redundancy is preferable after that.

All three problems formulated above can be transformed by logarithm transformation and by defining new 0-1 decision variables. This transformation linearizes the problems and allows for the use of integer programming algorithms. For each of these methods, however, no mixture of component types or redundancy strategies is allowed within any of the subsystems.

In addition, Ref [6] investigates the problem of where to allocate a spare in a k-out-of-n: F system of dependent components through minimal standby redundancy; and Ref [110] studies the allocation of one active redundancy when it differs based on the component with which it is to be allocated. Ref [101] considers the problem of optimally allocating a fixed number of s-identical multi-functional spares for a deterministic or stochastic mission time. In spite of some sufficiency conditions for optimality, the proposed algorithm can be easily implemented even for large systems.

1.3.1.2 Fault-Tolerance Mechanism

Fault tolerance is the ability of a system to continue performing its intended function in spite of faults. System designs with fault-tolerance mechanisms are particularly important for some computer-based systems with life-critical applications, since they must behave like a non-repairable system within each mission, and maintenance activities are performed only when the system is idle [3].

Ref [2] maximizes the reliability of systems subjected to imperfect fault-coverage. It generalizes that the reliability of such a system decreases with an increase in redundancy after a particular limit. The results include the effect of common-cause failures and the maximum allowable spare limit. The models considered include parallel, parallel-series, series-parallel, k-out-of-n and k-out-of-($2k$-1) systems.

Similarly to imperfect fault-coverage, Ref [3] later assumes the redundancy configurations of all subsystems in a non-series-parallel system are fixed except the k-out-of-n: G subsystem being analyzed. The analysis leads to n^*, the optimal number of components maximizing the reliability of this subsystem, which is shown to be necessarily greater than, or equal to, the optimal number required to maximize the reliability of the entire system. It also proves that n^* offers exactly the maximal system reliability if the subsystem being analyzed is in series with the rest of the system. These results can even be extended to cost minimization problems.

Ref [87] considers software component testing resource allocation for a system with single or multiple applications, each with a pre-specified reliability requirement. Given the coverage factors, it can also include fault-tolerance mechanisms in the problem formulation. The relationship between the component failure rates of and the cost of decreasing this rate is modeled by various types of reliability-growth curves.

For software systems, Ref [79] presents a UGF & GA based algorithm that selects the set of versions and determines the sequence of their execution, such that the system reliability (defined as the probability of obtaining the correct output within a specified time) is maximized subject to cost constraints. The software system is built from fault-tolerant *NVP* and *RB* components.

All of these optimization models mentioned above have been developed for hardware-only or software-only systems. Ref [124] first considers several simple configurations of fault-tolerant embedded systems (hardware and software) including *NVP/0/1, NVP/1/1,* and *RB/1/1*, where failures of software units are not necessarily statistically independent. A real-time embedded system is used to demonstrate and validate the models solved by a simulated annealing optimization algorithm. Moreover, Ref [80] generally takes into account fault-tolerant systems with series architecture and arbitrary number of hardware and software versions without common cause failures. An important advantage of the presented algorithm lies in its ability to evaluate both system reliability and performance indices.

1.3.2 Percentile Life Optimization Problem (P_2)

Many diversified models and solution methods, where reliability is used as the system performance measure, have been proposed and developed since the 1960s. However, this is not an appropriate choice when mission time cannot be clearly specified or a system is intended for use as long as it functions. Average life is also not reliable, especially when the implications of failure are critical or the variance in the system life is high. Percentile life is considered to be a more appropriate measure, since it incorporates system designer and user risk. When using percentile life as the objective function, the main difficulty is its mathematical inconvenience, because it is hard to find a closed analytical form of percentile life in the decision variables.

Ref [11] solves redundancy allocation problems for series-parallel systems where the objective is to maximize a lower percentile of the system time-to-failure (*TTF*) distribution. Component *TTF* has a Weibull distribution with known deterministic parameters. The proposed algorithm uses a genetic algorithm to search the prospective solution-space and a bisection search to evaluate t' in $R(t', \mathbf{x} : \mathbf{k}) = 1 - \alpha$. It is demonstrated that the solution that maximizes the reliability is not particularly effective at maximizing system percentile life at any level, and the recommended design configurations are very different depending on the level. Later in the literature, Ref [14] addresses similar problems where Weibull shape parameters are accurately estimated but scale parameters are random variables following a uniform distribution.

Ref [103] develops a lexicographic search methodology that is the first to provide exact optimal redundancy allocations for percentile life optimization problems. The continuous relaxed problem, solved by Kuhn-Tucker conditions and a two-stage hierarchical search, is considered for obtaining an upper bound, which is used iteratively to effectively reduce search space. This algorithm is general for any continuous increasing lifetime distribution.

Three important results are presented in [39] which describe the general relationships between reliability and percentile life maximizing problems.

- S_2 equals S_1 given $\alpha_t = 1 - R_s(t, \mathbf{x}_t^*)$, where $\mathbf{x}_t^* \in S_1$;
- S_1 equals S_2 given $t_{\alpha,\mathbf{x}_\alpha^*}$, where $\mathbf{x}_\alpha^* \in S_2$;
- Let $\psi(t)$ be the optimal objective value of P_1. For a fixed α, $t_\alpha = \inf\{t \geq 0 : \Psi(t) \leq 1 - \alpha\}$ is the optimal objective value of P_2.

Based on these results, a methodology for P_2 is proposed to repeat solving P_1 under different mission times satisfying $\psi(t_0) \geq 1 - \alpha \geq \psi(t_2)$ until $\mathrm{x}_{t=t_0}^* = \mathrm{x}_{t=t_2}^*$ and $t_0 - t_2$ is within a specified tolerance. It is reported to be capable of settling many unsolved P_2s using existing reliability-optimization algorithms. Without the necessity of an initial guess, this method is much better than [103] at reaching the exact optimal solution in terms of execution time.

1.3.3 MSS Optimization (P_3)

MSS is defined as a system that can unambiguously perform its task at different performance levels, depending on the state of its components which can be characterized by nominal performance rate, availability and cost. Based on their physical nature, multi-state systems can be classified into two important types: Type I MSS (e.g. power systems) and Type II MSS (e.g. computer systems), which use capacity and operation time as their performance measures, respectively.

In the literature, an important optimization strategy, combining UGF and GA, has been well developed and widely applied to reliability optimization problems of renewable MSS. In this strategy, there are two main tasks:

According to the system structure and the system physical nature, obtain the system UGF from the component UGFs;

Find an effective decoding and encoding technique to improve the efficiency of the GA.

Ref [52] first uses an UGF approach to evaluate the availability of a series-parallel multi-state system with relatively small computational resources. The essential property of the U-transform enables the total U-function for a MSS with components connected in parallel, or in series, to be obtained by simple algebraic operations involving individual component U-functions. The operator Ω_ω is defined by (1) - (3).

$$\begin{aligned}\Omega_\omega(U_1(z),U_2(z)) &= \Omega_\omega\left[\sum_{i=1}^{I} p_{1i} z^{g_{1i}}, \sum_{j=1}^{J} p_{2j} z^{g_{2j}}\right] \\ &= \sum_{i=1}^{I}\sum_{j=1}^{J} p_{1i} p_{2j} z^{\omega(g_{1i},g_{2j})}\end{aligned} \tag{1}$$

$$\begin{aligned}&\Omega_\omega(U_1(z),\cdots,U_k(z),U_{k+1}(z),\cdots U_n(z)) \\ &= \Omega_\omega(U_1(z),\cdots,U_{k+1}(z),U_k(z),\cdots U_n(z))\end{aligned} \tag{2}$$

$$\begin{aligned}&\Omega_\omega(U_1(z),\cdots,U_k(z),U_{k+1}(z),\cdots U_n(z)) \\ &= \Omega_\omega(\Omega_\omega(U_1(z),\cdots,U_k(z)),\Omega_\omega(U_{k+1}(z),\cdots U_n(z)))\end{aligned} \tag{3}$$

The function $\omega(\cdot)$ takes the form from (4) - (7).

For Type I MSS,

$$\omega_{s1}(g_1,g_2) = \min(g_1,g_2) \tag{4}$$

$$\omega_{p1}(g_1,g_2) = g_1 + g_2 \tag{5}$$

For Type II MSS,

$$\omega_{s2}(g_1,g_2) = \frac{g_1 g_2}{g_1 + g_2} \tag{6}$$

$$\omega_{p2}(g_1,g_2) = g_1 + g_2 \tag{7}$$

Later, Ref [55] combines importance and sensitivity analysis and Ref [75] extends this UGF approach to MSS with dependent elements. Table 3 summarizes the application of UGF to some typical MSS structures in optimal reliability design.

With this UGF & GA strategy, Ref [84] solves the structure optimization of a multi-state system with time redundancy. TRS can be treated as a Type II MSS, where the system and its component performance are measured by the processing speed. Two kinds of systems are considered: systems with hot reserves and systems with work sharing between components connected in parallel.

Table 3. Application of UGF approach

Series-Parallel System	[52], [54], [55], [59], [62], [65], [72], [73], [74], [92]
Bridge System	[53], [56], [71], [84]
LMSSWS	[64], [70], [78]
WVS	58], [66]
ATN	[63], [69]
LMCCS	[67], [68]
ACCN	[61]

Ref [57] applies the UGF & GA strategy to a multi-state system consisting of two parts:

- RGS including a number of resource generating subsystems;
- MPS including elements that consume a fixed amount of resources to perform their tasks.

Total system performance depends on the state of each subsystem in the RGS and the maximum possible productivity of the MPS. The maximum possible productivity of the MPS is determined by an integer linear programming problem related to the states of the RGS.

Ref [59] develops an UGF & GA strategy for multi-state series-parallel systems with two failure modes: open mode and closed mode. Two optimal designs are found to maximize either the system availability or the proximity of expected system performance to the desired levels for both modes. The function $\omega(\cdot)$ and the conditions of system success for both two modes are shown as follows.

For Type I MSS,

$$\omega_s^O(g_1, g_2) = \omega_s^C(g_1, g_2) = \min(g_1, g_2) \tag{8}$$

$$\omega_p^O(g_1, g_2) = \omega_p^C(g_1, g_2) = g_1 + g_2 \tag{9}$$

$$F_C(G_C, W_C) = G_C - W_C \geq 0 \tag{10}$$

$$F_O(G_O, W_O) = W_O - G_O \geq 0 \tag{11}$$

For Type II MSS,

$$\omega_s^O(g_1, g_2) = \min(g_1, g_2) \tag{12}$$

$$\omega_p^O(g_1,g_2)=\max(g_1,g_2) \tag{13}$$

$$\omega_s^C(g_1,g_2)=\max(g_1,g_2) \tag{14}$$

$$\omega_p^C(g_1,g_2)=\min(g_1,g_2) \tag{15}$$

$$F_C(G_C,W_C)=W_C-G_C\geq 0 \tag{16}$$

$$F_O(G_O,W_O)=W_O-G_O\geq 0 \tag{17}$$

Thus, the system availability can be denoted by

$$A_s(t)=1-\Pr\{F_C(G_C(t),W_C)<0\}-\Pr\{F_O(G_O(t),W_O)<0\} \tag{18}$$

Later [65, 71] introduces a probability parameter of 0.5 for both modes and [71] even extends this technique to evaluate the availability of systems with bridge structures.

To describe the ability of a multi-state system to tolerate both internal failures and external attacks, survivability, instead of reliability, is proposed in Refs [56, 60, 67, 72-74]. Ref [56] considers the problem of how to separate the elements of the same functionality between two parallel bridge components in order to achieve the maximal level of system survivability, while an UGF &GA strategy in [60] is used to solve the more general survivability optimization problem of how to separate the elements of a series-parallel system under the constraint of a separation cost. Ref [72] considers the problem of finding structure of series-parallel multi-state system (including choice of system elements, their separation and protection) in order to achieve a desired level of system survivability by minimal cost. To improve system's survivability, Ref [73, 74] further applies a multi level protection to its subsystems and the choice of structure of multi-level protection and choice of protection methods are also included. Other than series-parallel system, Ref [67] provides the optimal allocation of multi-state LCCS with vulnerable nodes. It should be noted that the solution that provides the maximal system survivability for a given demand does not necessarily provide the greatest system expected performance rate and that the optimal solutions may be different when the system operates under different vulnerabilities.

In addition to a GA, Ref [92] presents an ant colony method that combines with a UGF technique to find an optimal series-parallel power structure configuration.

Besides this primary UGF approach, a few other methods have been proposed for MSS reliability optimization problems. Ref [105] develops a heuristic algorithm RAMC for a Type I multi-state series-parallel system. The availability of each subsystem is determined by a binomial technique, and, thus, the system availability can be obtained in a straightforward manner from the product of all subsystem availabilities without using UGF. Nevertheless, this algorithm can only adapt to relatively simple formulations, including those with only two-state component behavior and no mixing of functionally equivalent components within a particular subsystem.

A novel continuous-state system model, which may represent reality more accurately than a discrete-state system model, is presented in Ref [86]. Given the system utility function and the component state probability density functions, a neural network approach is developed to approximate the objective reliability function of this continuous MSS optimal design problem.

1.3.4 Multi-Objective Optimization (P_4)

In the previous discussion, all problems were single-objective. Rarely does a single objective with several hard constraints adequately represent a real problem for which an optimal design is required. When designing a reliable system, as formulated by P_4, it is always desirable to simultaneously optimize several opposing design objectives such as reliability, cost, even volume and weight. For this reason, a recently proposed multi-objective system design problem deserves a lot of attention. The objectives of this problem are to maximize the system reliability estimates and minimize their associated variance while considering the uncertainty of the component reliability estimations. A Pareto optimal set, which includes all of the best possible trade-offs between the given objectives, rather than a single optimal solution, is usually identified for multi-objective optimization problems.

When considering complex systems, the reliability optimization problem has been modeled as a fuzzy multi-objective optimization problem in Ref [106], where linear membership functions are used for all of the fuzzy goals. With the Bellman & Zadeh model, the decision is defined as the intersection of all of the fuzzy sets represented by the objectives.

$$\mu_{\tilde{D}}(x) = (\mu_{\tilde{f}_1}(x) * \cdots * \mu_{\tilde{f}_i}(x) * \cdots * \mu_{\tilde{f}_m}(x)) \tag{19}$$

The influence of various kinds of aggregators, such as the *product* operator, *min* operator, *arithmetic mean* operator, fuzzy, and a convex combination of the *min* and the *max* operator and γ-operator on the solution is also studied primarily to learn each advantage over the non-compensatory *min* operator. It was found that in some problems illustrated in this paper, the fuzzy *and* the convex combination of the *min* and the *max* operator yield efficient solutions.

Refs [114, 115] solve multi-objective reliability- redundancy allocation problems using similar linear membership functions for both objectives and constraints. By introducing 0-1 variables and by using an *add* operator to obtain the weighted sum of all the membership functions, the problem is transformed into a bi-criteria single-objective linear programming problem with Generalized Upper Bounding (GUB) constraints. The proposed hybrid GA makes use of the GUB structure and combines it with a heuristic approach to improve the quality of the solutions at each generation.

With a weighting technique, Ref [28] also transfers P_4 into a single-objective optimization problem and proposes a GA-based approach whose parameters can be adjusted with the experimental plan technique. Ref [7] develops a multi-objective GA to obtain an optimal system configuration and inspection policy by considering every target as a separate objective. Both problems have two objectives: maximization of the system reliability and minimization of the total cost, subject to resource constraints.

P_4 is considered for series-parallel systems, *RB*/1/1, and bridge systems in [16] with multiple objectives to maximize the system reliability while minimizing its associated variance when the component reliability estimates are treated as random variables. For series-parallel systems, component reliabilities of the same type are considered to be dependent since they usually share the same reliability estimate from a pooled data set. The system variance is straightforwardly expressed as a function in the higher moments of the component unreliability estimates [38]. For *RB*/1/1, the hardware components are considered identical and statistically independent, while even independently developed software versions are found to have related faults as presented by the parameters *Prv* and *Pall*. Pareto optimal solutions are found by solving a series of weighted objective problems with incrementally varied weights. It is worth noting that significantly different designs are obtained when the formulation incorporates estimation uncertainty or when the component reliability estimates are treated as statistically dependent. Similarly, [91] utilizes a multi-objective GA to select an optimal network design that balances the dual objectives of high reliability and low uncertainty in its estimation. But the latter exploits Monte Carlo simulation as the objective function evaluation engine.

Ref [131] presents an efficient computational methodology to obtain the optimal system structure of a static transfer switch, a typical power electronic device. This device can be decomposed into several components and its equivalent reliability block diagram is obtained by the minimal cut set method. Because of the existence of unit-to-unit variability, each component chosen from several off-the-shelf types is considered with failure rate uncertainty, which is modeled by a normal, or triangular, distribution. The simulation of the component failure rate distributions is performed using the Latin Hypercube Sampling method, and a simulated annealing algorithm is finally applied to generate the Pareto optimal solutions.

Ref [116] illustrates the application of the ant colony optimization algorithm to solve both continuous function and combinatorial optimization problems in reliability engineering. The single or multi-objective reliability optimization problem is analogized to Dorigo's TSP problem, and a combinatorial algorithm, which includes a global search inspired by a GA coupled with a pheromone-mediated local search, is proposed. After the global search, the pheromone values for the newly created solution are calculated by a weighted average of the pheromone values of the corresponding parent solutions. A trial solution for conducting the local search is selected with a probability proportional to its current pheromone trial value. A two-step strength Pareto fitness assignment procedure is combined to handle multi-objective problems. The advantage of employing the ant colony heuristic for multi-objective problems is that it can produce the entire set of optimal solutions in a single run.

Ref [118] tests five simulated annealing-based multi-objective algorithms – SMOSA, UMOSA, PSA, PDMOSA and WMOSA. Evaluated by 10 comparisons, Measure *C* is introduced to gauge the coverage of two approximations for the real non-dominated set. From the analysis, the computational cost of the WMOSA is the lowest, and it works well even when a large number of constraints are involved, while the PDMOSA consumes more computational time and may not perform very well for problems with too many variables.

1.4 Developments in Optimization Techniques

This section reviews recent developments of heuristic algorithms, metaheuristic algorithms, exact methods and other optimization techniques in optimal reliability design. Due to their robustness and feasibility, metaheuristic algorithms, especially GAs, have been widely and successfully applied. To improve computation efficiency or to avoid premature convergence, an important part of this work has been devoted in recent years to developing hybrid genetic algorithms, which usually combine a GA with heuristic algorithms, simulation annealing methods, neural network techniques or other local search methods. Though more computation effort is involved, exact methods are particularly advantageous for small problems, and their solutions can be used to measure the performance of the heuristic or meta-heuristic methods [45]. No obviously superior heuristic method has been proposed, but several of them have been well combined with exact or meta-heuristic methods to improve their computation efficiency.

1.4.1 Meta-Heuristic Methods

Meta-heuristic methods inspired by natural phenomena usually include the genetic algorithm, tabu search, simulated annealing algorithm and ant colony optimization method. ACO has been recently introduced into optimal reliability design, and it is proving to be a very promising general method in this field. Ref [1] provides a comparison of meta-heuristics for the optimal design of computer networks.

1.4.1.1 Ant Colony Optimization Method

ACO is one of the adaptive meta-heuristic optimization methods developed by M. Dorigo for traveling salesman problems in [21] and further improved by him in [22-26]. It is inspired by the behavior of real life ants that consistently establish the shortest path from their nest to food. The essential trait of the ACO algorithm is the combination of *a priori* information about the structure of a promising solution with *posteriori* information about the structure of previously obtained good solutions [88].

Ref [81] first develops an ant colony meta-heuristic optimization method to solve the reliability-redundancy allocation problem for a k-out-of-n: G series system. The proposed ACO approach includes four stages:

1. Construction stage: construct an initial solution by selecting component j for subsystem i according to its specific heuristic η_{ij} and pheromone trail intensity τ_{ij}, which also sets up the transition probability mass function P_{ij}.
2. Evaluation stage: evaluate the corresponding system reliability and penalized system reliability providing the specified penalized parameter.
3. Improvement stage: improve the constructed solutions through local search
4. Updating stage: update the pheromone value online and offline given the corresponding penalized system reliability and the controlling parameter for pheromone persistence.

Ref [95] presents an application of the ant system in a reliability optimization problem for a series system, with multi-choice constraints incorporated at each subsystem, to maximize the system reliability subject to the system budget. It also combines a local search algorithm and a specific improvement algorithm that uses the remaining budget to improve the quality of a solution.

The ACO algorithm has also been applied to a multi-objective reliability optimization problem [116] and to the optimal design of multi-state series-parallel power systems [92].

1.4.1.2 Hybrid Genetic Algorithm

GA is a population-based directed random search technique inspired by the principles of evolution. Though it provides only heuristic solutions, it can be effectively applied to almost all complex combinatorial problems, and, thus, it has been employed in a large number of references as shown in Table 1. Ref [29] provides a state-of-the-art survey of GA-based reliability design.

To improve computational efficiency, or to avoid premature convergence, numerous researchers have been inspired to seek effective combinations of GAs with heuristic algorithms, simulation annealing methods, neural network techniques, steepest decent methods or other local search methods. The combinations are generally called hybrid genetic algorithms, and they represent one of the most promising developmental directions in optimization techniques.

Considering a complex system with a known system structure function, Ref [132] provides a unified modeling idea for both active and cold-standby redundancy optimization problems. The model prohibits any mixture of component types within subsystems. Both the lifetime and the cost of redundancy components are considered as random variables, so stochastic simulation is used to estimate the system performance, including the mean lifetime, percentile lifetime and reliability. To speed up the solution process, these simulation results become the training data for training a neural network to approximate the system performance. The trained neural network is finally embedded into a genetic algorithm to form a hybrid intelligent algorithm for solving the proposed model. Later [133] uses random fuzzy lifetimes as the basic parameters and employs a random fuzzy simulation to generate the training data.

Ref [48] develops a two-phase NN-hGA in which NN is used as a rough search technique to devise the initial solutions for a GA. By bounding the broad continuous search space with the NN technique, the NN-hGA derives the optimum robustly. However, in some cases, this algorithm may require too much computational time to be practical.

To improve the computation efficiency, Ref [49] presents a NN-flcGA to effectively control the balance between exploitation and exploration which characterizes the behavior of GAs. The essential features of the NN-flcGA include:

- combination with a NN technique to devise initial values for the GA
- application of a fuzzy logic controller when tuning strategy GA parameters dynamically
- incorporation of the revised simplex search method

Later, [50] proposes a similar hybrid GA called f-hGA for the redundancy allocation problem of a series-parallel system. It is based on

- application of a fuzzy logic controller to automatically regulate the GA parameters;
- incorporation of the iterative hill climbing method to perform local exploitation around the near optimum solution.

Ref [33] considers the optimal task allocation strategy and hardware redundancy level for a cycle-free distributed computing system so that the system cost during the period of task execution is minimized. The proposed hybrid heuristic combines the GA and the steepest decent method. Later [34] seeks similar optimal solutions to minimize system cost under constraints on the hardware redundancy levels. Based on the GA and a local search procedure, a hybrid GA is developed and compared with the simple GA. The simulation results show that the hybrid GA provides higher solution quality with less computational time.

1.4.1.3 Tabu Search

Though [45] describes the promise of tabu search, Ref [41] first develops a TS approach with the application of NFT [10] for reliability optimization problems. This method uses a subsystem-based tabu entry and dynamic length tabu list to reduce the sensitivity of the algorithm to selection of the tabu list length. The definition of the moves in this approach offers an advantage in efficiency, since it does not require recalculating the entire system reliability, but only the reliability of the changed subsystem. The results of several examples demonstrate the superior performance of this TS approach in terms of efficiency and solution superiority when compared to that of a GA.

1.4.1.4 Other Meta-Heuristic Methods

Some other adaptive meta-heuristic optimization methods inspired by activities in nature have also been proposed and applied in optimal reliability design. Ref [9] develops an immune algorithms-based approach inspired by the natural immune system of all animals. It analogizes antibodies and antigens as the solutions and objection functions, respectively. Ref [109] proposes a cellular evolutionary approach combining the multimember evolution strategy with concepts from Cellular Automata [125] for the selection step. In this approach, the parents' selection is performed only in the neighborhood in contrast to the general evolutionary strategy that searches for parents in the whole population. And a great deluge algorithm is extended and applied to optimize the reliability of complex systems in [107]. When both accuracy and speed are considered simultaneously, it is proven to be an efficient alternative to ACO and other existing optimization techniques.

1.4.2 Exact Methods

Unlike meta-heuristic algorithms, exact methods provide exact optimal solutions though much more computation complexity is involved. The development of exact methods, such as the branch-and-bound approach and lexicographic search, has recently been concentrated on techniques to reduce the search space of discrete optimization methods.

Ref [119] considers a reliability-redundancy allocation problem in which multiple-choice and resource constraints are incorporated. The problem is first transformed into a bi-criteria nonlinear integer programming problem by introducing 0-1 variables. Given a good feasible solution, the lower reliability bound of a subsystem is determined by the product of the maximal component reliabilities of all the other subsystems in the solution, while the upper bound is determined by the maximal amount of available sources of this subsystem. A branch-and-bound procedure, based on this reduced solution space, is then derived to search for the global optimal solution. Later, Ref [120] even combines the lower and upper bounds of the system reliability, which are obtained by variable relaxation and Lagrangean relaxation techniques, to further reduce the search space.

Also with a branch-and-bound algorithm, Ref [20] obtains the upper bound of series-parallel system reliability from its continuous relaxation problem. The relaxed problem is efficiently solved by the greedy procedure described in [19], combining heuristic methods to make use of some slack in the constraints obtained from rounding down. This technique assumes the objective and constraint functions are monotonically increasing.

Ref [30] presents an efficient branch-and-bound approach for coherent systems based on a 1-neighborhood local maximum obtained from the steepest ascent heuristic method. Numerical examples of a bridge system and a hierarchical series-parallel system demonstrate the advantages of this proposed algorithm in flexibility and efficiency.

Apart from the branch-and-bound approach, Ref [102] presents a partial enumeration method for a wide range of complex optimization problems based on a lexicographic search. The proposed upper bound of system reliability is very useful in eliminating several inferior feasible or infeasible solutions as shown in either big or small numerical examples. It also shows that the search process described in [95] does not necessarily give an exact optimal solution due to its logical flows.

Ref [83] develops a strong Lin & Kuo heuristic to search for an ideal allocation through the application of the reliability importance. It concludes that, if there exists an invariant optimal allocation for a system, the optimal allocation is to assign component reliabilities according to B-importance ordering. This Lin & Kuo heuristic can provide an exact optimal allocation.

Assuming the existence of a convex and differential reliability cost function $C_i(y_{ij})$, $y_{ij} = \log(1-r_{ij})$ for all component j in any subsystem i, Ref [27] proves that the components in each subsystem of a series-parallel system must have identical reliability for the purpose of cost minimization. The solution of the corresponding unconstrained problem provides the upper bound of the cost, while a doubly minimization problem gives its lower bound. With these results, the algorithm ECAY, which can provide either exact or approximate solutions depending on different stop criteria, is proposed for series-parallel systems.

1.4.3 Other Optimization Techniques

For series-parallel systems, Ref [104] formulates the reliability-redundancy optimization problem with the objective of maximizing the minimal subsystem reliability.

Problem 5 (P_5):

$$\max_{\mathbf{x}_1,\mathbf{x}_2,\ldots} (\min_i (1-\prod_j (1-r_{ij})^{x_{ij}}))$$

s.t.

$$g_i(\mathbf{x}) \le b_i, \qquad \text{for } i = 1,\ldots m$$

$$\mathbf{x} \in \mathbf{X}$$

Assuming linear constraints, an equivalent linear formulation of P_5 [36] can be obtained through an easy logarithm transformation, and, thus, the problem can be solved by readily available commercial software. It can also serve as a surrogate for traditional reliability optimization problems accomplished by sequentially solving a series of max-min subproblems.

Ref [51] presents a comparison between the Nakagawa and Nakashima method [43] and the max-min approach used by Ramirez-Marquez from the standpoint of solution quality and computational complexity. The experimental results show that the max-min approach is superior to the Nakagawa and Nakashima method in terms of solution quality in small-scale problems, but the analysis of its computational complexity demonstrates that the max-min approach is inferior to other greedy heuristics.

Ref [129] develops a heuristic approach inspired by the greedy method and a GA. The structure of this algorithm includes:

1. randomly generating a specified population size number of minimum workable solutions;
2. assigning components either according to the greedy method or to the random selection method;
3. improving solutions through an inner-system and inter-system solution revision process.

Ref [97] applies a hybrid dynamic programming/ depth-first search algorithm to redundancy allocation problems with more than one constraint. Given the tightest upper bound, the knapsack relaxation problem is formulated with only one constraint, and its solution $f_1(b)$ is obtained by a dynamic programming method. After choosing a small specified parameter e, the depth-first search technique is used to find all near-optimal solutions with objectives between $f_1(b)$ and $f_1(b) - e$. The optimal solution is given by the best feasible solution among all of the near-optimal solutions.

Ref [127] also presents a new dynamic programming method for a reliability-redundancy allocation problem in series-parallel systems where components must be chosen among a finite set. This pseudo-polynomial YCC algorithm is composed of two steps: the solution of the sub-problems, one for each subsystem, and the global resolution using previous results. It shows that the solutions converge quickly toward the optimum as a function of the required precision.

1.5 Comparisons and Discussions of Algorithms Reported in Literature

In this section, we provide a comparison of several heuristic or meta-heuristic algorithms reported in the literature. The compared numerical results are from the GA in Coit and Smith [10], the ACO in Liang and Smith [81], TS in Kulturel-Konak *et al.* [41], linear approximation in Hsieh [36], the IA in Chen and You [9] and the heuristic method in You and Chen [129]. The 33 variations of the Fyffe *et al.* problem, as devised by Nakagawa and Miyazaki [96], are used to test their performance, where different types are allowed to reside in parallel. In this problem set, the cost constraint is maintained at 130 and the weight constraint varies from 191 to 159.

As shown in Table 4, ACO [81], TS [41], IA [36] and heuristic methods [129] generally yield solutions with a higher reliability. When compared to GA [10],

- ACO [81] is reported to consistently perform well over different problem sizes and parameters and improve on GA's random behavior;
- TS [41] results in a superior performance in terms of best solutions found and reduced variability and greater efficiency based on the number of objective function evaluations required;
- IA [9] finds better or equally good solutions for all 33 test problems, but the performance of this IA-based approach is sensitive to value-combinations of the parameters, whose best values are case-dependent and only based upon the experience from preliminary runs. And more CPU time is taken by IAs;
- The best solutions found by heuristic methods [129] are all better than, or as good as, the well-known best solutions from other approaches. With this method, the average CPU time for each problem is within 8 seconds;
- In terms of solution quality, the proposed linear approximation approach [36] is inferior. But it is very efficient and the CPU time for all of the test problems is within one second;
- If a decision-maker is considering the max-min approach as a surrogate for system reliability maximization, the max-min approach [104] is shown to be capable of obtaining a close solution (within 0.22%), but it is unknown whether this performance will continue as problem sizes become larger.

For all the optimization techniques mentioned above, it might be hard to discuss about which tool is superior because in different design problems or even in a same problem with different parameters, these tools will perform variously.

Table 4. Comparison of several algorithms in the literature. Each for the test problems form [96]

W	System Reliability					
	GA [10]	ACO [81]	TS [41]	Hsieh[36]	IA [9]	Y&C [129]
191	0.98670	0.9868	0.98681	0.98671	0.98681	0.98681
190	0.98570	0.9859	0.98642	0.98632	0.98642	0.98642
189	0.98560	0.9858	0.98592	0.98572	0.98592	0.98592
188	0.98500	0.9853	0.98538	0.98503	0.98533	0.98538
187	0.98440	0.9847	0.98469	0.98415	0.98445	0.98469
186	0.98360	0.9838	0.98418	0.98388	0.98418	0.98418
185	0.98310	0.9835	0.98351	0.98339	0.98344	0.98350
184	0.98230	0.9830	0.98300	0.9822	0.9827	0.98299
183	0.98190	0.9822	0.98226	0.98147	0.98221	0.98226
182	0.98110	0.9815	0.98152	0.97969	0.98152	0.98152
181	0.98020	0.9807	0.98103	0.97928	0.98103	0.98103
180	0.97970	0.9803	0.98029	0.97833	0.98029	0.98029
179	0.97910	0.9795	0.97951	0.97806	0.97951	0.97950
178	0.97830	0.9784	0.97840	0.97688	0.97821	0.97840
177	0.97720	0.9776	0.97747	0.9754	0.97724	0.97760
176	0.97640	0.9765	0.97669	0.97498	0.97669	0.97669
175	0.97530	0.9757	0.97571	0.9735	0.97571	0.97571
174	0.97435	0.9749	0.97479	0.97233	0.97469	0.97493
173	0.97362	0.9738	0.97383	0.97053	0.97376	0.97383
172	0.97266	0.9730	0.97303	0.96923	0.97303	0.97303
171	0.97186	0.9719	0.97193	0.9679	0.97193	0.97193
170	0.97076	0.9708	0.97076	0.96678	0.97076	0.97076
169	0.96922	0.9693	0.96929	0.96561	0.96929	0.96929
168	0.96813	0.9681	0.96813	0.96415	0.96813	0.96813
167	0.96634	0.9663	0.96634	0.96299	0.96634	0.96634
166	0.96504	0.9650	0.96504	0.96121	0.96504	0.96504
165	0.96371	0.9637	0.96371	0.95992	0.96371	0.96371
164	0.96242	0.9624	0.96242	0.9586	0.96242	0.96242
163	0.96064	0.9606	0.95998	0.95732	0.96064	0.96064
162	0.95912	0.9592	0.95821	0.95555	0.95919	0.95919
161	0.95803	0.9580	0.95692	0.9541	0.95804	0.95803
160	0.95567	0.9557	0.9556	0.95295	0.95571	0.95571
159	0.95432	0.9546	0.95433	0.9508	0.95457	0.95456

Generally, if computational efficiency is of most concern to designer, linear approximation or heuristic methods can obtain competitive feasible solutions within a very short time (few seconds), as reported in [36, 129]. The proposed linear approximation [36] is also easy to implement with any LP software. But the main limitation of those reported approaches is that the constraints must be linear and separable.

Due to their robustness and feasibility, meta-heuristic methods such as GA and recently developed TS and ACO could be successfully applied to almost all NP-hard reliability optimization problems. However, they can not guarantee the optimality and sometimes can suffer from the premature convergence situation of their solutions because they have many unknown parameters and they neither use a prior knowledge nor exploit local search information. Compared to traditional meta-heuristic methods, a set of promising algorithms, hybrid GAs [33-34, 48-50, 132-133], are attractive since they retain the advantages of GAs in robustness and feasibility but significantly improve their computational efficiency and searching ability in finding global optimum with combining heuristic algorithms, neural network techniques, steepest decent methods or other local search methods.

For reliability optimization problems, exact solutions are not necessarily desirable because it is generally difficult to develop exact methods for reliability optimization problems which are equivalent to methods used for nonlinear integer programming problems [45]. However, exact methods may be particularly advantageous when the problem is not large. And more importantly, such methods can be used to measure the performance of heuristic or meta-heuristic methods.

1.6 Conclusions and Discussions

We have reviewed the recent research on optimal reliability design. Many publications have addressed this problem using different system structures, performance measures, problem formulations and optimization techniques.

The systems considered here mainly include series-parallel systems, k-out-of-n: G systems, bridge networks, n-version programming architecture, recovery block architecture and other unspecified coherent systems. The recently introduced *NVP* and *RB* belong to the category of fault tolerant architecture, which usually considers both software and hardware.

Reliability is still employed as a system performance measure in a majority of cases, but percentile life does provide a new perspective on optimal design without the requirement of a specified mission time. Availability is primarily used as the performance measure of renewable multi-state systems whose optimal design has been emphasized and well developed in the past 10 years. Optimal design problems are generally formulated to maximize an appropriate system performance measure under resource constraints, and more realistic problems involving multi-objective programming are also being considered.

When turning to optimization techniques, heuristic, meta-heuristic and exact methods are significantly applied in optimal reliability design. Recently, many advances in meta-heuristics and exact methods have been reported. Particularly, a new meta-heuristic method called ant colony optimization has been introduced and demonstrated to be a very promising general method in this field. Hybrid GAs may be the most important recent development among the optimization techniques since they retain the advantages of GAs in robustness and feasibility but significantly improve their computational efficiency.

Optimal reliability design has attracted many researchers, who have produced hundreds of publications since 1960. Due to the increasing complexity of practical engineering systems and the critical importance of reliability in these complex systems, this still seems to be a very fruitful area for future research.

Compared to traditional binary-state systems, there are still many unsolved topics in MSS optimal design including

- using percentile life as a system performance measure;
- involving cold-standby redundancy;
- nonrenewable MSS optimal design;
- applying optimization algorithms other than GAs, especially hybrid optimization techniques.

From the view of optimization techniques, there are opportunities for improved effectiveness and efficiency of reported ACO, TS, IA and GDA, while some new meta-heuristic algorithms such as Harmony Search Algorithm and Particle Swarm Optimization may offer excellent solutions for reliability optimization problems. Hybrid optimization techniques are another very promising general developmental direction in this field. They may combine heuristic methods, NN or some local search methods with all kinds of meta-heuristics to improve computational efficiency or with exact methods to reduce search space. We may even be able to combine two meta-heuristic algorithms such as GA and SA or ACO.

The research dealing with the understanding and application of reliability at the nano level has also demonstrated its attraction and vitality in recent years. Optimal system design that considers reliability within the uniqueness of nano-systems has seldom been reported in the literature. It deserves a lot more attention in the future. In addition, uncertainty and component dependency will be critical areas to consider in future research on optimal reliability design.

Acknowledgement

This work was partly supported by NSF Projects DMI-0429176. It is republished with the permission of IEEE Transactions on Systems, Man, Cybernetics, Part A: Systems and Humans.

References

[1] Altiparmak F, Dengiz B, Smith AE (2003) Optimal design of reliable computer networks: a comparison of meta heuristics. Journal of Heuristics 9: 471-487

[2] Amari SV, Dugan JB, Misra RB (1999) Optimal reliability of systems subject to imperfect fault-coverage. IEEE Transactions on Reliability 48: 275-284

[3] Amari SV, Pham H, Dill G (2004) Optimal design of k-out-of-n: G subsystems subjected to imperfect fault-coverage. IEEE Transactions on Reliability 53: 567-575

[4] Andrews JD, Bartlett LA (2005) A branching search approach to safety system design optimization. Reliability Engineering & System Safety 87: 23-30

[5] Aneja YP, Chandrasekaran R, Nair KPK (2004) Minimal-cost system reliability with discrete-choice sets for components. IEEE Transactions on Reliability 53:71-76

[6] Bueno VC (2005) Minimal standby redundancy allocation in a k-out-of-n: F system of dependent components. European Journal of Operation Research 165: 786-793

[7] Busacca PG, Marseguerra M, Zio E (2001) Multi-objective optimization by genetic algorithms: application to safety systems. Reliability Engineering & System Safety 72: 59-74

[8] Chern MS (1987) On the computational complexity of reliability redundancy allocation in a series system. Operation Research Letter SE-13: 582-592

[9] Chen TC, You PS (2005) Immune algorithms-based approach for redundant reliability problems with multiple component choices. Computers in Industry 56: 195-205

[10] Coit DW, Smith AE (1996) Reliability optimization of series-parallel systems using a genetic algorithm. IEEE Transactions on Reliability 45: 254-260

[11] Coit DW, Smith AE (1998) Redundancy allocation to maximize a lower percentile of the system time-to-failure distribution. IEEE Transactions on Reliability 47: 79-87
[12] Coit DW, Liu J (2000) System reliability optimization with k-out-of-n subsystems. International Journal of Reliability, Quality and Safety Engineering 7: 129-142
[13] Coit DW (2001) Cold-standby redundancy optimization for nonrepairable systems. IIE Transactions 33: 471-478
[14] Coit DW, Smith AE (2002) Genetic algorithm to maximize a lower-bound for system time-to-failure with uncertain component Weibull parameters. Computer & Industrial Engineering 41: 423-440
[15] Coit DW (2003) Maximization of system reliability with a choice of redundancy strategies. IIE Transactions 35: 535-543
[16] Coit DW, Jin T, Wattanapongsakorn N (2004) System optimization with component reliability estimation uncertainty: a multi-criteria approach. IEEE Transactions on Reliability 53: 369-380
[17] Cui L, Kuo W, Loh HT, Xie M (2004) Optimal allocation of minimal & perfect repairs under resource constraints. IEEE Transactions on Reliability 53: 193-199
[18] Dimou CK, Koumousis VK (2003) Competitive genetic algorithms with application to reliability optimal design. Advances in Engineering Software 34: 773-785
[19] Djerdjour M (1997) An enumerative algorithm framework for a class of nonlinear programming problems. European Journal of Operation Research 101: 101-121
[20] Djerdjour M, Rekab K (2001) A branch and bound algorithm for designing reliable systems at a minimum cost. Applied Mathematics and Computation 118: 247-259
[21] Dorigo M (1992) Optimization learning and natural algorithm. Ph.D. Thesis, Politecnico di Milano, Italy
[22] Dorigo M, Maniezzo V, Colorni A (1996) Ant system: optimization by a colony of cooperating agents. IEEE Transactions on System, Man, and Cybernetics-part B: Cybernetics 26: 29-41
[23] Dorigo M, Gambardella LM (1997) Ant colonies for traveling salesman problem. BioSystem 43: 73-81
[24] Dorigo M, Gambardella LM (1997) Ant colony system: a cooperative learning approach to the traveling salesman problem. IEEE Transactions on Evolutionary computation 1: 53-66
[25] Dorigo M, Caro G (1999) The ant colony optimization meta heuristic. In: Corne D, Dorigo M, Glover F (eds.) New Ideas in Optimization. McGraw-Hill, pp 11-32.
[26] Dorigo M, Caro GD, Gambardella LM (1999) Ant algorithm for discrete optimization. Artificial Life 5: 137-172
[27] Elegbede C, Chu C, Adjallah K, Yalaoui F (2003) Reliability allocation through cost minimization. IEEE Transactions on Reliability 52: 106-111

[28] Elegbede C, Adjallah K (2003) Availability allocation to repairable systems with genetic algorithms: a multi-objective formulation Reliability Engineering and System Safety 82: 319-330

[29] Gen M, Kim JR (1999) GA-based reliability design: state-of-the-art survey. Computer & Industrial Engineering 37: 151-155

[30] Ha C, Kuo W (2006) Reliability redundancy allocation: an improved realization for nonconvex nonlinear programming problems. European Journal of Operational Research 171: 24-38

[31] Ha C, Kuo W (accepted) Multi-paths iterative heuristic for redundancy allocation: the tree heuristic. IEEE Transactions on Reliability

[32] Ha C, Kuo W (2005) Multi-path approach for Reliability-redundancy allocation using a scaling method. Journal of Heuristics, 11: 201-237

[33] Hsieh CC, Hsieh YC (2003) Reliability and cost optimization in distributed computing systems. Computer & Operations Research 30: 1103-1119

[34] Hsieh CC (2003) Optimal task allocation and hardware redundancy policies in distributed computing system. European Journal of Operational Research 147: 430-447

[35] Hsieh YC, Chen TC, Bricker DL (1998) Genetic algorithm for reliability design problems. Microelectronics Reliability 38: 1599-1605

[36] Hsieh YC (2002) A linear approximation for redundant reliability problems with multiple component choices. Computer & Industrial Engineering 44: 91-103

[37] Huang J, Zuo MJ, Wu Y (2000) Generalized multi-state k-out-of-n: G systems. IEEE Transactions on Reliability 49: 105-111

[38] Jin T, Coit DW (2001) Variance of system-reliability estimates with arbitrarily repeated components. IEEE Transactions on Reliability 50: 409-413

[39] Kim KO, Kuo W (2003) Percentile life and reliability as a performance measures in optimal system design. IIE Transactions 35: 1133-1142

[40] Kulturel-Kotz S, Lai CD, Xie M (2003) On the effect of redundancy for systems with dependent components. IIE Transactions 35: 1103-1110

[41] Kulturel-Konak S, Smith AE, Coit DW (2003) Efficiently solving the redundancy allocation problem using tabu search. IIE Transactions 35: 515-526

[42] Kumar UD, Knezevic J (1998) Availability based spare optimization using renewal process. Reliability Engineering and System Safety 59: 217-223

[43] Kuo W, Hwang CL, Tillman FA (1978) A note on heuristic method for in optimal system reliability. IEEE Transactions on Reliability 27: 320-324

[44] Kuo W, Chien K, Kim T (1998) Reliability, Yield and Stress Burn-in. Kluwer

[45] Kuo W, Prasad VR (2000) An annotated overview of system-reliability optimization. IEEE Transactions on Reliability 49: 487-493

[46] Kuo W, Prasad VR, Tillman FA, Hwang CL (2001) Optimal Reliability Design: Fundamentals and Applications. Cambridge University Press

[47] Kuo W, Zuo MJ (2003) Optimal Reliability Modeling: Principles and Applications. John Wiley & Sons

[48] Lee CY, Gen M, Kuo W (2002) Reliability optimization design using hybridized genetic algorithm with a neural network technique. IEICE Transactions

on Fundamentals of Electronics, Communications and Computer Sciences: E84-A: 627-637
[49] Lee CY, Yun Y, Gen M (2002) Reliability optimization design using hybrid NN-GA with fuzzy logic controller. IEICE Transactions on Fundamentals of Electronics, Communications and Computer Sciences E85-A: 432-447
[50] Lee CY, Gen M, Tsujimura Y (2002) Reliability optimization design for coherent systems by hybrid GA with fuzzy logic controller and local search. IEICE Transactions on Fundamentals of Electronics, Communications and Computer Sciences E85-A: 880-891
[51] Lee H, Kuo W, Ha C (2003) Comparison of max-min approach and NN method for reliability optimization of series-parallel system. Journal of System Science and Systems Engineering 12: 39-48
[52] Levitin G, Lisnianski A, Ben-Haim H, Elmakis D (1998) Redundancy optimization for series-parallel multi-state systems. IEEE Transactions on Reliability 47: 165-172
[53] Levitin G, Lisnianski A (1998) Structure optimization of power system with bridge topology. Electric Power Systems Research 45: 201-208
[54] Levitin G, Lisnianski A (1999) Joint redundancy and maintenance optimization for multistate series-parallel systems. Reliability Engineering and System Safety 64: 33-42
[55] Levitin G, Lisnianski A (1999) Importance and sensitivity analysis of multistate systems using the universal generating function method. Reliability Engineering and System Safety 65: 271-282
[56] Levitin G, Lisnianski A (2000) Survivability maximization for vulnerable multi-state systems with bridge topology. Reliability Engineering and System Safety 70: 125-140
[57] Levitin G (2001) Redundancy optimization for multi-state system with fixed resource-requirements and unreliable sources. IEEE Transactions on Reliability 50: 52-59
[58] Levitin G, Lisnianski A (2001) Reliability optimization for weighted voting system. Reliability Engineering and System Safety 71: 131-138
[59] Levitin G, Lisnianski A (2001) Structure optimization of multi-state system with two failure modes. Reliability Engineering and System Safety 72: 75-89
[60] Levitin G, Lisnianski A (2001) Optimal separation of elements in vulnerable multi-state systems. Reliability Engineering and System Safety 73: 55-66
[61] Levitin G (2001) Reliability evaluation for acyclic consecutively connected network with multistate elements. Reliability Engineering and System Safety 73: 137-143
[62] Levitin G, Lisnianski A (2001) A new approach to solving problems of multi-state system reliability optimization Quality and Reliability Engineering International 17: 93-104
[63] Levitin G (2002) Optimal allocation of multi-state retransmitters in acyclic transmission networks. Reliability Engineering and System Safety 75: 73-82
[64] Levitin G (2002) Optimal allocation of elements in a linear multi-state sliding window system. Reliability Engineering and System Safety 76: 245-254

[65] Levitin G (2002) Optimal series-parallel topology of multi-state system with two failure modes. Reliability Engineering and System Safety 77: 93-107
[66] Levitin G (2002) Evaluating correct classification probability for weighted voting classifiers with plurality voting. European Journal of Operational Research 141: 596-607
[67] Levitin G (2003) Optimal allocation of multi-state elements in linear consecutively connected systems with vulnerable nodes. European Journal of Operational Research 150: 406-419
[68] Levitin G (2003) Optimal allocation of multistate elements in a linear consecutively-connected system. IEEE Transactions on Reliability 52: 192-199
[69] Levitin G (2003) Reliability evaluation for acyclic transmission networks of multi-state elements with delays. IEEE Transactions on Reliability 52: 231-237
[70] Levitin G (2003) Linear multi-state sliding-window systems. IEEE Transactions on Reliability 52: 263-269
[71] Levitin G (2003) Reliability of multi-state systems with two failure-modes. IEEE Transactions on Reliability 52: 340-348
[72] Levitin G, Lisnianski A (2003) Optimizing survivability of vulnerable series-parallel multi-state systems. Reliability Engineering & System Safety 79: 319-331
[73] Levitin G (2003) Optimal multilevel protection in series-parallel systems. Reliability Engineering and System Safety, 81: 93-102
[74] Levitin G, Dai Y, Xie M, Poh KL (2003) Optimizing survivability of multi-state systems with multi-level protection by multi-processor genetic algorithm. Reliability Engineering and System Safety. 82: 93-104
[75] Levitin G (2004) A universal generating function approach for the analysis of multi-state systems with dependent elements. Reliability Engineering and System Safety 84: 285-292
[76] Levitin G (2004) Consecutive k-out-of-r-from-n system with multiple failure criteria. IEEE Transactions on Reliability 53: 394-400
[77] Levitin G (2004) Reliability optimization models for embedded systems with multiple applications. IEEE Transactions on Reliability 53: 406-416
[78] Levitin G (2005) Uneven allocation of elements in linear multi-state sliding window system. European Journal of Operational Research 163: 418-433
[79] Levitin G (2005) Optimal structure of fault-tolerant software systems. Reliability Engineering and System Safety 89: 286-295
[80] Levitin G (accepted) Reliability and performance analysis of hardware-software systems with fault-tolerant software components. Reliability Engineering and System Safety
[81] Liang YC, Smith AE (2004) An ant colony optimization algorithm for the redundancy allocation problem. IEEE Transactions on Reliability 53: 417-423
[82] Lin F, Kuo W, Hwang F (1999) Structure importance of consecutive-k-out-of-n systems. Operations Research Letters 25: 101-107
[83] Lin F, Kuo W (2002) Reliability importance and invariant optimal allocation. Journal of Heuristics 8: 155-172

[84] Lisnianski A, Levitin G, Ben-Haim H (2000) Structure optimization of multi-state system with time redundancy. Reliability Engineering and System Safety 67: 103-112

[85] Lisianski A, Levitin G (2003) Multi-state System Reliability, Assessment, Optimization and Applications. World Scientific

[86] Liu PX, Zuo MJ, Meng MQ (2003) Using neural network function approximation for optimal design of continuous-state parallel-series systems. Computers & Operations Research 30: 339-352

[87] Lyu MR, Rangarajan S, Moorsel APA (2002) Optimal allocation of test resources for software reliability growth modeling in software development. IEEE Transactions on Reliability 51: 183-192

[88] Maniezzo V, Gambardella LM, De Luigi F. (2004) Ant Colony Optimization. In: Onwubolu GC, Babu BV (eds) New Optimization Techniques in Engineering. Springer-Verlag Berlin Heidelberg, pp 101-117

[89] Marseguerra M, Zio E (2000) Optimal reliability/availability of uncertain systems via multi-objective genetic algorithms. 2000 Proceedings Annual Reliability and Maintainability Symposium, pp 222-227

[90] Marseguerra M, Zio E (2004) System design optimization by genetic algorithm. IEEE Transactions on Reliability 53: 424-434

[91] Marseguerra M, Zio E, Podofillini L, Coit DC (2005) Optimal design of reliable network systems in presence of uncertainty. IEEE Transactions on Reliability 54: 243-253

[92] Massim Y, Zeblah A, Meziane R, Benguediab M, Ghouraf A (2005) Optimal design and reliability evaluation of multi-state series-parallel power systems. Nonlinear Dynamics 40: 309-321

[93] Misra KB (1986) On optimal reliability design: a review. System Science 12: 5-30

[94] Misra KB (1991) An algorithm to solve integer programming problems: an efficient tool for reliability design. Microelectronics and Reliability 31: 285-294

[95] Nahas N, Nourelfath M (2005) Ant system for reliability optimization of a series system with multiple-choice and budget constraints. Reliability Engineering and System Safety 87: 1-12

[96] Nakagawa Y, Miyazaki S (1981) Surrogate constraints algorithm for reliability optimization problems with two constraints. IEEE Transactions on Reliability R-30: 175-180

[97] Kevin YKNG, Sancho NGF (2001) A hybrid dynamic programming/depth-first search algorithm with an application to redundancy allocation. IIE Transactions 33: 1047-1058

[98] Nordmann L, Pham H (1999) Weighted voting systems. IEEE Transactions on Reliability 48: 42-49

[99] Nourelfath M, Dutuit Y (2004) A combined approach to solve the redundancy optimization problem for multi-state systems under repair policies. Reliability Engineering and System Safety 84: 205-213

[100] Prasad VR, Raghavachari M (1998) Optimal allocation of interchangeable components in a series-parallel system. IEEE Transactions on Reliability 47: 255-260

[101] Prasad VR, Kuo W, Kim KO (2000) Optimal allocation of s-identical multi-functional spares in a series system. IEEE Transactions on Reliability 48: 118-126

[102] Prasad VR, Kuo W (2000) Reliability optimization of coherent systems. IEEE Transactions on Reliability 49: 323-330

[103] Prasad VR, Kuo W, Kim KO (2001) Maximization of a percentile life of a series system through component redundancy allocation IIE Transactions 33: 1071-1079

[104] Ramirez-Marquez JE, Coit DW, Konak A (2004) Redundancy allocation for series-parallel systems using a max-min approach. IIE Transactions 36: 891-898

[105] Ramirez-Marquez JE, Coit DW (2004) A heuristic for solving the redundancy allocation problem for multi-state series-parallel system. Reliability Engineering and System Safety 83: 341-349

[106] Ravi V, Reddy PJ, Zimmermann HJ (2000) Fuzzy global optimization of complex system reliability. IEEE Transactions on Fuzzy Systems 8: 241-248

[107] Ravi V (2004) Optimization of complex system reliability by a modified great deluge algorithm. Asia-Pacific Journal of Operational Research 21: 487-497

[108] Ren Y, Dugan JB (1998) Design of reliable system using static & dynamic fault trees. IEEE Transactions on Reliability 47: 234-244

[109] Rocco CM, Moreno JA, Carrasquero N (2000) A cellular evolution approach applied to reliability optimization of complex systems. 2000 Proceedings Annual Reliability and Maintainability Symposium, pp 210-215

[110] Romera R, Valdés JE, Zequeira RI (2004) Active-redundancy allocation in systems. IEEE Transactions on Reliability 53: 313-318

[111] Royset JO, Polak E (2004) Reliability-based optimal design using sample average approximations. Probabilistic Engineering Mechanics 19: 331-343

[112] Royset JO, Kiureghian AD, Polak E (2001) Reliability-based optimal design of series structural systems. Journal of Engineering Mechanics 127: 607-614

[113] Ryoo HS (2005) Robust meta-heuristic algorithm for redundancy optimization in large-scale complex systems. Annals of Operations Research 133: 209-228

[114] Sasaki M, Gen M (2003) A method of fuzzy multi-objective nonlinear programming with GUB structure by Hybrid Genetic Algorithm. International Journal of Smart Engineering Design 5: 281-288

[115] Sasaki M, Gen M (2003) Fuzzy multiple objective optimal system design by hybrid genetic algorithm. Applied Soft Computing 2: 189-196

[116] Shelokar PS, Jayaraman VK, Kulkarni BD (2002) Ant algorithm for single and multiobjective reliability optimization problems. Quality and Reliability Engineering International 18: 497-514

[117] Stocki R, Kolanek K, Jendo S, Kleiber M (2001) Study on discrete optimization techniques in reliability-based optimization of truss structures. Computers and Structures 79: 2235-2247

[118] Suman B (2003) Simulated annealing-based multi-objective algorithm and their application for system reliability. Engineering Optimization 35: 391-416

[119] Sung CS, Cho YK (1999) Branch-and-bound redundancy optimization for a series system with multiple-choice constraints. IEEE Transactions on Reliability 48: 108-117

[120] Sung CS, Cho YK (2000) Reliability optimization of a series system with multiple-choice and budget constraints. European Journal of Operational Research 127: 158-171

[121] Taguchi T, Yokota T, Gen M (1998) Reliability optimal design problem with interval coefficients using hybrid genetic algorithms. Computers & Industrial Engineering 35: 373-376

[122] Taguchi T, Yokota T (1999) Optimal design problem of system reliability with interval coefficient using improved genetic algorithms. Computers & Industrial Engineering 37: 145-149

[123] Tillman FA, Hwang CL, Kuo W (1977) Optimization techniques for system reliability with redundancy- a review. IEEE Transactions on Reliability R-26: 148-152

[124] Wattanapongsakorn N, Levitan SP (2004) Reliability optimization models for embedded systems with multiple applications. IEEE Transactions on Reliability 53: 406-416

[125] Wolfram S (1984) Cellular automata as models of complexity. Nature, 3H

[126] Yalaoui A, Chu C, Châtelet E (2005) Reliability allocation problem in a series-parallel system. Reliability Engineering and System Safety 90: 55-61

[127] Yalaoui A, Châtelet E, Chu C (2005) A new dynamic programming method for reliability & redundancy allocation in a parallel-series system. IEEE Transactions on Reliability 54: 254-261

[128] Yang JE, Hwang MJ, Sung TY, Jin Y (1999) Application of genetic algorithm for reliability allocation in nuclear power plants. Reliability Engineering and System Safety 65: 229-238

[129] You PS, Chen TC (2005) An efficient heuristic for series-parallel redundant reliability problems. Computer & Operations Research 32: 2117-2127

[130] Yun WY, Kim JW (2004) Multi-level redundancy optimization in series systems. Computer and Industrial Engineering 46: 337-346

[131] Zafiropoulos EP, Dialynas EN (2004) Reliability and cost optimization of electronic devices considering the component failure rate uncertainty. Reliability Engineering and System Safety 84: 271-284

[132] Zhao R, Liu B (2003) Stochastic programming models for general redundancy-optimization problems. IEEE Transactions on Reliability 52: 181-191

[133] Zhao R, Liu B (2004) Redundancy optimization problems with uncertainty of combining randomness and fuzziness. European Journal of Operation Research 157: 716-735

Multiobjective Metaheuristic Approaches to Reliability Optimization

Sadan Kulturel-Konak

Management Information Systems, Penn State Berks, Reading, PA

Abdullah Konak

Information Sciences and Technology, Penn State Berks, Reading, PA

David W. Coit

Department of Industrial and Systems Engineering, Rutgers University, NJ

2.1 Introduction

Many engineering system design and reliability optimization problems involve several design criteria or objectives that are in conflict with each other, e.g., minimize cost, maximize reliability, minimize weight, and maximize performance. In most cases, the objectives are non-commensurable, and the relative importance of the objectives and the tradeoffs between them are not fully understood by decision makers, especially for design problems with several objectives. In its general form, a multiobjective problem can be stated as follows [9]:

$$\left.\begin{array}{lll} \text{Optimize} & f_k(\mathbf{x}), & k=1,2,\ldots,K; \\ \text{Subject to} & g_j(\mathbf{x})\geq 0, & j=1,2,\ldots,J; \\ & h_l(\mathbf{x})=0, & l=1,2,\ldots,L; \\ & \mathbf{x}\in\mathbf{S}; & \end{array}\right\} \qquad (1)$$

where $\mathbf{x}$ is a solution vector of n decision variables, $\mathbf{x}=(x_1, x_2, \ldots, x_n)$ and $\mathbf{S}$ is the solution space. There are K conflicting objective functions, and each objective function can be minimization or maximization type. Optimizing $\mathbf{x}$ with respect to a single objective often results in unacceptable results

K.-K. Sadan et al.: *Multiobjective Metaheuristic Approaches to Reliability Optimization*, Computational Intelligence in Reliability Engineering (SCI) **39**, 37–62 (2007)
www.springerlink.com

with respect to the other objectives. Therefore, a perfect multiobjective solution **x** that simultaneously optimizes each objective function is almost impossible. A reasonable approach to a multiobjective problem is to investigate a set of solutions, each of which satisfies the objectives at an acceptable level without being dominated by any other solution.

Assuming that all objectives are minimization type in problem (1), a feasible solution **x** is said to be dominated by another feasible solution **y**, if and only if $f_k(\mathbf{y}) \leq f_k(\mathbf{x}), \forall k = 1,...,K$ and $f_k(\mathbf{y}) < f_k(\mathbf{x})$ for at least one objective k. A solution is said to be *Pareto optimal* if it is not dominated by another solution in the feasible solution space. The set of all feasible non-dominated solutions is referred to as the *Pareto optimal set*. The curve formed by joining all Pareto optimal solutions in the objective space is called the *Pareto optimal front*.

Multiobjective optimization and decision making have been also applied in designing reliability systems. Earlier work on multiobjective approaches to reliability optimization was reported in [30], [31], [47], [48], [49], [50], [51], and [52]. Sakawa [47; 48; 51] formulated the redundancy allocation problem of a series-parallel system with two objectives, minimizing system cost and maximizing system reliability, under multiple constraints. Sakawa developed the surrogate worth tradeoff method to solve the formulated multiobjective problems. In this approach, the first step involves determining a trade-off surface using a Lagrangien relaxation approach, and in the second step Lagrangien multipliers are updated based on the decision makers articulated preferences. Inagaki et al. [31] proposed a multiobjective decision making procedure based on interactive optimization and mathematical programming to design reliability systems with minimum cost and weight and maximum reliability. Hwang et al. [30] used multiobjective optimization to determine the replacement ages of critical items of a complex system considering three decision criteria, minimum replacement cost-rate, maximum availability, and a lower-bound on mission reliability. Kuo et al. [38] provided a survey of traditional multiobjective reliability problems and solutions approaches.

In the last two decades, metaheuristics have been the primary tool to solve multiobjective optimization problems [32]. This chapter surveys metaheuristic approaches to reliability optimization problems and presents two example applications, a multiobjective tabu search approach to the redundancy allocation problem and a multiobjective genetic algorithm approach to the economic design of reliable networks.

2.2 Metaheuristics and Multiobjective Optimization

2.2.1 Metaheuristics

In many optimization problems, some decision variables have only discrete values. The term *combinatorial optimization* refers to finding optimal solutions to such problems. Finding an optimal solution for a combinatorial optimization problem is a daunting task due to the combinatorial explosion of possible solutions to the problem with the number of discrete decision variables. Classical approaches such as enumeration (implicit enumeration, brand-and-bound and dynamic programming), Lagrangian relaxation, decomposition, and cutting plane techniques or their combinations may not be computationally feasible or efficient to solve a combinatorial optimization problem of a practical size. Therefore, researchers focus on *heuristic techniques* that seek a good solution to a complex combinatorial problem within a reasonable time. A metaheuristic is usually defined as a top-level general (intelligent) strategy which guides other heuristics to search for good feasible solutions to a difficult problem. Examples of metaheuristics are Simulated Annealing (SA), Tabu Search (TS), Genetic Algorithms (GA), Ant Colony Approaches (ACO), and memetic algorithms.

Although there are many different metaheuristic techniques, all metaheuristic techniques can be depicted as a cycle of creating a set of candidate solutions from a set of current solutions using some forms of new solution generation operator and updating the set of current solutions based on a selection mechanism (Fig. 1).

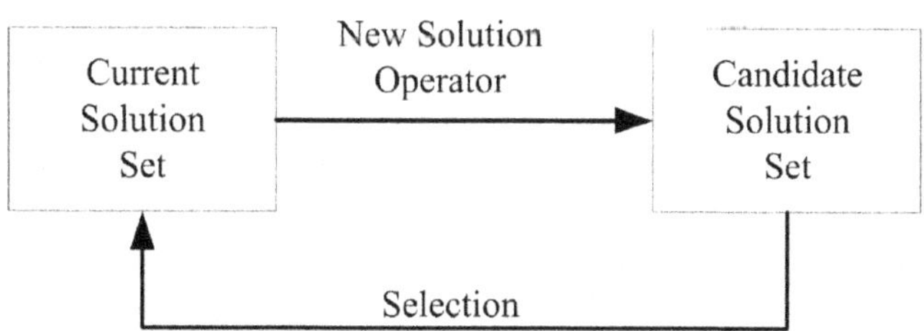

Fig. 1. Basic principle of metaheuristics.

The intelligence of a metaheuristic is usually embedded into its selection mechanism. For example, TS uses the objective function value and a long term memory to determine the new current solution. The new solution creation operator can be random or systematic. For example, in GA, a crossover operator randomly recombines multiple solutions to create new solutions and a mutation operator randomly perturbs solutions. In SA and TS, on the other hand, a move operator systematically generates a set of neighborhood solutions from the current solution.

Although a correct classification of different metaheuristics is a difficult task, based on the size of the current solution set, metaheuristics can be grouped into two general groups, population-based approaches, such as GA, which maintain and use multiple current solutions to generate new solutions and single-solution approaches, such as SA and TS, in which a single current solution is used.

A common property of metaheuristics is their ability to avoid being trapped in local optima. In a single objective minimization problem, a solution $\mathbf{x}$ is called the *global optimum* solution if $f(\mathbf{x}) \leq f(\mathbf{y})$ for every feasible solution $\mathbf{y} \in \mathbf{S}$. A solution $\mathbf{x}$ is called *local optimum* if it is the best solution within a neighboring set of solutions, i.e., $f(\mathbf{x}) \leq f(\mathbf{y})$ for every feasible solution $\mathbf{y} \in N(\mathbf{x})$. A local steepest descent heuristic searches a defined neighborhood $N(\mathbf{x})$ of current solution $\mathbf{x}$ and selects the best solution in $N(\mathbf{x})$ as the new current solution to continue. Therefore, such a heuristic is trapped after finding the first local optimum solution (see Fig. 2).

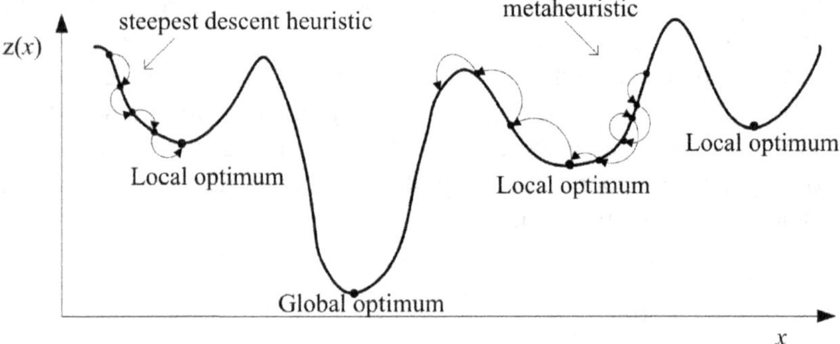

Fig. 2. Search patterns of steepest descent search and metaheuristics.

On the other hand, metaheuristics implement various techniques to move away from local optimum, such as noise in selection of the current solution, random perturbations to create new solutions, and a memory of recently visited solutions. As seen in Fig. 2, the objective function value of current solution may deteriorate during the search process; however, this enables a metaheuristic to move away from local attraction points in order to explore different regions of the solution space where global optimum may reside.

2.2.2 Metaheuristic Approaches to Multiobjective Optimization

Most multiobjective metaheuristics have been derived from their single-objective counterparts using three different approaches, weighted sum of objectives, alternating objectives, and Pareto ranking. In the weighted sum of objectives approach, a weight w_i is assigned to each normalized objective function $f_i'(\mathbf{x})$ so that the problem is converted into a single objective problem with a scalar objective function as follows

$$\min z = w_1 f_1'(\mathbf{x}) + w_2 f_2'(\mathbf{x}) + \ldots + w_K f_K'(\mathbf{x}) \tag{2}$$

where $\sum_{i=1}^{K} w_i = 1$. Then, a single objective metaheuristic can be used to solve the problem. The weighted sum approach is straightforward and easy to implement. However, a single solution is obtained for a given weight vector $\mathbf{w} = (w_1, w_2, \ldots, w_K)$, and if multiple solutions are desired, the problem must be solved multiple times with different weight combinations. The main difficulty with this approach is selecting a weight vector in each run.

The alternating objectives approach is another straightforward extension of a single objective metaheuristic to solve multiobjective problems. In this approach, only a single objective function, which is randomly selected in each iteration, is used to evaluate candidate solutions in the selection of new current solutions. Therefore, a minimum modification is required in the original single objective metaheuristic. Non-dominated solutions discovered during the search are stored in a list, and new solutions are compared to existing non-dominated solutions for possible updates.

The Pareto-ranking approach explicitly utilizes the concept of Pareto dominance in the selection. This approach is mainly used by population based multiobjective evolutionary algorithms. The population is ranked according to a dominance rule, and then solutions are selected based on their ranks.

The ultimate goal in multiobjective optimization is to investigate the Pareto optimal set. For many multiobjective problems, however, the size of the Pareto optimal set is very large. In addition, it is usually impossible to prove global optimality in combinatorial optimization problems. Therefore, the output of a multiobjective metaheuristic is called the *best-known Pareto set.* An effective multiobjective metaheuristic should achieve the following three goals [62]:

- Final solutions should be as close as possible to the true Pareto front.

- Final solutions should be uniformly distributed and diverse over the final Pareto front.
- The extreme ends of the true Pareto front should be investigated.

2.3 Multiobjective Tabu Search and Reliability Optimization

TS is a local (neighborhood) search guided by a selection function which is used to evaluate candidate solutions [21; 22]. To avoid being trapped in local optima, TS depends on the tabu list. A new solution (candidate solution) is produced by the move operator slightly perturbing a current solution. The set of all candidate solutions produced by the move operator is called the neighborhood of the current solution. The best candidate solution is the one with the best objective function value in the neighborhood. To avoid cycling, some aspect(s) of recent moves are classified as forbidden and stored on a tabu list for a certain number of iterations. For canonical TS, in each iteration, the algorithm is forced to select the best move which is not tabu. In some cases, however, if a tabu move improves upon the best solution found so far, then that move can be accepted. This is called an aspiration criterion. After the neighborhood of the current solution is investigated and the best candidate is determined, the tabu list is updated, the best candidate is assigned to the current solution, and the entire process starts again. The search continues until a predetermined stopping condition is satisfied.

Although TS has been frequently implemented to solve single objective combinatorial optimization problems, it has rarely been the primary focus of multiobjective optimization approaches. Multiobjective TS were proposed in [1], [3], [4], [25], and [27]. In Kulturel-Konak et al. [35], a new multiobjective TS approach was proposed, called the multinomial TS, to solve the multiobjective series-parallel system redundancy allocation problem.

2.3.1 The Multinomial Tabu Search Algorithm to Solve Redundancy Allocation Problem

2.3.1.1 Multiobjective System Redundancy Allocation Problem

In this section, a multinomial TS (MTS) approach of Kulturel-Konak et al. [35] is presented to solve the multiobjective series-parallel system redundancy allocation problem (RAP). For MTS, the series-parallel system RAP is as follows: determine the optimal design configuration for given objectives and constraints when there are multiple component choices available for each of several *k*-out-of-*n*:*G* subsystems. While the RAP has been extensively studied as a single objective problem in the literature (refer to [36] for a complete set of citations) to maintain computational tractability, the design problem is actually more realistically a multiobjective problem. In this case, the problem is formulated as determining the optimal design configuration to maximize system reliability (*R*) and to minimize system cost (*C*) and weight (*W*) when there are multiple component choices available for each of several *k*-out-of-n:*G* subsystems. The mathematical formulation of the problem is given below.

$$\left.\begin{array}{l} \max\,(R=\prod_{i=1}^{s}R_i(\mathbf{x}_i\,|\,k_i)),\ \min\,(C=\sum_{i=1}^{s}C_i(\mathbf{x}_i)),\ \min\,(W=\sum_{i=1}^{s}W_i(\mathbf{x}_i)) \\ \text{subject to} \\ \qquad k_i \le \sum_{j=1}^{m_i} x_{ij} \le n_{\max,i} \qquad \forall i=1,2,\ldots,s \end{array}\right\} \tag{3}$$

where

s	number of subsystems
$\mathbf{x}_i$	$(x_{i1}, x_{i2}, \ldots, x_{i,m_i})$
x_{ij}	quantity of jth component in subsystem i
n_i	total number of components used in subsystem i ($n_i = \sum_{j=1}^{m_i} x_{ij}$)
$n_{\max,i}$	user assigned maximum number of components in parallel used in subsystem i
k_i	minimum number of components in parallel required for subsystem i
$R_i(\mathbf{x}_i\|\, k_i)$	reliability of subsystem i, given k_i
$C_i(\mathbf{x}_i)$	total cost of subsystem i
$W_i(\mathbf{x}_i)$	total weight of subsystem i

2.3.1.2 Multinomial Tabu Search Algorithm

Multinomial TS (MTS) uses a multinomial probability mass function to select an objective to become active in each iteration, i.e., one of the objectives, maximizing *R*, minimizing *C*, or minimizing *W* is used to identify the best candidate. The idea of alternating objectives in multiobjective optimization has been previously applied in the area of evolutionary computation by Kursawe [39] and Schaffer [55]. This idea is very general and easy to implement. It can accommodate two or more objectives. There are no weights to set and no scaling adjustment to be made. The procedure of MTS is given as follows:

Step 1. **Initial Solution:** To obtain an initial feasible solution, for each subsystem i, s integers between k_i and $n_{\max,i}-3$ (inclusive) are chosen according to a discrete uniform distribution to represent the number of components in parallel (n_i). Then, the n_i components are randomly and uniformly assigned to the m_i different types. If feasible, it becomes the initial CURRENT solution. Initialize the tabu list and the non-dominated (ND) solutions list as empty.

Step 2. **Select an objective:** Randomly choose an objective - maximizing R, minimizing C, or minimizing W - as the objective to actively evaluate candidate solutions using equal probabilities of $p_R = p_C = p_W = 1/3$ (i.e., $p_R + p_C + p_W = 1$).

Step 3. **Moves:** Moves operate on each subsystem. The first type of move changes the number of a particular component type by adding or subtracting one (i.e., $x_{ij} \leftarrow x_{ij}+1$, or $x_{ij} \leftarrow x_{ij} - 1$). These moves are considered individually for all available component type j within each subsystem i. The second type of move simultaneously adds one component to a certain subsystem and deletes another component, of a different type, within the same subsystem ($x_{ij} \leftarrow x_{ij}+1$, and $x_{ik} \leftarrow x_{ik} - 1$ for $j \neq k$, enumerating all possibilities). The two types of moves are performed independently on the CURRENT solution to create CANDIDATE solutions. An important advantage of these types of moves is that they do not require recalculating the entire system reliability. Note that subsystems are changed one-at-a-time. Hence a CANDIDATE solution is different from the CURRENT solution only at a single subsystem. Therefore, only the reliability of the changed subsystem is recalculated, and system reliability is updated accordingly.

Step 4. **Update ND and BEST CANDIDATE:** Compare each feasible CANDIDATE solution with the current ND solutions list as follows. If a CANDIDATE solution dominates some current ND solutions, remove these dominated solutions from the ND solutions list and add the candidate to the ND solutions list. If a candidate solution is not dominated by any current ND solution, add it to the ND solutions list. Choose the non-tabu (or if it is tabu, but dominates any solution in the ND solutions list) CANDIDATE solution with the best objective function value according to the objective chosen in Step 3 as the BEST CANDIDATE solution.

Step 5. **Update Tabu List:** The structure of the subsystem that has been changed in the accepted move is stored in the tabu list. The dynamic tabu list length is changed according to a uniform integer random number between [s, $3s$].

Step 6. **Diversification:** A diversification scheme based on restart is used. If the list of ND solutions has not been updated in the last (stopping criterion/4) moves, one of the ND solutions found during the search is uniformly randomly selected as the new CURRENT solution, the tabu list is reset to empty, and the search restarts from this solution (i.e., return to Step 3).

Step 7. **Stopping Criterion:** The stopping criterion is defined as the maximum number of iterations conducted without updating the ND solutions list and is set to 1,000. If the stopping criterion is satisfied, stop and return the ND solution list. Otherwise, go to Step 3.

2.3.1.3 Computational Experiments

The test problem considered herein uses the input parameters of the problem originally proposed by Fyffe et al. [20]. This problem includes fourteen 1-out-of-n subsystems connected in series. For each subsystem i, $n_{\max,I} = 8$ and the number of available component types (m_i) varies between 3 and 4. In the original problem, the upper bounds on the system cost and weight were specified as 130 and 170 units, respectively. However, these constraint limits are not used in this work. The Tabu Search for Multiobjective Combinatorial Optimization algorithm (TAMOCO) of Hansen [25] was used for comparison. Both algorithms were coded in C and run using an Intel Pentium IV with 2.2 GHz processor and 1 GB RAM. To equally compare the MTS algorithm with TAMOCO, all general TS parameters, such as encoding, stopping criterion, tabu list structure and length, were set to the same values in both algorithms. For TAMOCO, the set of current solutions, S, is set to 10, and the DRIFT criterion is set after 250 iterations for each current solution, similar to the diversification of Step 6 given above.

Wu and Azarm [58] provides a survey of the metrics for quality assessment of final Pareto fronts. In the multiobjective optimization literature, two final Pareto fronts P_A and P_B found by two different algorithms A and B are usually compared using a coverage measure as follows [65]:

$$C(P_A, P_B) = \frac{|\{\mathbf{y} \in P_B ; \exists \mathbf{x} \in P_A : \mathbf{x} \succ \mathbf{y}\}|}{|P_B|} \tag{4}$$

$C(P_A, P_B)$ gives the ratio of the solutions in Pareto front P_B which are dominated by at least one solution in Pareto front P_A. Note that $C(P_A, P_B) \neq C(P_B, P_A)$. Therefore, both $C(P_A, P_B)$ and $C(P_B, P_A)$ should be provided to compare two Pareto fronts P_A and P_B.

$C(P_A, P_B)$ metric provides information about coverage but does not address range of objective function values. An ideal Pareto front set should spread over the maximum theoretical range of the objective space. Overall Pareto Spread measure (*OS*) can be used to compare two final Pareto fronts P_A and P_B as follows [58]:

$$OS(P_A, P_B) = \frac{\prod_{i=1}^{K} |\max_{\mathbf{x} \in P_A}(f_i(\mathbf{x})) - \min_{\mathbf{x} \in P_A}(f_i(\mathbf{x}))|}{\prod_{i=1}^{K} |\max_{\mathbf{x} \in P_B}(f_i(\mathbf{x})) - \min_{\mathbf{x} \in P_B}(f_i(\mathbf{x}))|} \tag{5}$$

If $OS(P_A, P_B) > 1$, P_A is said to be preferred to P_B with respect to the overall spread. One can calculate the spread measure with respect to each individual objective and compare final Pareto fronts based on each objective's spread.

In the TAMOCO algorithm, the set of current solutions, *S*, was initially set to 10. However, since the number of moves and CPU seconds are higher in the MTS algorithm, a larger *S*, 50, was tried. This increase did not improve the Pareto front identified by TAMOCO. Table 1 summarizes the results. The number of Pareto optimal solutions of TAMOCO is much less than those of MTS, and all solutions found by the former are dominated by the solutions found by the latter, that is $C(P_{MTS}, P_{TAMOCO})=1$ and $C(P_{TAMOCO}, P_{MTS})=0$ for $S=10$ and $S=50$. MTS also is much better with respect to the spread of the final Pareto fronts, $OS(P_{MTS}, P_{TAMOCO})=9.2$ for $S=10$ and $OS(P_{MTS}, P_{TAMOCO})=5.9$ for $S=50$.

Table 1. The Summary of the Results of the RAP with Three Objectives

	TAMOCO		MTS
	$S=10$	$S=50$	
Number of moves	143,810	557,100	424,576
Number of diversifications	-	-	1,454
Number of DRIFT	550	1,000	-
$\|P\|$	1,134	1,059	5,230
CPU time (seconds)	151.34	258.59	354.00

MTS compared very favorably to TAMOCO. MTS is simple, general and tractable, and results show that it is quite effective at identifying a wide Pareto front as well.

2.4 Multiobjective Genetic Algorithms

GA was first developed by Holland and his colleagues in the 1960s and 1970s. GA is inspired by the evolutionist theory explaining the origin of species [28]. GA operates with a collection of solutions, called the population. Initially, the population includes randomly generated solutions. As the search evolves, the population includes fitter and fitter solutions and eventually it converges, meaning that it is dominated by a single solution. Holland presented the schema theorem as the proof of convergence to the global optimum where a solution is represented by a binary vector. GA has been proven to be an effective search technique for many combinatorial optimization problems.

GA has two operators to generate new solutions from existing ones: crossover and mutation. In crossover, generally two solutions, called parents, are combined together to form new solutions, called offspring. The parents are selected from the population with preference towards good solutions so that offspring are expected to inherit good features that make the parents desirable. By repeatedly applying crossover, features of good solutions are expected to appear more frequently in the population, eventually leading to convergence to an overall good solution.

Mutation introduces random changes into characteristics of solutions. Mutation also plays a critical role in GA. As discussed earlier, crossover leads the population to converge by making the solutions in the population alike. Mutation reintroduces genetic diversity back into the population and helps the search avoid early convergence to local optima. Selection specifically determines solutions for the next generation. In the most general case, the fitness of a solution determines the probability of its survival for the next generation.

Being a population based approach, GA is well suited to find multiple solutions to a multiobjective problem in a single run. In addition, a single objective GA can be easily modified to solve multiobjective problems. Therefore, the majority of all metaheuristics approaches to multiobjective optimization are based on evolutionary approaches [32]. The first multiobjective GA, called Vector Evaluated Genetic Algorithms (or VEGA), was proposed by Schaffer [55]. Afterward, several multiobjective evolutionary algorithms were developed including: Multiobjective Genetic Algorithm

(MOGA) [19], Niched Pareto Genetic Algorithm (NPGA) [29], Weight-Based Genetic Algorithm (WBGA) [24], Random Weighted Genetic Algorithm (RWGA) [46], Nondominated Sorting Genetic Algorithm (NSGA) [57], Strength Pareto Evolutionary Algorithm (SPEA) [64], Improved SPEA (SPEA2) [63], Pareto-Archived Evolution Strategy (PAES) [34], Pareto Envelope-based Selection Algorithm (PESA) [8], Region-based Selection in Evolutionary Multiobjective Optimization (PESA-II) [7], Fast Non-dominated Sorting Genetic Algorithm (NSGA-II) [11], Multiobjective Evolutionary Algorithm (MEA) [53], Micro-GA [6], Rank-Density Based Genetic Algorithm (RDGA) [41], and Dynamic Multiobjective Evolutionary Algorithm (DMOEA) [60].

2.4.1 Multiobjective GA Approaches to Reliability Optimization

Deb et al. [10] formulated the placement of electronic components on a printed circuit board as a bi-objective optimization problem where the objectives are minimizing the total wire-length and minimizing the overall component failure rate due to the heat generated by components. In this problem, the total wire-length can be minimized by placing components closer. However, closely placed components generate heat that may cause component and overall system failures. The formulated problem was solved by the NSGA-II. The final solutions were analyzed to understand the trade-offs between reliability and printed circuit board design.

The reliable network design problem has also been studied using multiobjective GA. Kumar et al. [37] presented a multiobjective GA approach to design telecommunication networks while simultaneously minimizing network performance and design costs under a reliability constraint. Kim and Gen [33] studied bicriteria spanning tree networks with minimum cost and maximum reliability.

Elegbede and Adjallah [18] formulized a multiobjective problem to simultaneously optimize the availability and the cost of repairable parallel-series systems. The formulated problem was transformed into a single objective optimization problem using the weighted sum technique and penalty approach, and the reduced problem was solved with a GA. Marseguerra et al. [42] determined optimal surveillance test intervals using a multiobjective GA with the goal of improving reliability and availability. Their research implemented a multiobjective GA which explicitly accounts for the uncertainties in the parameters. The objectives considered were the inverse of the expected system failure probability and the inverse of its variance. These are used to drive the genetic search toward solutions which are guaranteed to give optimal performance with high assurance, i.e., low estimation variance.

Martorell et al. [44] studied the selection of technical specifications and maintenance activities at nuclear power plants to increase reliability, availability and maintainability (RAM) of safety-related equipment. However, to improve RAM, additional limited resources (e.g., budget and work force) are required, posing a multiobjective problem. They demonstrated the viability and significance of their proposed approach using multiobjective GA for an emergency diesel generator system. Additionally, Martorell et al. [43] considered the optimal allocation of more reliable equipment, testing and maintenance activities to assure high RAM levels for safety-related systems. For these problems, the decision-maker encounters a multiobjective optimization problem where the parameters of design, testing and maintenance are decision variables. Solutions were obtained by using both single-objective GA and multiobjective GA, which were demonstrated to solve the problem of testing and maintenance optimization based on unavailability and cost criteria.

Sasaki and Gen [54] introduced a multiobjective problem which had fuzzy multiple objective functions and constraints with a Generalized Upper Bounding (GUB) structure. They solved this problem using a new hybridized GA. This approach leads to a flexible optimal system design by applying fuzzy goals and constraints. A new chromosome representation was introduced in their work. To demonstrate the effectiveness of their method, a large-scale optimal system reliability design problem was analyzed.

Busacca et al. [5] proposed a multiobjective GA based on raking of population. They applied this GA to determine optimal test intervals of the components of a complex system in a nuclear plant.

Reliability allocation to minimize total plant costs, subject to an overall plant safety goal, was presented by Yang et al. [59]. For their problem, design optimization was needed to improve the design, operation and safety of new and/or existing nuclear power plants. They presented an approach to determine the reliability characteristics of reactor systems, subsystems, major components and plant procedures that are consistent with a set of top-level performance goals. To optimize the reliability of the system, the cost for improving and/or degrading the reliability of the system was also included in the reliability allocation process creating a multiobjective problem. GA was applied to the reliability allocation problem of a typical pressurized water reactor.

2.5 Multiobjective Optimization of Telecommunication Networks Considering Reliability

Recent developments in the telecommunications technology, such as high capacity fiber-optic cables, have provided economical benefits in terms of capacity concentration. As a result, modern telecommunication networks tend to have sparse topologies with a few alternative paths between communication centers. In such networks, catastrophic events, such as the total lost of links or nodes, will have profound effects on performance and connectivity of the network. Therefore, network survivability and reliability have become important concerns while designing telecommunication networks. This section presents an implementation of a multiobjective GA to the network design problem considering survivability and reliability. The multiobjective network design problem is formulated as follows:

Objectives:

- To design a network with minimum cost and maximum all-terminal reliability

Given:

- Geographical distribution of communication nodes
- Cost of installing a link between nodes *i* and *j*

Constraints:

- 2-node connectivity

Decision Variables:

- Network topology

Assumptions:

- Links can be either in two states, operative or failed, with known probabilities.
- Link failures are independent.
- No parallel link exists between nodes.
- Nodes are assumed to be perfectly reliable.
- No repair is considered.

2.5.1 Network Survivability and Reliability

The term “network survivability” refers to as the ability of a network to continue network services in case of component failures. Traditionally, the survivability of a network is achieved by imposing redundant paths between communication centers. *K*-link and *k*-node connectivity measures are frequently used in practice to express survivability of networks. A network topology is said to be k-link (k-node) connected if the network is still connected after removing of any k-1 links (k-1 nodes) from the network. For many real-life network applications, a level of redundancy that provides connectivity, in the case of a single link or node failure, is sufficient [23].

Providing connectivity is the ultimate function of a network. If the failure probabilities of the components of a network are known, then the probability that the network provides the connectivity service is an important design criterion. In network reliability analysis, a network with unreliable components is generally modeled as a stochastic binary system of which each component can be independently either in two states: operative or failed. Then, network reliability is defined as the probability that the desired connectivity is achieved. For example, *all-terminal reliability*, which is defined as the probability that all node pairs are communicating, is often used to evaluate the ability of networks to provide connectivity service. Consider a network G(V, E) with node set V and link set E. Let x_{ij} denote the state of link (i, j) such that $x_{(i,j)}=1$ if link (i, j) is operative with probability p_{ij} and $x_{(i,j)}=0$ with probability $1-p_{ij}$, otherwise. The probability of observing a particular network state $\mathbf{X}=\{x_{ij}:(i,j)\in E\}$ is given as

$$\Pr\{\mathbf{X}\}=\prod_{(i,j)\in E}[1-p_{ij}+x_{(i,j)}(2p_{ij}-1)] \tag{6}$$

Then, the all-terminal reliability of a network is given as follows:

$$R=\sum_{\mathbf{X}}\Phi(\mathbf{X})\Pr\{\mathbf{X}\} \tag{7}$$

where the structure function Φ for all-terminal reliability maps a given state vector $\mathbf{X}$ on the state of the network as follows: $\Phi(\mathbf{X})=1$ if all nodes of the network are connected to each other in state $\mathbf{X}$, or $\Phi(\mathbf{X})=0$, otherwise. Computing the exact reliability is known to be an NP-hard problem [2] and requires substantial computational effort even for small size networks. Therefore, simulation and other approximation techniques are frequently used to estimate network reliability. In this study, the Sequential Construction Method of Easton and Wong [17], which is an efficient simulation approach to network reliability, is used to estimate all-terminal reliability.

2.5.2 Multiobjective Elitist GA with Restricted Mating

The multiobjective GA approach described in this section is called Multiobjective Elitist GA with Restricted Mating (MEGA-RM) because of the distinct features as follows:

- The population includes only nondominated solutions.
- A single solution is generated at a time.
- The population size is dynamic and limited by an upper bound (μ_{max}).
- Multiple copies of the same solution are not allowed in the population.
- In crossover, a mating restriction is enforced. A solution is not allowed to mate with solutions outside its neighborhood which is defined in the objective space.
- The crowding distance measure of the NSGA-II [11] is used as the diversification mechanism.

2.5.2.1 Problem Encoding

The first step in any GA implementation is to develop an encoding of the problem. GA's crossover and mutation operators operate on the encoding of the problem, rather than the problem itself. A node adjacency matrix representation is used to represent solutions. In this representation, a network is stored in an $n \times n$ matrix, $\mathbf{x}=\{x_{ij}\}$, where n is the number of nodes, such that $x_{ij}=1$ if there exists a link between nodes i and j and $x_{ij}=0$, otherwise.

2.5.2.2 Crossover Operator

The function of crossover is to generally combine two solutions, called parents, together to form new solutions, called offspring. The parents are selected from the population with preference towards good solutions so that offspring is expected to inherit good features that make the parents desirable.

A uniform crossover operator is used. Two existing solutions $\mathbf{x}$ and $\mathbf{y}$ are randomly selected from the population without replacement, and offspring $\mathbf{z}$ randomly inherits links from either of the parents, one at a time, as follows: for $i=1,...,n$ and $j=i+1,...,n$ if $U[0,1] > 0.5$ then $z_{ij} = x_{ij}$, else $z_{ij} = y_{ij}$. As a result of uniform crossover, it is possible that the offspring is not 2-node connected or not even connected at all. A repair algorithm is used to produce an acceptable topology. The repair algorithm identifies the node cut sets that violate the connectivity requirements, and then the shortest link across such cuts are added to the offspring for repair.

2.5.2.3 Mutation

As discussed earlier, crossover leads the population to converge by making the solutions in the population alike. Mutation reintroduces genetic diversity back into the population and helps the search avoid early convergence to local optima. The MEGA-RM has three mutation operators perturbing an offspring without violating the connectivity constraint. The *add-a-link* mutation operator simply adds a non-existing link to a solution. The *remove-a-link* operator removes a link from a solution without violating the connectivity constraint. To achieve this, a random cycle C is found and a link (a, b) between the nodes of cycle C (excluding the ones on the cycle) is removed. The *swap-a-link* operator removes link (a, b) and adds another link (c, d) between the nodes of a randomly selected cycle C.

2.5.2.4 Overall Algorithm

The MEGA-RM starts with μ_0 non-dominated solutions randomly generated using a procedure based on the two-tree heuristic [45] as follows:

Step 1. Find a random spanning tree of the nodes.
Step 2. Form a second random spanning tree spanning only the extreme nodes of the first spanning tree (nodes with a single link).
Step 3. Randomly add links that have not been used in Steps 1 and 2 to the solution with a probability p which is uniformly selected from 0 to 0.5. Steps 1 and 2 generate a variety of sparse 2-node connected random networks. In this step, additional links are added to obtain random non-dominated solutions with different density. This way, a wide spectrum of the initial non-dominated solutions might be sampled.

In each generation of the MEGA-RM, a new solution is produced by crossover and/or mutation. The crowding distance measure of the NSGA-II is used to ensure a diverse set of final solutions. Since the population of the MEGA-RM includes only ND solutions, the solutions are always maintained in the ascending order of cost (which is also the descending order of reliability). Therefore, the crowding distance of the ith solution in the population ($\mathbf{x}_{[i]}$) can easily be calculated as follows:

$$\mathbf{x}_{[i]}.cd = \frac{C(\mathbf{x}_{[i+1]}) - C(\mathbf{x}_{[i-1]})}{C(\mathbf{x}_{[\mu]}) - C(\mathbf{x}_{[0]})} + \frac{R(\mathbf{x}_{[i+1]}) - R(\mathbf{x}_{[i-1]})}{R(\mathbf{x}_{[\mu]}) - R(\mathbf{x}_{[0]})} \tag{8}$$

where $\mathbf{x}_{[0]}.cd = +\infty$ and $\mathbf{x}_{[\mu]}.cd = +\infty$. When a new non-dominated solution is found, it is inserted into the population while maintaining the order of the solutions. Therefore, the crowding distance is calculated only for the new dominated solution and its immediate predecessor and successor.

The first parent for crossover is selected using a binary tournament selection based on the crowding distance, and the second parent is selected from the neighborhood of the first solution. In the sorted population, the neighborhood of the ith solution in the population is defined as follows:

$$N(\mathbf{x}_{[i]}) = \{\mathbf{x}_{[j]} : \max\{0, i-a\} \le j \le \min\{\mu, i+a\}, i \ne j\} \tag{9}$$

where a is an integer number defining the size of the neighborhood. Hajela and Lin [24] suggested that such restricted mating in multiobjective GA usually results in improved performance. The overall algorithm of the MEGA-RM is given as follows:

Step 1. Randomly generate μ_0 initial solutions.

Step 2. **Crossover:**

- Step 2.1. Randomly select two solutions and set the solution with the higher crowding distance as the first parent. Let i be the index of the first parent.
- Step 2.2. Randomly and uniformly select a solution from neighborhood $N(\mathbf{x}_{[i]})$.
- Step 2.3. Apply uniform crossover to generate offspring $\mathbf{z}$ and repair it if necessary.

Step 3. **Mutation:**

- Step 3.1. Generate UNI, a uniformly distributed random variable between 0 and 1. If UNI > *mutation rate,* skip mutation and go to Step 4.
- Step 3.2. Randomly and uniformly select a mutation operator and apply that mutation operator on offspring $\mathbf{z}$.

Step 4. Evaluate offspring using simulation [17] and compare it to the population as follows.

- Step 4.1. If offspring $\mathbf{z}$ is dominated by an existing solution or the same solution exists in the population, discard it.
- Step 4.2. If offspring $\mathbf{z}$ dominates a set of solutions, remove those solutions and insert $\mathbf{z}$ into the population, calculate its and its immediate neighbors' crowding distances using equation (8).
- Step 4.3. If offspring $\mathbf{z}$ neither dominates nor is dominated, add offspring to the population and calculate its and its immediate neighbors' crowding distances using equation (6).

Step 5. If $\mu > \mu_{max}$, remove the solution with the smallest crowding distance from the population, update the crowding distances of the immediate neighbors of the removed solution.

Step 6. $g = g+1$. If $g > g_{max}$, stop. Otherwise, go to Step 2.

2.5.3 Computational Experiments

The MEGA-RM was compared to the NSGA-II on three problems with 15, 20, and 25 nodes from the literature [12; 13]. The NSGA-II used the same crossover and mutation operators described above for the MEGA-RM. For each problem, both algorithms were run five times, each with a different random number seed. The parameters for the MEGA-RM are: $g_{max}=50{,}000$, $\mu_0=5$, $\mu_{max}=50$ for the 15-node problem and $\mu_0=10$, $\mu_{max}=100$ for the 20 and 25-node problems, and mutation rate=0.50. The same parameters are also used for the NSGA-II, μ_{max} being the population size.

The final Pareto fronts found by MEGA-RM in five replications were compared to ones found by the NSGA-II using the coverage measure given in equation (4). Therefore, this approach resulted in 25 comparisons for each problem. In Table 2, $(a; b)_{ij}$ represents the calculated coverage values $(C(P_{MEGA\text{-}RM}, P_{NSGA}); C(P_{NSGA}, P_{MEGA\text{-}RM}))$ for the ith run of MEGA-RM and the jth run of the NSGA-II.

Table 2. Comparison of the NSGA-II and the MEGA-RC on the test problems.

15-node	NSGA				
MEGA-RM	1	2	3	4	5
1	(0.71; 0.00)	(0.24; 0.06)	(0.61; 0.00)	(0.06; 0.14)	(0.15; 0.10)
2	(0.67; 0.00)	(0.26; 0.08)	(0.65; 0.00)	(0.04; 0.14)	(0.15; 0.12)
3	(0.78; 0.00)	(0.17; 0.08)	(0.65; 0.04)	(0.02; 0.16)	(0.13; 0.08)
4	(0.64; 0.02)	(0.15; 0.08)	(0.43; 0.04)	(0.02; 0.12)	(0.02; 0.12)
5	(0.69; 0.00)	(0.22; 0.04)	(0.61; 0.00)	(0.02; 0.08)	(0.15; 0.06)
20-node	NSGA				
MEGA-RM	1	2	3	4	5
1	(1.00; 0.00)	(1.00; 0.00)	(1.00; 0.00)	(1.00; 0.00)	(1.00; 0.00)
2	(1.00; 0.00)	(0.99; 0.00)	(1.00; 0.00)	(1.00; 0.00)	(1.00; 0.00)
3	(1.00; 0.00)	(0.99; 0.00)	(0.99; 0.00)	(0.94; 0.01)	(1.00; 0.00)
4	(1.00; 0.00)	(0.99; 0.00)	(1.00; 0.00)	(0.98; 0.00)	(1.00; 0.00)
5	(1.00; 0.00)	(1.00; 0.00)	(1.00; 0.00)	(0.99; 0.00)	(1.00; 0.00)
25-node	NSGA				
MEGA-RM	1	2	3	4	5
1	(1.00; 0.00)	(1.00; 0.00)	(1.00; 0.00)	(1.00; 0.00)	(1.00; 0.00)
2	(1.00; 0.00)	(1.00; 0.00)	(1.00; 0.00)	(1.00; 0.00)	(1.00; 0.00)
3	(1.00; 0.00)	(1.00; 0.00)	(0.98; 0.00)	(1.00; 0.00)	(1.00; 0.00)
4	(0.95; 0.02)	(0.95; 0.01)	(1.00; 0.00)	(1.00; 0.00)	(1.00; 0.00)
5	(0.67; 0.16)	(0.80; 0.07)	(0.96; 0.00)	(1.00; 0.00)	(1.00; 0.00)

Note that $C(P_A, P_B) \neq C(P_B, P_A)$. Therefore, both $C(P_A, P_B)$ and $C(P_B, P_A)$ are provided to compare two Pareto fronts P_A and P_B.

As seen in Table 2, the MEGA-RM is clearly superior to the NSGA-II. Especially, for the 20 and 25-node problems, the final Pareto fronts found by the NSGA-II are almost fully covered by the Pareto fronts of the MEGA-RM. In terms of distribution of the final Pareto fronts, MEGA-RM was also superior. However, overall spread results are not given herein for sake of briefity.

2.6 Other Metaheuristic Techniques to Multiobjective Reliability Optimization

2.6.1 Ant Colony Optimization

Ant colony optimization (ACO) was first proposed by Dorigo et al. [15; 16] and inspired by the behavior of real ants while searching for food. Since then, ACO approaches have been successfully applied to many combinatorial optimization problems, including reliability optimization [40] [56]. ACO is different than other metaheuristics approaches in the way that candidate solutions are created. In GA, TS, or SA, candidate solutions are created from current solutions using various operators while in ACO, entire candidate solutions are randomly constructed from scratch at each iteration. Each ant constructs a new single solution, and a local search is used to improve these new solutions. Therefore, ACO can be considered as a construction metaheuristic rather than improvement one. Shelokar et al. [56] proposed a multiobjective ACO to determine solutions to the redundancy allocation problem (RAP). In this approach to RAP, each ant of the colony constructs a solution by visiting subsystems (s=1,...,N) and randomly selecting the number of redundant components in each subsystem with a probability distribution p_{is} which denotes the probability of selecting total i components in subsystem s. After evaluating the performance of each ant in the colony, the selection probabilities are updated such that p_{is} values are increased for the selection decisions made by the best performing ants and reduced for the decisions made by low performing and infeasible ants. This probability update procedure mimics pheromone trails deposited by ants to determine shortest path to a food source (i.e., the shortest route will have the highest pheromone concentration). To solve a multiobjective problem in [56], the ants were ranked using the strength Pareto fitness assignment approach [64].

2.6.2 Simulated Annealing

Simulated annealing (SA) is based on ideas from physical annealing of metals and has been successfully applied to many difficult combinatorial optimization problems. Similar to TS, a neighborhood of the current solution is produced by slightly perturbing a current solution. While selecting the new current solution in the neighborhood, inferior solutions are allowed based on a probability function in form of $e^{-\Delta/t}$ where Δ is the change in the objective function due to accepting a new solution and t is the temperature parameter. Initially, the temperature is set to a high value for more stochastic selection and it is gradually reduced for more deterministic selection toward the end of the search.

Zafiropoulos and Dialynas [61] used a multiobjective SA to design electronic devices where component failure rates are uncertain. In this approach, a single objective SA was modified to simultaneously maximize reliability and minimize cost. Hao-Zhong et al. [26] proposed a hybrid multiobjective GA/SA for electric power network planning considering reliability and cost.

Dhingra [14] presented an application of goal programming to find Pareto optimal solutions for a reliability apportionment problem with fuzzy objectives and constraints. Fuzzy sets were used to model incomplete and imprecise information regarding the problem parameters.

2.6.3 Conclusions

Metaheuristic models have been particularly effective for the practical solution of multiobjective system reliability optimization problems. For many reliability design engineering problems, the decision maker has to solve multiobjective problems to find at least one feasible solution that can be implemented in the system design. Multiobjective problems are often solved by consolidating the objective functions into equivalent single objective problems using pre-defined weights or utility functions, or by the determination or approximation of a Pareto optimal set. Methods involving a single consolidated objective function can be problematic because assigning appropriate numerical values (i.e., weights) to an objective function can be challenging for many practitioners. Alternatively, methods such as genetic algorithms and tabu search provide attractive alternatives to yield non-dominated Pareto optimal solutions.

Multiobjective formulations are realistic models for many complex engineering optimization problems. The objectives under consideration often conflict with each other, and optimizing a particular solution with respect to a single objective can result in unacceptable results with respect to the other objectives. A reasonable solution to a multiobjective problem is to investigate a set of solutions, each of which satisfies the objectives at an acceptable level without being dominated by any other solution. Pareto optimal solution sets are often preferred to single solutions because they can be practical when considering real-life problems since the final solution of the decision maker is always a trade-off. Pareto optimal sets can be of varied sizes. Determination of an entire Pareto optimal solution set or a representative subset can be readily achieved using available metaheuristics such as GA, TS, SA, ACO, etc. The multiple objective variations of these metaheuristics differ primarily by using specialized fitness functions and introducing methods to promote solution diversity.

References

1. Alves MJ, Climaco J (2000) An interactive method for 0-1 multiobjective problems using simulated annealing and tabu search. Journal of Heuristics 6:385-403.
2. Ball M (1980) Complexity of Network Reliability Calculation. Networks 10:153-165.
3. Baykasoglu A (2001) Goal programming using multiple objective tabu search. The Journal of the Operational Research Society 52:1359-69.
4. Baykasoglu A (2001) MOAPPS 1.0: aggregate production planning using the multiple-objective tabu search. International Journal of Production Research 39:3685-702.
5. Busacca PG, Marseguerra M, Zio E (2001) Multiobjective optimization by genetic algorithms: Application to safety systems. Reliability Engineering and System Safety 72:59-74.
6. Coello CAC (2000) An updated survey of GA-based multiobjective optimization techniques. ACM Computing Surveys 32:109-43.
7. Corne D, Jerram NR, Knowles J, Oates J (2001) PESA-II: Region-based Selection in Evolutionary Multiobjective Optimization. In: Proceedings of the Genetic and Evolutionary Computation Conference (GECCO-2001) San Francisco, California, pp 283-290.
8. Corne DW, Knowles JD, Oates MJ (2000) The Pareto envelope-based selection algorithm for multiobjective optimization. In: Proceedings of 6th Interna-

tional Conference on Parallel Problem Solving from Nature, 18-20 Sept. 2000, Springer-Verlag, Paris, France, pp 839-48.
9. Deb K (2001) Multi-Objective Optimization using Evolutionary Algorithms. John Wiley & Sons, Ltd.
10. Deb K, Jain P, Gupta NK, Maji HK (2004) Multiobjective placement of electronic components using evolutionary algorithms. IEEE Transactions on Components and Packaging Technologies 27:480-92.
11. Deb K, Pratap A, Agarwal S, Meyarivan T (2002) A fast and elitist multiobjective genetic algorithm: NSGA-II. IEEE Transactions on Evolutionary Computation 6:182-97.
12. Dengiz B, Altiparmak F, Smith AE (1997) Efficient optimization of all-terminal reliable networks, using an evolutionary approach. IEEE Transactions on Reliability 46:18-26.
13. Dengiz B, Altiparmak F, Smith AE (1997) Local search genetic algorithm for optimal design of reliable networks. IEEE Transactions on Evolutionary Computation 1:179-88.
14. Dhingra AK (1992) Optimal apportionment of reliability and redundancy in series systems under multiple objectives. IEEE Transactions on Reliability 41:576-82.
15. Dorigo M, Di Caro G (1999) Ant colony optimization: a new meta-heuristic. In: Proceedings of the 1999. Congress on Evolutionary Computation-CEC99, 6-9 July 1999, IEEE, Washington, DC, USA, pp 1470-7.
16. Dorigo M, Di Caro G, Gambardella LM (1999) Ant algorithms for discrete optimization. Artificial Life 5:137-72.
17. Easton MC, Wong CK (1980) Sequential destruction method for Monte Carlo evaluation of system reliability. IEEE Transactions on Reliability R-29:27-32.
18. Elegbede C, Adjallah K (2003) Availability allocation to repairable systems with genetic algorithms: a multi-objective formulation. Reliability Engineering & System Safety 82:319-30.
19. Fonseca CM, Fleming PJ (1993) Multiobjective genetic algorithms. In: IEE Colloquium on 'Genetic Algorithms for Control Systems Engineering' (Digest No. 1993/130), 28 May 1993, IEE, London, UK, pp 6-1.
20. Fyffe DE, Hines WW, Lee NK (1968) System Reliability Allocation and A Computational Algorithm. IEEE Transactions on Reliability 7:74-79.
21. Glover F (1989) Tabu search I. ORSA Journal on Computing 1:190-206.
22. Glover F (1990) Tabu search II. ORSA Journal on Computing 2:4-32.
23. Grötschel M, Monma CL, Stoer M (1995). Design of survivable networks. In: Network Models. Elsevier Science B.V., Amsterdam, pp 617-672.
24. Hajela P, Lin C-Y (1992) Genetic search strategies in multicriterion optimal design. Structural Optimization 4:99-107.
25. Hansen MP (2000) Tabu search for multiobjective combinatorial optimization: TAMOCO. Control and Cybernetics 29:799-818.
26. Hao-zhong C, Ci-wei G, Ze-liang M (2004) General optimized model of multi-objective electric power network planning. Journal of Shanghai Jiaotong University 38:1229-32.

27. Hertz A, Jaumard B, Ribeiro CC, Formosinho Filho WP (1994) A multi-criteria tabu search approach to cell formation problems in group technology with multiple objectives. RAIRO Recherche Operationelle 28:303-28.
28. Holland JH (1975) Adaptation in Natural and Artificial Systems. University of Michigan Press, Ann Arbor.
29. Horn J, Nafpliotis N, Goldberg DE (1994) A niched Pareto genetic algorithm for multiobjective optimization. In: Proceedings of the First IEEE Conference on Evolutionary Computation. IEEE World Congress on Computational Intelligence, 27-29 June 1994, IEEE, Orlando, FL, USA, pp 82-7.
30. Hwang CL, Tillman FA, Wei WK, Lie CH (1979) Optimal scheduled-maintenance policy based on multiple-criteria decision-making. IEEE Transactions on Reliability R-28:394-9.
31. Inagaki T, Inoue K, Akashi H (1978) Interactive optimization of system reliability under multiple objectives. IEEE Transactions on Reliability R-27:264-7.
32. Jones DF, Mirrazavi SK, Tamiz M (2002) Multiobjective meta-heuristics: an overview of the current state-of-the-art. European Journal of Operational Research 137:1-9.
33. Kim JR, Gen M (1999) Genetic algorithm for solving bicriteria network topology design problem. In: Proceedings of the 1999. Congress on Evolutionary Computation-CEC99, 6-9 July 1999, IEEE, Washington, DC, USA, pp 2272-9.
34. Knowles J, Corne D (1999) The Pareto archived evolution strategy: a new baseline algorithm for Pareto multiobjective optimisation. In: Proceedings of the 1999. Congress on Evolutionary Computation-CEC99, 6-9 July 1999, IEEE, Washington, DC, USA, pp 98-105.
35. Kulturel-Konak S, Norman BA, Coit DW, Smith AE (2004) Exploiting tabu search memory in constrained problems. INFORMS Journal on Computing 16:241-54.
36. Kulturel-Konak S, Smith AE, Coit DW (2003) Efficiently solving the redundancy allocation problem using tabu search. IIE Transactions 35:515-26.
37. Kumar R, Parida PP, Gupta M (2002) Topological design of communication networks using multiobjective genetic optimization. In: Proceedings of 2002 World Congress on Computational Intelligence - WCCI'02, 12-17 May 2002, IEEE, Honolulu, HI, USA, pp 425-30.
38. Kuo W, Prasad VR, Tillman FA, Hwang CL (2001) Optimal Reliability Design Cambridge University Press, Cambridge, U.K.
39. Kursawe F (1991) A variant of evolution strategies for vector optimization. In: Parallel Problem Solving from Nature. 1st Workshop, PPSN 1 Proceedings, 1-3 Oct. 1990, Springer-Verlag, Dortmund, West Germany, pp 193-7.
40. Liang Y-C, Smith AE (2004) An ant colony optimization algorithm for the redundancy allocation problem (RAP). IEEE Transactions on Reliability 53:417-23.
41. Lu H, Yen GG (2002) Rank-density based multiobjective genetic algorithm. In: Proceedings of 2002 World Congress on Computational Intelligence - WCCI'02, 12-17 May 2002, IEEE, Honolulu, HI, USA, pp 944-9.

42. Marseguerra M, Zio E, Podofillini L (2004) Optimal reliability/availability of uncertain systems via multi-objective genetic algorithms. IEEE Transactions on Reliability 53:424-34.
43. Martorell S, Sanchez A, Carlos S, Serradell V (2004) Alternatives and challenges in optimizing industrial safety using genetic algorithms. Reliability Engineering & System Safety 86:25-38.
44. Martorell S, Villanueva JF, Carlos S, Nebot Y, Sanchez A, Pitarch JL, Serradell V (2005) RAMS+C informed decision-making with application to multi-objective optimization of technical specifications and maintenance using genetic algorithms. Reliability Engineering and System Safety 87:65-75.
45. Monma CL, Shallcross DF (1989) Methods for designing communications networks with certain two-connected survivability constraints. Operations Research 37:531-41.
46. Murata T, Ishibuchi H (1995) MOGA: multi-objective genetic algorithms. In: Proceedings of 1995 IEEE International Conference on Evolutionary Computation, 29 Nov.-1 Dec. 1995, IEEE, Perth, WA, Australia, pp 289-94.
47. Sakawa M (1978) Multiobjective optimisation by the Surrogate Worth trade off method. IEEE Transactions on Reliability R-27:311-14.
48. Sakawa M (1978) Multiobjective reliability and redundancy optimization of a series-parallel system by the Surrogate Worth Trade-off method. Microelectronics and Reliability 17:465-7.
49. Sakawa M (1980) Decomposition approaches to large-scale multiobjective reliability design. Journal of Information & Optimization Sciences 1:103-20.
50. Sakawa M (1980) Multiobjective optimization for a standby system by the surrogate worth trade-off method. Journal of the Operational Research Society 31:153-8.
51. Sakawa M, Arata K (1978) Optimal reliability design of a series-parallel system with multiple objectives. Memoirs of the Faculty of Engineering, Kobe University 13-20.
52. Sakawa M, Arata K (1980) Reliability design of a series-parallel system by a large-scale multiobjective optimization method. Transactions of the Institute of Electronics and Communication Engineers of Japan, Section E (English) E63:252.
53. Sarker R, Liang K-H, Newton C (2002) A new multiobjective evolutionary algorithm. European Journal of Operational Research 140:12-23.
54. Sasaki M, Gen M (2003) A method of fuzzy multi-objective nonlinear programming with GUB structure by Hybrid Genetic Algorithm. International Journal of Smart Engineering System Design 5:281-288.
55. Schaffer JD (1985) Multiple Objective optimization with vector evaluated genetic algorithms. In: Proceedings of International Conference on Genetic Algorithm and their applications pp 93-100.
56. Shelokar PS, Jayaraman VK, Kulkarni BD (2002) Ant algorithm for single and multiobjective reliability optimization problems. Quality and Reliability Engineering International 18:497-514.

57. Srinivas N, Deb K (1994) Multiobjective Optimization Using Nondominated Sorting in Genetic Algorithms. Journal of Evolutionary Computation 2:221-248.
58. Wu J, Azarm S (2001) Metrics for Quality Assessment of a Multiobjective Design Optimization Solution Set Transactions of ASME 23:18-25.
59. Yang J-E, Hwang M-J, Sung T-Y, Jin Y (1999) Application of genetic algorithm for reliability allocation in nuclear power plants. Reliability Engineering & System Safety 65:229-38.
60. Yen GG, Lu H (2003) Dynamic multiobjective evolutionary algorithm: adaptive cell-based rank and density estimation. IEEE Transactions on Evolutionary Computation 7:253-74.
61. Zafiropoulos EP, Dialynas EN (2004) Reliability and cost optimization of electronic devices considering the component failure rate uncertainty. Reliability Engineering & System Safety 84:271-84.
62. Zitzler E, Deb K, Thiele L (2000) Comparison of multiobjective evolutionary algorithms: empirical results. Evolutionary Computation 8:173-95.
63. Zitzler E, Laumanns M, Thiele L (2001) SPEA2: Improving the strength Pareto evolutionary algorithm. Swiss Federal Institute Techonology, TIK-Rep 103.
64. Zitzler E, Thiele L (1999) Multiobjective evolutionary algorithms: a comparative case study and the strength Pareto approach. IEEE Transactions on Evolutionary Computation 3:257-71.
65. Zitzler E, Thiele L, Laumanns M, Fonseca CM, Da Fonseca VG (2003) Performance assessment of multiobjective optimizers: An analysis and review. IEEE Transactions on Evolutionary Computation 7:117-132.

Genetic Algorithm Applications in Surveillance and Maintenance Optimization

Sebastián Martorell, Sofía Carlos, José F. Villanueva

Department of Chemical and Nuclear Engineering, Polytechnic University of Valencia, Spain

Ana Sánchez

Department of Statistics and Operational Research, Polytechnic University of Valencia, Spain

3.1 Introduction

Safety of industrial installations has been a matter of concern to many experts for more than five decades (McCormick 1981). The relevance of this topic in the normal operation of industrial installations can be realized for example analyzing the cost associated with their safety-related systems devoted to prevent or mitigate accidents, which ranges from 2-5% to 30-40%, the later for example at nuclear and chemical plants (Martorell et al. 2004).

Many studies have been developed in this period within the nuclear field aimed at improving safety systems, with the main focus on developing designs that use more reliable and redundant equipment (e.g. reliability allocation) (Yang et al. 1999) and implementing surveillance and maintenance policies (e.g. testing and maintenance optimization) more effective in terms of risk control and resource expenditures (Vesely 1999).

In particular, optimization of surveillance tests and maintenance activities (T&M) is being considered in the framework of several on-going programs at Nuclear Power Plants (NPPs) (e.g. risk-informed technical specifications, reliability centered maintenance, maintenance rule implementation, etc.), with the main focus on the improvement of the operation and maintenance of safety-related equipment (SRE) to allow both flexibility and safety in the operation of NPPs (Harunuzzaman and Aldemir 1996).

S. Martorell et al.: *Genetic Algorithm Applications in Surveillance and Maintenance Optimization*, Computational Intelligence in Reliability Engineering (SCI) **39**, 63–99 (2007)
www.springerlink.com

Nowadays, it can be said that T&M optimization of SREs of NPPs has to be developed in the framework of the so called RAMS+C (Reliability, Availability, Maintainability, Safety and Costs) informed applications (Martorell et al. 2005). Thus, the optimization goal is to make T&M activities more risk-effective and less costly by using all or partially the RAMS+C information to focus better on what is risk and cost important.

There are several T&M activities in plant associated with controlling risk or with satisfying requirements, such as Technical Specification (TS) including Surveillance Test Intervals (STI) and Allowed Outage Times (AOT), In-Service Test (IST), In-Service Inspection (ISI), Condition Based and Time-Directed Preventive Maintenance (CBPM and TDPM respectively), etc., which are candidate to be evaluated for their resource effectiveness in RAMS+C-informed optimization problems.

Optimization of T&M normally enters two types of problems: Single-objective Optimization Problem (SOP) and Multi-objective Optimization Problem (MOP). In the former, the optimization problem can be formulated in terms of an objective function, to be either minimized or maximized, under constraints that apply for a given scope (i.e. component, system or plant performance). The MOP involves the simultaneous optimization of the global set of normally conflicting objective functions subject to a set of constraints. In general, an MOP admits multiple solutions. Therefore, the decision-maker must ultimately make a value judgment among the options to arrive at a particular decision. The solutions of a MOP problem are always situated in its Pareto optimal (non-dominated) set. The SOP corresponds to a particular case of an MOP that provides a unique solution.

In optimizing T&M based on RAMS+C information, like in many engineering optimization problems (e.g. design, etc.), one normally faces multi-modal and non-linear objective functions and a variety of both linear and non-linear constraints. In addition, quantification of such objective functions and constraints requires the development of analytical or numerical models which often depend on time. This results in a complex and discrete search space with regions of feasible and unfeasible solutions for a discontinuous objective function that eventually presents local optima. For such problems, it is desired to find global optima not violating any constraint. In addition, requirements such as continuity and differentiability of both normally conflicting objective and constraints functions add yet another element to the decision process.

Methods based on Evolutionary Algorithms (EAs) have been investigated in order to try to solve such multi-modal non-linear problems that may exhibit numerical and combinatorial features at the same time. Genetic Algorithms (GAs) are very likely to be the most widely known type of Evolutionary Algorithms. GAs are adaptive methods that can be used in searching and optimization, which possess many characteristics desirable to face above optimization problems, both SOP and MOP.

The introduction of the use of GAs in the field of T&M optimization based on RAMS+C information dates less than a decade. Muñoz et al. 1997 were the first to propose the use of GAs as an optimization tool for surveillance and maintenance scheduling. Here the GAs are firstly introduced. Since then, the number of papers covering this topic has grown exponentially during the last years. Among their relevant characteristics it is worth highlighting once more that they do not require the computation of derivatives to perform their search. This is particularly important when the objective functions are computed via simulation. In addition, the development of improved GAs has allowed formulating more complex MOP and improving the reliability of the obtained solutions and at the same time reducing the computational effort.

This chapter presents a state of the art review of the use of Genetic Algorithms in surveillance and maintenance optimization based on quantifying their impact on RAMS+C using both analytical and simulation models. It begins with the analysis of published papers covering this area of research. It follows an overview of surveillance and maintenance activities, that is, types of activities and relevant characteristics to be taken into account in optimization problems. Next, the underlying multiple-criteria decision-making is introduced that is then formulated in terms of an optimization problem as either a SOP or MOP. Based on such a classification, the most commonly adopted criteria for the decision-making are reviewed from the side of the models used, that is, analytical and numerical models (time-dependent or not). Next, the chapter provides an overview of the evolution and the fundamentals of the GAs applied to solve such problems, which are divided into two main groups SOGA (Single Objective GA) and MOGA (Multiple Objectives GA). The chapter concludes with the introduction of future research and application areas.

Acronyms and notation

A	Availability
A[U]+C	Availability, or unavailability, and Cost
AOT	Allowed Outage Time
CBPM	Condition-Based Preventive Maintenance
CBRM	Condition-Based Replacement Maintenance
C	Cost
CM	Corrective Maintenance
EA	Evolutionary Algorithm

EDG	Emergency Diesel Generator
GA	Genetic Algorithm
ISI	In-Service Inspection
IST	In-Service Test
LCO	Limiting Condition for Operation
M	Maintainability
MC	Monte Carlo procedure
MC-GA	Monte Carlo procedure coupled with a Genetic Algorithm
MCDM	Multiple Criteria Decision Making
MOGA	Multiple Objectives GA
MOP	Multi-objective Optimization Problem
NPP	Nuclear Power Plant
OFRM	On-Failure Replacement Maintenance
PdM	Predictive Maintenance
PM	Preventive Maintenance
PRA	Probabilistic Risk Analysis
R	Reliability
RAM	Reliability, Availability, Maintainability
RAM+C	Reliability, Availability, Maintainability and Cost
RAMS+C	Reliability, Availability, Maintainability, Safety and Cost
RM	Replacement Maintenance
S	Safety
SOGA	Single Objective GA
SOP	Single-objective Optimization Problem
SR	Surveillance Requirement
SRE	Safety Related Equipment
STI	Surveillance Test Interval
T&M	Surveillance Testing and Maintenance
TDPdM	Time-Directed Predictive Maintenance
TDPM	Time-Directed Preventive Maintenance
TS	Technical Specification
U	Unavailability

Symbols

α	ageing rate
τ	downtime for testing
σ	downtime for preventive maintenance
γ	downtime for replacement maintenance
μ	downtime for corrective maintenance
ρ	on demand failure probability

λ_0	residual standby failure rate
λ_D	degradation rate
w	component age
D	allowed outage time
f	fraction of failures coming after degradation
T	test interval
M	time-directed preventive maintenance interval
L	time-directed replacement maintenance interval
$Pr\{\text{Degr}\}$	Degradation probability
$Pr\{\text{Fail}\}$	Failure probability
oF	= {CM, RM}
oD	= {PM, RM}

3.2 Analysis of Published Research

Muñoz et al. 1997 were the first to propose the use of GAs as an optimization tool for testing and maintenance scheduling. Since then, the number of papers covering this topic has grown exponentially during the last years. Fig. 1 summarizes a survey of published research in the area of GA applications in testing and maintenance optimization within the most significant scientific journals linked to the nuclear field.

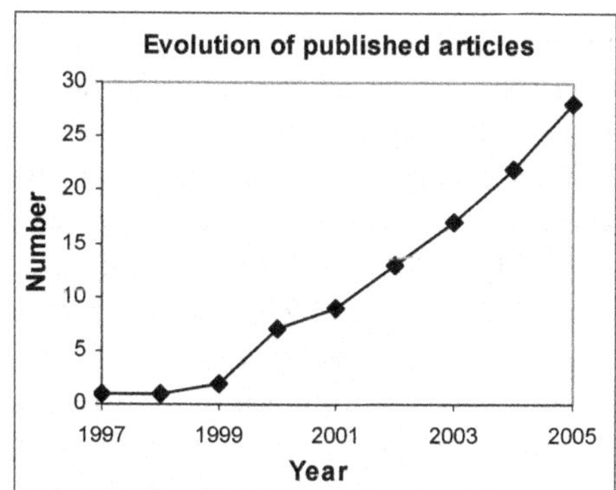

Fig. 1. Evolution of articles published on GA applications in T&M optimization

Fig. 2.A summarizes the published research grouped according to the particular application area with respect to testing and maintenance activities. Thus, almost 55% or the articles covers the topic of optimizing surveillance test of safety related equipment and other 50% are devoted to time-directed preventive maintenance optimization. Both represent classical application areas. On the other hand, inspection and condition-based preventive maintenance, the latter linked also to predictive maintenance, represent a lower percentage that is expected to be increased in the coming years as there is a current trend at NPP to replace TDPM with CBPM when applicable. In addition, corrective maintenance (CM) is another classical optimization area since optimization of CM very often enters simultaneously optimization problems of testing and preventive maintenance activities and it is rarely accounted for individually.

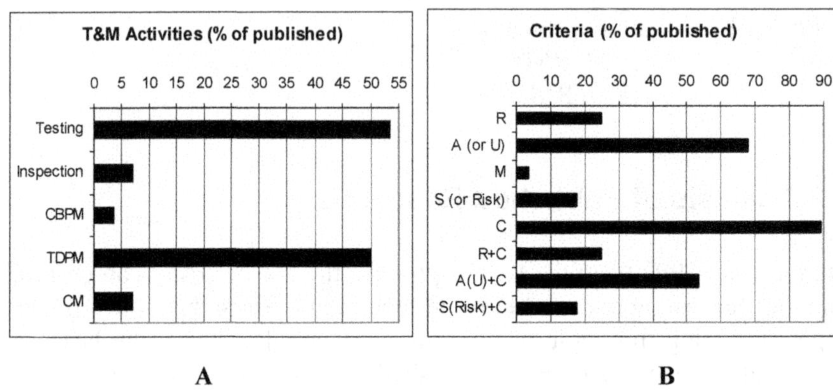

Fig. 2. Percentage of published work.
A: according to T&M activities; B: according to decision criteria

Figure 2.B summarizes the published research grouped according to the RAMS+C optimization criteria adopted for the decision-making. One can observe how almost 90% considers T&M related costs (C). Decision-making is normally based on multiple criteria that involve reliability (R) or availability/unavailability (A/U) in addition to costs as normally conflicting decision criteria. Thus, A+C and R+C combinations of decision criteria are considered in near the 55% and 25% of the articles respectively.

It is also worthy to note how safety (S) (or risk) criterion enters the formulation of surveillance and maintenance optimization problems by GA in recent years to join the cost criterion in the multi-criteria decision-making process. Thus, for potentially hazardous and risky industries, such as the nuclear one, decision-making for system design and maintenance optimization must account also for risk attributes, which integrate the effects of testing and maintenance choices on the system as a whole by including both the likelihood of hazardous events and their expected consequences, e.g. damages to environment and public health, degradation of plant performance, loss of economic income, etc. Correspondingly, the decision-making problem focuses on risk to environmental and public safety, to plant or system performance, to economic risk, etc.

On the other hand, maintainability (M) criterion is taken into account normally embedded into the availability criterion and it is rarely accounted for separately. An example of its use alone is the implementation of the maintenance rule at NPPs, which limits the maintainability of the safety related equipment very significant for the plant safety.

Figure 3 summarizes the published research grouped according to the type of model used to quantify the RAMS+C criteria adopted for the decision-making. One can observe how most of articles adopt analytical models (75%) to quantify RAMS+C criteria and in addition they often do not depend on time (about 60%).

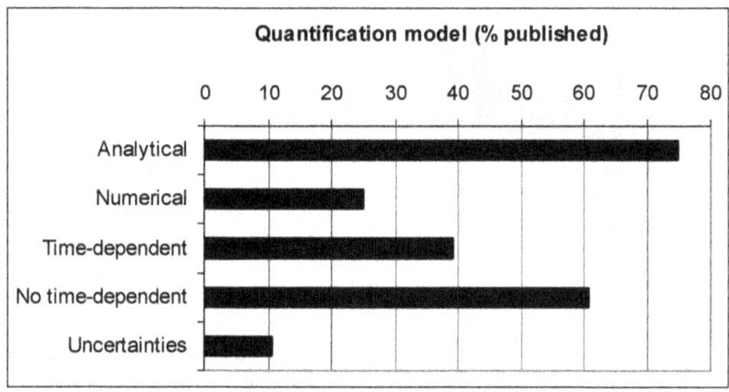

Fig. 3. Percentage of published work according to quantification models

Numerical and time-dependent models normally allow estimating in a more realistic way how the changes on T&M policies affect RAMS+C attributes. However, such models also entail more complexity in solving the optimization problem, which normally requires coupling a Monte Carlo (MC) simulation model embedded into a GA (MC-GA). Even so, almost 40% and 25% of the articles adopt time-dependent models and numerical models respectively.

In Bunea and Bedford 2002, it is shown how the impact of model uncertainties on testing and maintenance optimization can be significant. Therefore, it is important for the decision-maker to be able to estimate how uncertain the results are and how the uncertainty in the model parameters is propagated in the decision model based on RAMS+C criteria. Otherwise, the prediction of the system behaviour on the basis of the parameters' best estimates may be scarcely significant, if not accompanied by some measure of the associated uncertainty. However, the adoption of such models that consider uncertainties also entails more complexity in solving the optimization problem. Only 10% of the articles consider uncertainties in the optimization of T&M by GA. This is however a very promising area which requires further research.

The published research work can be also grouped according to the type of optimization problem being solved: SOP or MOP. In Fig. 4, one can observe how the early GA applications to testing and maintenance optimization faced resolution of single-objective optimization problems. However, in recent years a half of even more of the GA applications to testing and maintenance optimization faces the resolution of multiple objective optimization problems directly.

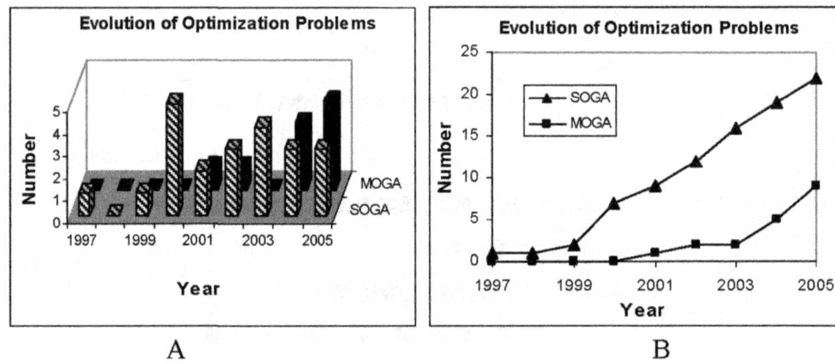

Fig. 4. Evolution of published works. A: according to the optimization Problem; B: according to GA type (accumulated evolution)

As said, decision-making has been normally based on multiple criteria since the early applications, where for example A+C and R+C combinations have been considered in near the 55% and 25% of the articles respectively. Consequently, the formulation of the optimization problem as a MOP seems the most natural way of facing decision-making. However, the progression of the development of GAs has largely influenced the transformation of the original MOP into SOP in a number of ways that will be discussed later on in this chapter.

Thus, the original GAs introduced after Holland 1975 were essentially unconstrained search techniques, which required the assignment of a scalar measure of quality, or fitness, to candidate solutions to a single-objective optimization problem. However, the early 1990 saw a growing effort for applying GAs to constrained optimization problems as most of engineering optimization problems are based on multiple criteria and often present their solution constrained by a number of restrictions imposed on the decision variables and criteria. In parallel, the middle 1990 saw the development of the first generation of multiple objectives GAs. However, the last generation of multiple objectives GAs released in the early 2000's has brought the possibility of formulating and solving directly more complex and constrained MOP as they improve the reliability of the obtained solutions and at the same time reduce the computational effort. This is the case of GAs application to T&M optimization based on RAMS+C. In addition, the delay between the development of new GAs and their applications to T&M optimization problems has reduced drastically in the last years.

3.3 Overview of Testing and Maintenance

3.3.1 RAMS and the role of T&M at NPP

The basic concepts and the role of Testing and Maintenance relative to RAMS of SRE at NPP are illustrated in Fig. 5 (Martorell et al. 2005).

Reliability of SRE represents its capability to respond and sustain operation under specified conditions without failure during a given period or to a given age. Thus, failure is defined as an interruption of its functional capability or loss of performance below the threshold defined in the functional specification. In turn, degradation is defined as the loss of performances of characteristics in the limits of the specifications as time or age passes, which results in a failure just below the failure point defined in the functional specifications.

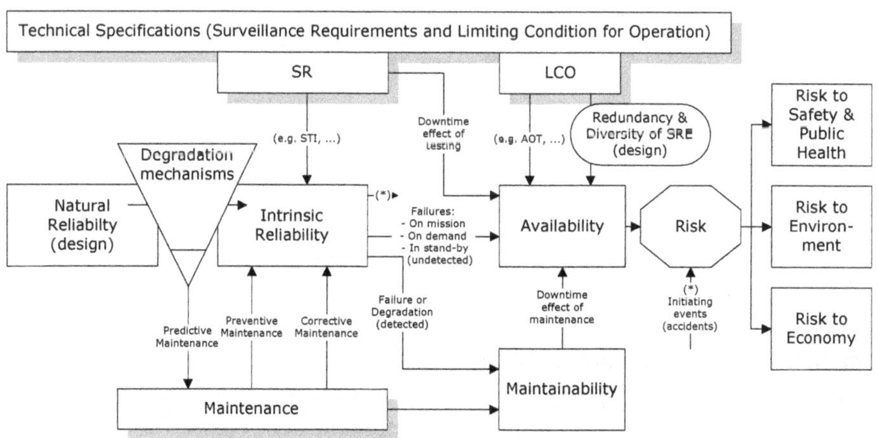

Fig. 5. Role of TSM and RAMS criteria

Natural reliability is the reliability of the equipment with no maintenance at all, which directly depends on its physical characteristics or design, while the intrinsic reliability is the value (in principle higher than natural) obtained with a normal amount of quality maintenance (usually preventive maintenance).

Maintenance represents all activities performed on equipment in order to assess, maintain or restore its operational capabilities. The goal of maintenance is to reach and maintain a generally implicit target for actual reliability, since a high intrinsic reliability results in minimal maintenance needs, while a low intrinsic reliability places unreasonable demands on maintenance process.

As shown in Fig. 6 one can distinguish two main types of maintenance at NPP, named corrective maintenance and preventive maintenance (PM) respectively. CM shall restore the operational capability of the failed or degraded equipment while PM shall increase the intrinsic reliability of non-failed equipment beyond the natural reliability, for example, controlling its degradation below the failure point.

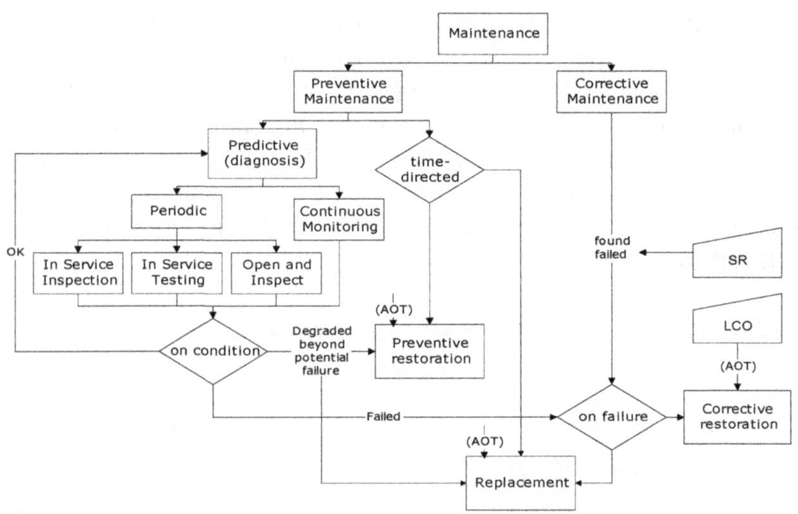

Fig. 6. Types of testing and maintenance activities at NPP

In addition, PM is normally divided into two more sub-categories, named time-directed and condition-based preventive maintenance respectively. The TDPM represents PM that is performed according to a pre-established schedule (depending on equipment age, elapsed time, level of production, etc.) regardless of equipment condition. However, CBPM represents PM that is initiated as a result of a pre-determined type of criteria that normally measures equipment deterioration. Condition is measured using appropriate techniques that fall into the group named predictive maintenance (diagnosis). Depending on the condition assessed, a preventive restoration is to be performed if the equipment is found degraded beyond potential failure, while a corrective restoration is to be performed on component failure, and no action is required otherwise (condition is OK). On the other hand, most of the SRE at NPP are normally in stand-by; therefore, CM may be also lunched if the equipment is found in a failed condition after performing an operational test following a Surveillance Requirement (SR) established by NPP Technical Specifications. In addition, a replacement activity is also considered for critical equipment which is normally launched on a time base, on condition or following failure finding activities.

Maintenance used to be mainly of concern of NPP staff, which supposed a wider margin in adopting changes on them. However, the US Nuclear Regulatory Commission Maintenance Rule released in July 1991 and subsequently amended (10 CFR 50.65) has significant impact on how nuclear power plants perform and document their maintenance.

Maintenance was traditionally carried out during shutdown, but now increasing amounts, in particular of preventive maintenance, are being scheduled during power operation due to different reasons. It is performed using the Limiting Condition for Operation (LCO) in TS, either by means of voluntarily declaring equipment inoperable to perform preventive maintenance activities or forced as a consequence of failure or excessive degradations that may require preventive or corrective restoration or even replacement. Therefore, the duration of this on-line maintenance is limited by the Allowed Outage Time (AOT).

As introduce above, predictive maintenance (PdM) represents monitoring and diagnostic tasks that relay on variations in operation parameters. There are several types of predictive maintenance activities: monitoring, testing, inspections and open and inspect (overhaul).

Monitoring tasks include operator rounds (based on human senses), maintenance personnel walk-downs (based on human senses or parameter measurements), operational parameters monitoring from inside the control room (continuos monitoring). Examples are monitoring of large rotating machinery, motor-operated valves' monitoring, lost parts monitoring, lube oil analysis, etc.

In-Service Inspection represents visual or non-destructive examination of equipment, applied for example for the assessment of the condition of massive pressure vessels, associated boundaries, pipes, tanks, heat exchangers, turbine blades, etc. The current In-service Inspection requirements for major components at NPP are found in *"Codes and Standards"* 10 CFR 50.55a and the *General Design Criteria* listed in Appendix A to 10 CFR 50. These requirements are throughout the General Design Criteria, such as in

Criterion I, "Overall Requirements," Criterion II, "Protection by Multiple Fission Product Barriers," Criterion III, "Protection and Reactivity Control Systems," and Criterion IV, "Fluid Systems."

In-Service Testing represents tasks measuring equipment readiness, such as full-size performance test to demonstrate the capability of the equipment to be operable, verification of the proper adjustment of equipment instrumentation and control systems, etc. Current In-Service Testing programs at NPP are performed in compliance with the requirements of 10 CFR 50.55a (f), which are requirements for all plants.

Overhaul represents tasks involving opening, dismantling and inspecting equipment that are generally very expensive.

Also, replacement is a planned activity for singular equipment, for example equipment under a graded quality program, which is replaced on a time base. On the contrary, other equipment is not replaced along the NPP operational life. Between both extreme situations, partial or total replacement maintenance (RM) may be planned for equipment in case of degradation being beyond a given threshold or on equipment failure.

Maintenance introduces two types of positive aspects. Firstly, corrective maintenance restores the operational capability of the failed or degraded equipment and secondly preventive maintenance increases the intrinsic reliability of non-failed equipment beyond the natural reliability, for example, controlling its degradation below the failure point. Although the equipment is subjected to preventive and corrective maintenance it may degrade over age depending on the working conditions of the equipment and the effectiveness of the maintenance being applied (so called imperfect maintenance). So that, several activities are scheduled to control evolution of degradation mechanisms that fall into the categories of continuous monitoring or periodic predictive maintenance, which are responsible for launching on condition preventive or corrective activities when necessary.

On the contrary, maintenance also introduces adverse effects, called the downtime effect that represents the time the equipment is out of service to overcome maintenance (corrective, preventive, repair, overhaul, etc.). Thus, the adverse effect depends on the maintainability characteristics of the equipment.

Maintainability represents the capability of the equipment or systems to be maintained under specified conditions during a given period, which depends to some extent on their physical characteristics or design among other factors, such as maintenance task force, spare parts, etc.

Equipment **availability** represents its capability to be in a state to perform its intended function under specified conditions without failure during a given period. To be ready means not to be out of service or failed. Thus, the availability, or more directly the unavailability of the equipment not only depends on the downtime effect, but also depends on the probability of failure to perform its intended function (unreliability effect). A failure can occur while the equipment or system is performing its intended function (mission failure), at the moment of the demand to operate (on demand) or before the demand (in stand-by), the later associated only with safety-related equipment normally in stand-by, which can experience failures in such period of time that will remain undetected until what ever becomes first a true demand to operate or a given operational test.

NPP TS, throughout their SR, establishes how and when performing such operational tests of equipment normally in stand-by to limit the unreliability effect, providing the STI. Also TS, throughout their LCO, establishes the maximum AOT of equipment to limit the downtime effect.

In particular, **Surveillance Requirements and Limiting Conditions for Operation** have been paid more attention over the past two decades in risk-informed application of the NPP-specific Probabilistic Risk Analysis (PRA) to support changes on these requirements. Fig. 7 shows a flowchart of normal contents in LCO and SR of typical TS. Herein, LCO and SR for AC Sources of a PWR NPP during normal operation (TS 3/4.8.1) have been chosen to illustrate the fundamentals of each section. In particular, Emergency Diesel Generators (EDG) are taken to guide the following explanation.

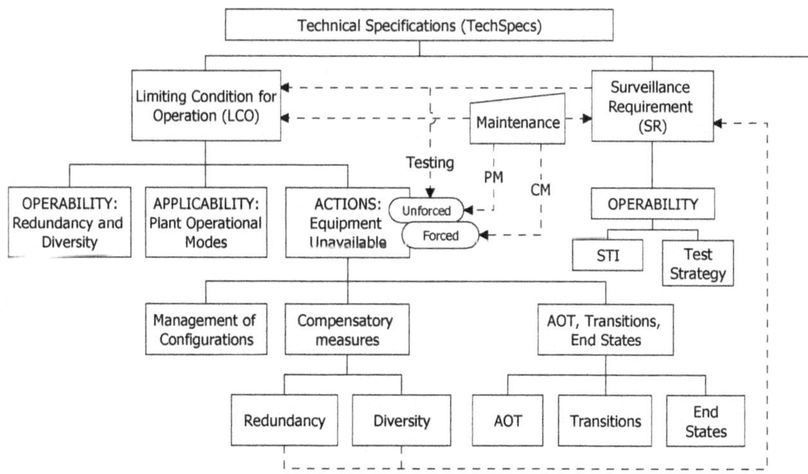

Fig. 7. Testing and maintenance with regard to Technical Specifications at NPP

LCO are the lowest functional capability or performance levels of equipment required for safe operation of the NPP. Thus, for the example of application, LCO establishes the operability of redundant AC Sources of diverse nature (external supply, EDG, etc.), which is of application to NPP

Operational Modes from full power to Shutdown. Redundancy of EDG is stated as follows: Two emergency diesel generators physically independent and able to supply demanded electricity to Class 1E systems (section 3.8.1.1.b) in TS). When a LCO of a NPP is not met, the licensee shall shut down the reactor or follow any remedial action permitted by the technical specifications until the condition can be met. Thus, for example, with one EDG inoperable, it is stated to manage risky configuration and introduce compensatory measures, such as demonstrate operability of external supply (action b).1) and operability of redundant EDG train (action b).3) by performing surveillance tests. Also, LCO establishes that the downed EDG train has to be restored to the operability condition before 72 hours (action b).4), which represents the AOT, or otherwise it is required to shut down the NPP providing guidance on the NPP transitions (power operation, hot standby, cold shutdown) to follow and the time available to reach a safer end state.

SR prescribe periodic surveillance tests for detecting faults, verifying the operability of safety equipment which define the type of surveillance test, test strategy and frequency, the last establishing the STI between two consecutive tests. For example, SR 4.8.1.1.2 establishes a STI of 31 days for the EDG under a staggered testing (testing one EDG train every fifteen days).

Finally, the attribute of **safety** of SRE can be defined as its capability to prevent or mitigate the consequences of postulated accidents at NPP Final Safety Assessment Report, which could cause undue risk to the health and safety of the public. Therefore, **risk** is often adopted as measure of NPP safety, which can be defined as the probability of causing damage to the health and safety of the public. In addition, environmental and financial risks can be considered as both also depend partially on the RAM attributes of SRE.

As shown in Fig. 5, NPP safety is based on the principles of redundancy, diversity and availability (reliability) of its SRE. Thus, safety-related systems consist of a number of redundant trains to perform the safety function. In addition, NPP safety is based on diversity and independency of SRE to perform such safety functions; even more, different trains of the same system are of different in nature in some cases. Consequently, NPP risk increases with redundancy loss and unavailability (unreliability) of its SRE. NPP TS, throughout their LCO, establishes the lowest functional capability or performance levels of equipment required for safe operation of the facility (SRE herein). When a LCO of a nuclear reactor is not met, the licensee shall shut down the reactor or follow any remedial action permitted by the technical specifications until the condition can be met (the time available to do it is limited by the AOT). Thus, LCO establishes the number and type of redundant trains or SRE and their level of availability and reliability; and in this way it limits NPP risk.

3.3.2 Failure types and T&M activities

Failures of SRE at NPP are linked to applicable types of T&M activities. Fig. 8 shows a possible categorization of such failures types linked to applicable types of T&M activities.

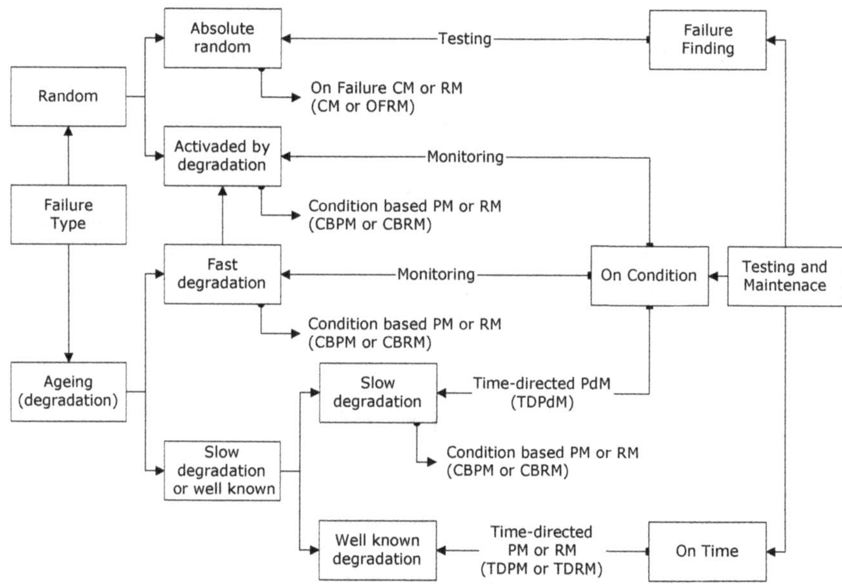

Fig. 8. Testing and maintenance activities depending on the failure mode

Thus, planned tasks, either scheduled or done as a result of unacceptable equipment condition are performed to predetermined criteria and prior to failure with the purpose of preventing unanticipated failure.

Some of the pre-defined maintenance activities are directly concerned with dealing with the results of degradation or ageing. Degradation can not only accumulate progressively (i.e. slow degradation) as in case of wear-out, e.g. normal wear from corrosion, but also can occur randomly and with no warning (i.e. fast degradation), e.g. damage due to a foreign object in the piping. When the degradation reaches the point the component can no longer meet it intended function, it is a functional failure. Condition monitoring is necessary to handle fast degradation while slow degradation and well known degradation mechanisms can be managed through

Time-Directed PdM (TDPdM) and TDPM respectively. In addition, monitoring and TDPdM have to support CBPM or Condition-Based RM (CBRM) when necessary, i.e. intervention when the degradation reaches the predefine degradation point.

However, most simple safety-related components have a low and constant failure rate during most of their life from random failures. In addition, complex components contain many pieces which can cause failure, all with different lifetimes and ages. An skid failure rate supposedly becomes essentially constant when all of the subparts are considered together. Constant failure rate means that failures are random and just as likely at any time. We can not prevent truly random events, but many of these events are randomly initiated short term wear-out modes (i.e. random failures activated by fast degradation or random degradation).

We can protect against absolute random failures only through performing failure finding activities (i.e. testing) and performing CM or On Failure RM (OFRM) if necessary. Protection against random degradation, that needs an initiator to get wear-out started, consists of performing frequent or continuous condition monitoring tasks to identify the onset degradation which could be duly managed through CBPM or CBRM.

3.4 Decision-making Based on RAMS+C

3.4.1 Basis for the RAMS+C informed decision-making

Current situation at NPP shows how different programs implemented at the plant aim to the improvement of particular T&M-related parameters (Martorell et al. 2005). According to the objectives established in these programs, it is possible to divide them into two different groups, one that focuses directly on risk (or safety) and the other that focuses manly on RAM (Reliability, Maintainability and Availability), being aware of the fact that both groups incorporate implicit or explicitly costs in the decision-making process.

The existence of such T&M improvement programs, performed at the same time, imposes the consideration of all their acceptance criteria simultaneously. In addition, the relationships that exist among the decision parameters involved in each program introduce synergic effects of the changes with regard to several criteria simultaneously. Thus, a change in a LCO, will affect the way how the maintenance can be carried out, for example, limiting the AOT, may require the establishment of compensatory measures or the management of the plant configuration. On the contrary, the improvement of the maintenance policy based on RAM criteria may lead to a wider margin to propose a change in the LCO or SR requirements based on safety criteria.

In general, it can be said that system reliability and availability optimization is classically based on quantifying the effects that design choices and testing and maintenance activities have on RAM attributes. A quantitative model is used to asses how design and maintenance choices affect the system RAM attributes and the involved costs. Thus, the design and maintenance optimization problem must be framed as a Multiple Criteria Decision Making (MCDM) problem where RAM+C attributes act as the conflicting decision criteria with the respect to which optimization is sought and the relevant design and maintenance parameters (e.g. redundancy configuration, component failure rates, maintenance periodicities, testing frequencies) act as decision variables.

In particular for potentially hazardous and risky industries, such as the nuclear one, decision-making for system design and maintenance optimization normally account also for risk attributes, which integrate the effects of design and maintenance choices on the system as a whole by including both the likelihood of hazardous events and their expected consequences. Then, testing and maintenance decision-making for NPP entails the simultaneous consideration of RAM+Safety (RAMS) criteria. For example, optimization of testing and maintenance activities of safety-related systems aim at increasing their RAM attributes which, in turn, yields an improved plant safety level. This, however, is obtained at the expense of an increased amount of resources invested (e.g. costs, task forces, etc.). Therefore, the multiple criteria decision-making task aims at finding the appropriate choices of reliability design, testing and maintenance procedures that optimally balance the conflicting RAMS and Costs attributes. To this aim, the different vector choices of the decision variables **x** are evaluated with respect to numerical objectives regarding reliability, availability, maintainability, risk/safety, but also cost attributes:

- $R(\boldsymbol{x})$ = System Reliability;
- $A(\boldsymbol{x})$ = System Availability ($U(\boldsymbol{x})$= system unavailability = 1 – $A(\boldsymbol{x})$);
- $\mathrm{M}(\boldsymbol{x})$ = System Maintainability, i.e. the unavailability contribution due to test and maintenance;
- $S(\boldsymbol{x})$ = System Safety, normally quantified in terms of the system risk, measure $Risk(\boldsymbol{x})$ (e.g. as assessed from a Probabilistic Risk Analysis);

- $C(\boldsymbol{x})$ = Cost required to implement the vector choice $\boldsymbol{x}$.

For different practical and research applications, a reduced decision-making process may suffice based on a subset of the RAMS+C criteria, such as for example the A[U]+C criteria.

In this general view, the vector of the decision variables $\boldsymbol{x}$ encodes the parameters related to the inherent equipment reliability (e.g. per demand failure probability, failure rate, etc.) and to the system logic configuration (e.g. number of redundant trains, etc.), which define the system reliability allocation, and those relevant to testing and maintenance activities (test intervals, maintenance periodicities, renewal periods, maintenance effectiveness, mean repair times, allowed downtimes, etc...) which govern the system availability and maintainability characteristics. Herein, the interest is only on those parameters related to testing and maintenance.

3.4.2 Quantification Models of RAMS+C

A quantitative model needs to be used to asses how testing and maintenance choices affect the system RAMS+C attributes. Thus, the relevant criteria are to be formulated in terms of the decision variables using appropriate models. These models have to be able to show explicitly the relationships among the criteria (decision criteria) and the variables of interest involved for the decision-making (decision variables).

In particular, PRA-based models and data are often used to assess RAMS criteria of SRE at NPP formulated in terms of the T&M-related parameters, the later will act as decision variables in the multi-objective optimization problem.

PRA uses analytical and time-independent models that need extension to account explicitly for the effects of testing and maintenance activities (Martorell et al. 2002).

In addition, averaged or yearly cost contributions are obtained based on analytical models of dependability (Harunuzzaman and Aldemir 1996, Busacca et al. 2001, Martorell et al. 2002, Vinod and Kushwaha 2004).

However, time-dependent and numerical models can better represent real life. One can find many examples in the literature of analytical and numerical time-dependent models of RAMS+C (Lapa et al. 2000, Bunea and Bedford 2002, Cepin 2002, Pereira and Lapa 2003, Lapa et al. 2003, Lapa and Pereira 2006, Martorell et al. 2006). In particular, a Monte Carlo based simulation model provides a flexible tool which enables one to describe many of the relevant aspects for testing and maintenance optimization such as aging, imperfect repair, obsolescence, renovation, which are not easily captured by analytical models (Borgonovo et al. 2000, Barata et al. 2002, Marseguerra et al. 2002).

The impact of model uncertainties on testing and maintenance optimization results can be significant. Then, it is required an appropriate representation of model and data uncertainties and a systematic procedure to estimate how uncertain the results are and how the uncertainty in the model parameters is propagated in the decision model based on RAMS+C criteria. (Rocco et al. 2000, Rocco 2002, Bunea and Bedford 2002, Marseguerra et al. 2004a, Coit et al. 2004). A probabilistic uncertainty representation and a Monte Carlo propagation procedure with its embedded sampling method, e.g. Simple Random Sampling or Latin Hypercube Sampling, are often adopted to quantify RAMS+C uncertainty.

This section does not attempt to provide a comprehensive presentation of all the models available in the literature to formulate the whole set of cause-effect relationships introduced in the previous section. Instead, Table 1 summarizes modeling requirements to address relationships among testing and maintenance activities, triggering events and RAM effects associated with several types of T&M.

In Table 1, T&M type represents one of the possible T&M activities. The Pre-T&M category is included to account for the fact that some T&M activities are preceded by another activity, which is normally performed to decide on the need of conducting the second one. Thus, for example, CM of stand-by related components is normally performed after the component has been found in a failed condition in a previous test. However, a TDPM does not require a previous activity as it is launched based on a time limit. The T&M Trigger category represents the condition that launches the second activity, e.g. a failed condition or a time limit in the previous examples respectively. As another example, a CBPM may start once the component has been found degraded in a previous predictive maintenance activity. The Post-T&M category represents the activity that follows a second activity. Thus, in the CM example, an operational test of the component may be performed once the CM has finished. The same Post-T&M activity may be found after performing a preventive maintenance. The Eq. Condition category represents the condition of the component before entering the activity. The R-effect, A-effect and M-effect categories represent how the activity impacts the reliability, availability and maintainability of the component respectively. Thus, for example the component is known to be available, or failed, after a test, but the test can also degrade the component's reliability and imposes unavailability as a consequence of the component downtime for testing.

Figure 9 shows a diagram with possible functional states of SRE at NPP, T&M activities and transitions, which accommodates to the previous model requirements.

Table 1. Example of modelling requirements of triggering and RAM-effects linked to T&M activities

Category	Testing	Predictive		Preventive	
T&M Type	-	Monitor-ing	TDPdM	TDPM	CBPM
Pre- T&M	None	None	None	None	Monitor-Predictive
Post T&M	None	None	None	Test	Test
T&M Trigger	Time limit	Continu-ous	Time or per-formance	Time or per-formance	Degraded con-dition
Equipment Condition	Un-known	Known	Unknown	Unknown	Degraded
R effect	Degradation	None	None	Imperfect maintenance	Imperfect maintenance
A effect	Available or failed	None	None	None (1)	None (1)
M effect	Down time	None	Down time	Down time	Down time

Category	Corrective	Replacement		
T&M Type	CM	TDRM	CDRM	OFRM
Pre- T&M	Testing	None	Monitor- Pre-dictive	Testing
Post-T&M	Test	None	None	None
T&M Trigger	Failed condition	Time or per-formance	Degraded con-dition	Failed condition
Equipment Condition	Failed	Unknown	Degraded	Failed
R effect	Bad As Old	Good As New	Good As New	Good As New
A effect	None (2)	Available	Available	Available
M effect	Down time	Down time	Down time	Down time

(1) Available if testing after PM (2) Available if testing after CM

Thus, SRE is normally operable in standby but it can fail at random or as a consequence of degradation or ageing. Thus, there is a probability for the component to be also in a degraded or failed state. Testing or real demands to operate are the only ways of discovering a failure of stand-by components. Testing is performed with frequency $1/T$, while τ represents the mean downtime for testing. Corrective maintenance follows testing once the component has been found failed in a previous test. Parameter μ represents the mean downtime for corrective maintenance. Preventive maintenance is intended to remove degradation from the component and this way improving the component inherent reliability, i.e. reducing the component failure rate. Preventive maintenance is normally performed on a time limit, TDPM, with frequency $1/M$, or alternatively following a degraded condition found in a previous predictive maintenance activity, which is performed with frequency $1/D$. Parameters σ and β represent the mean downtime for preventive and predictive maintenance respectively. Replacement of the whole or an important part of the component is conducted for some critical equipment for the NPP safety. It may be performed on a time limit, with frequency $1/L$, or on condition.

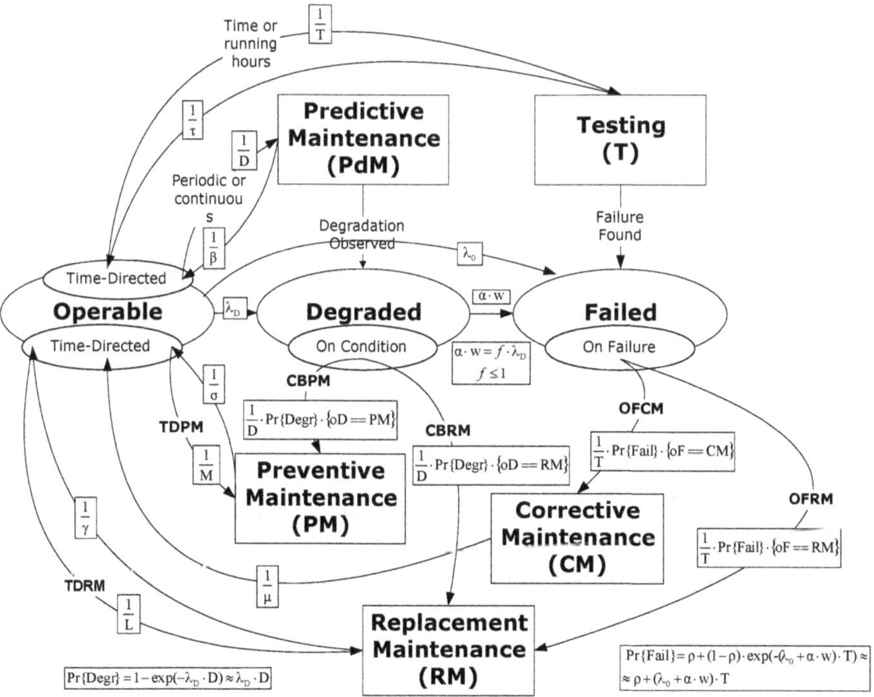

Fig. 9. Diagram of functional states, T&M activities and transitions for safety-related equipment

3.5 Optimization Problem

3.5.1 Problem Formulation

The commonly accepted manner to tackle the previously illustrated MCDM problem is to formulate it as a MOP. A general MOP considers a set of decision variables $\boldsymbol{x}$, a set of objective functions $\boldsymbol{f}(\boldsymbol{x})$ and a set of constraints $\boldsymbol{g}(\boldsymbol{x})$. Adapting the formal definition of Goldberg 1989 and Zitzler 1999, in the RAMS field the MOP regards the optimization of the vector of multi-objective functions

$$y = \boldsymbol{f}(\boldsymbol{x}) = (R(\boldsymbol{x}), U(\boldsymbol{x}), M(\boldsymbol{x}), Risk(\boldsymbol{x}), C(\boldsymbol{x})) \quad (1)$$

subject to the vector of constraints

$$\boldsymbol{g}(\boldsymbol{x}) = (R(\boldsymbol{x}) \geq R_L, U(\boldsymbol{x}) \leq U_L, M(\boldsymbol{x}) \leq M_L, Risk(\boldsymbol{x}) \leq Risk_L, C(\boldsymbol{x}) \leq C_L) \quad (2)$$

where

$$\boldsymbol{x} = \{x1, x2, ..., xds\} \in X$$
$$y = \{y1, y2, ..., yos\} \in Y$$

The quantities R_L, U_L, M_L, $Risk_L$, C_L represent constraining threshold values for the reliability, unavailability, maintainability, risk and cost objectives respectively. Martorell et al. 2005, gives typical threshold values for RAMS+C objectives. X is the decision space and Y is the objective space with dimensions ds and os respectively.

The vector of constraints $\boldsymbol{g}(\boldsymbol{x})$ limits the objective space, and indirectly the decision space, to seek acceptable values of the relevant parameters. These are denoted domain implicit constraints (Martorell et al. 2000). In addition, one can impose explicit constraints to limit directly the feasible values of the parameters included in the decision vector. Both types of constraints delineate, directly or indirectly, the set of acceptable values in the decision space, here denoted by Xp, and its image set in the objective space denoted by Yp.

3.5.2 Solution Approaches

There exist many works in the scientific literature devoted to solve the above MOP using different optimization techniques. Two types of approaches using GAs are here briefly summarized (Martorell et al. 2004).

In general, a MOP admits multiple solutions due to the conflicting nature of its attributes. Therefore, to arrive at a final decision, the decision maker must make a value judgment among the identified options, giving preference to some attributes at the expense of other, possibly conflicting, ones. The ways to impose such a value judgment can be broadly classified into three categories (Zitzler 1999): a) decision making before optimality search, in which the objectives of the MOP are aggregated into a single objective that implicitly includes a preference function, b) search before decision making, in which optimization is performed without introducing any a priori preference information to arrive at a Pareto optimal set of candidate solutions from which the final choice is made, and c) decision making during search, in which the decision maker articulates preferences during an interactive optimization process that guides the search. The way of establishing a preference function influences the optimization method used in the search of optimal solutions. In particular, depending on whether the decision making is performed before or after the search, the optimization problem to be solved may be transformed into a SOP or remain MOP.

The SOP is a particular case of a MOP when only one criterion is involved in the optimization process or, alternatively, many criteria are involved but only one criterion or a combination of some of them act as the single objective function, a priori defined, and the remaining criteria are implemented into the set of constraints. Thus, in the RAMS+C problem different solutions optimize the different RAMS+C attributes, see figure 2. Most of the standard optimization methods developed focus on solving SOP. Thus, for example in references Muñoz et al. 1997, Levitin and Lisnianski 1999, Joon-Eon et al. 2000, Rocco et al. 2000, Martorell et al. 2000, Bris and Chayelet 2003 and Tong et al. 2004, a SOP problem is solved using cost as objective function while using different functions as a constraint. Muñoz et al. 1997, Levitin and Lisnianski 1999, Martorell et al. 2000, and Bris and Chayelet 2003 adopt availability (unavailability) as constrain in the SOP problem, while Joon-Eon et al. 2000 and Tong et al. 2004 consider risk, in terms of core damage frequency, and Rocco et al. 2000 and Martorell et al. 2000 impose constraints on the ranges of the decision variables. The consideration of multiple criteria in the optimization problem has also been solved as a SOP approach, as for example in reference Elegbede and Adjallah 2003 and Martorell et al. 2004, where availability (unavailability) and cost are combined into a single objective function based on a weighting method. In addition, Elegbede and Adjallah 2003 impose constraints on the range of variation of the optimization variables adopting a penalization method, while Martorell et al. 2004 also impose restrictions on the values of the unavailability and cost functions. Finally, one can also find references of SOP problems in which reliability is involved in the objective function, as for example Levitin and Lisnianski 2000 where the reliability of a multi-state system is taken as objective function without imposing any constraint, or Tsai et al. 2001, where the aim is to optimize the ratio cost/reliability considering as constraint a desired value in the system reliability.

On the contrary, the direct resolution of a MOP amounts to performing the optimality search without taking decisions a priori. Since the MOP has no unique solution that can simultaneously optimize all objectives, some of which are conflicting, the search leads to a set of potential solutions, the so-called Pareto optimal (non-dominated) set, from which the final decision is drawn a posteriori. As a practical matter, the concept of Pareto optimality is of no direct help for selecting a single solution because there is no universal standard to judge the merits of a solution of the Pareto optimal sets and all are equally optimal in the multi-objective sense. The decision maker must make a choice, a posteriori, on the basis of subjective preference values and trade-offs with respect to the optimization objectives. On the other hand, different preferences and trade-offs are conveniently portrayed in the Pareto optimal set. For example, Busacca et al. 2001 face a MOP in which three criteria are considered simultaneously: availability, cost and exposure time, but without imposing any constraint, Marseguerra et al. 2002 adopt net profit and availability as multiple objective functions, and Marseguerra et al. 2004a and Marseguerra et al. 2004b consider the system unavailability and its variance as objective functions under cost constraints. In Martorell et al. 2004 and Martorell et al. 2005, the system unavailability and cost criteria are selected as multiple objective functions and constraints. In Podofillini et al. 2006 breakage probability and cost are considered as objective functions, and restrictions are also imposed on both criteria. Finally, Martorell et al. 2005 face the general MOP that consists of considering all the RAMS+C criteria as objective function in the optimization problem using constraints regarding all the RAMS+C criteria involved as well.

3.6 Genetic Algorithms

3.6.1 Origin, fundamentals and first applications

Genetic Algorithms are very likely the most widely known type of Evolutionary Algorithms (EAs) (Bäck 1996). GAs are adaptive methods used in searching and optimization problems, which work by imitating the principles of natural selection and genetics. GAs in their usual form were developed by John Holland in 1965 who summarized his work in Holland 1975. His work and a number of PhD dissertations by his students (De Jong

1975) were the only documents available on GAs until late seventies, which serve as starting point of nearly all known applications and implementations of GAs.

Fundamentals on how GAs work, GAs implementation and applications have been introduced in many papers, e.g. Beasley et al. 1993a, Beasley et al. 1993b, Fogel 1994, Marseguerra et al. 2006, in several books, e.g. Bäck 1996, Goldberg 1989, Davis 1991, Fogel 1995, Michalewicz 1996 and in PhD dissertations, e.g. Zitzler 1999.

Because of their simplicity, flexibility, easy operation, minimal requirements and global perspective, GAs had been successfully used in a wide variety of problems in several areas of engineering and life science, as described for example in Beasley et al. 1993b, Fogel 1994, Herrera and Verdegay 1996.

Middle 90's saw the first GAs applications in the field of optimizing equipment reliability, maintainability and availability (Painton and Campbell 1995, Kumar et al. 1995, Muñoz et al. 1997, Levitin and Lisnianski 1999).

3.6.2 Pioneering GA

The pioneering studies on evolutionary optimization (before 90's) were based on transforming the original MOPs into SOPs as discussed in the previous section. Fig. 10 summarizes the evolution of the development of the several generations of GA.

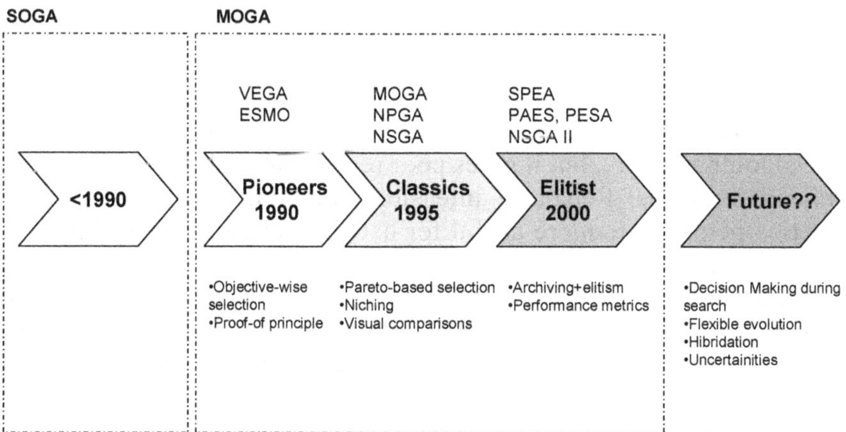

Fig. 10. Evolution of Genetic Algorithm based optimization approaches

The type of GA-based optimization approach that faces a SOP is known as SOGA. As one can observe in Figure 4, many recent applications of GA to optimizing surveillance and maintenances have been based on SOGA. However, a number of developments have been necessary to improve the performance of the initial GA to reach the capability of the SOGA used nowadays.

GAs are essentially unconstrained search techniques, which require the assignment of a scalar measure of quality, or fitness, to candidate solutions to the optimization problem. Then, it has been necessary a growing effort to apply GAs to general constrained optimization problems since most of engineering optimization problems often present their solution constrained by a number of restrictions imposed on the decision variables (Joines and Houck 1994, Michalewicz 1995, Carlson 1995, Carlson et al. 1995, Martorell et al. 1995, Leemis 1995).

In engineering optimization problems, constraints can be classified as explicit or implicit (Carlson 1995, Carlson et al. 1995). The former involve, for example, the parameter to be optimized (decision variables) that can only adopt a given set of values, where violation is directly revealed. However, an implicit constraint requires a simulation to check for a violation. For example, it is necessary to evaluate the cost or risk associated with given values of the decision variables to reveal a violation on either of them.

Many approaches have been considered in SOGA for handling solutions that violate one or more constraints, but no one general method has emerged. Approaches can be categorized as follows (Joines et al. 1994, Michalewicz 1995, Carlson 1995, Carlson et al. 1995): (1) eliminating infeasible solutions, (2) repairing infeasible solutions, (3) using modified genetic operators and (4) applying penalties for infeasible solutions. The first approach is used for example in Muñoz et al. 1997, Levitin and Lisnianski 1999, Goldberg 1989, but it gives poor results particularly in highly constrained problems. Repairing infeasible solutions and using modified genetic operators are more useful for handling explicit constraints since handling implicit constraints is more difficult an usually too time consuming. The last approach is the most prevalent technique for handling implicit constraints although it is also possible to handle explicit constraints. This technique transforms the constrained problem into an unconstrained problem by penalizing those solutions that are infeasible after evolution. Last generation of SOGA in the field of design, testing and maintenance optimization based on RAMS+C use mainly dynamic penalization to handle implicit constraints (Martorell et al. 2006).

3.6.3 Development of multi-objective GA

A MOGA faces the MOP directly. Both SOGA and MOGA approaches are currently in use for testing and maintenance optimization. However, in recent years a half of even more of the GA applications to testing and maintenance optimization use a MOGA approach (see Fig. 4), showing a growing trend that overpasses the use of SOGA.

The first MOGA was developed by Schaffer (Schaffer 1985) who proposed the VEGA (Vector Evaluation Genetic Algorithm). Later Kursawe developed the ESMO algorithm (Kursawe, 1991), which also addressed multiple objectives in a non-aggregating manner. These first approaches are known as Non-Pareto approaches.

Goldberg 1989 was the first to propose the Pareto-based approach that is behind of most of the current MOGA. The idea behind the Pareto-based approach is to assign equal probability of reproduction to all non-dominated individual in the population. The concept of Pareto dominance is well known and is described for example in Goldberg 1989, Zietler 1999 and Marseguerra et al. 2006.

Within the genetic approach, three major problems must be addressed when a genetic algorithm is applied to multi-objective optimization problems:

a) How to measure the fitness of a solution and how to perform selection in order to guide the search towards the Pareto-optimal set. This is accomplished by considering Pareto dominance.

b) How to maintain a diverse population in order to prevent premature convergence and achieve a well distributed and well spread non-dominated set. This is accomplished by density estimation.

c) How to avoid losing best individuals along the evolution process. This is accomplished through elitism.

In addition, what concerns constraints handling, explicit constraints are handled in a similar way to the case of single-objective GAs, i.e. by testing whether, in the course of the population creation and replacement procedures, the candidate solution fulfills the constraints. Also, a dynamic penalization approach similar to that described for implicit constraints handling within the single-objective GA can be adopted to degrade the fitness of those individuals which violate constraints on any of the objectives. Other methods based on constrained dominance are proposed.

A number of well-known MOGA were developed as the first generation of Pareto-based approaches: MOGA (Multiobjective Genetic Algorithm) (Fonseca and Fleming 1993), NSGA (Non-dominated Sorting Genetic Algorithm) (Srinivas and Deb 1994), and Niched Pareto Genetic Algorithm (NPGA) (Horn et al. 1994). These algorithms have common properties such as that the solutions are ranked according to their dominance in the population and diversity is maintained using niching strategies.

Second generation of GA represents an evolution of the first generation of Pareto-based approaches, which implement the use of elitism to improve convergence (e.g. reliability and speed up). The SPEA (Strength Pareto Evolutionary Algorithm) (Zitzler 1999), SPEA2 (Zitzler et al. 2001), NSGA-II (Deb et al. 2002), Pareto Archived Evolution Strategy (PAES) (Knowles and Corne 1999) are example of this generation of GA.

Concerning the fitness assignment task, there are three typical dominance-based ranking methods:

1) Dominance rank which is based on assessing by how many individuals an individual is dominated (Fonseca and Fleming 1993).

2) Dominance depth which is based on assessing which dominance front an individual belongs to (Deb et al. 2002).

3) Dominance count which is based on assessing how many individuals an individual dominates (Zitzler et al. 2001).

To maintain diversity in evolving the GA, the most frequently used techniques in multi-objective optimization are based on density estimation. Fitness sharing is, for example, used to this purpose (Fonseca and Fleming 1993, Deb et al. 2002). Alternatively, a technique is used in which the density at any point in the solution space is a decreasing function of the distance to the k-th nearest data point (Zitzler et al. 2001).

An archiving procedure allows implementation of the concept of elitism in the multi-objective GA (Deb et al. 2002, Zitzler et al. 2001).

In parallel, the use of performance metrics to support comparison of GA performance has been enhanced. Quantitative performance metrics are introduced to assess the efficiency of the multi-objective GA concerning both accuracy, i.e. approximation to the optimal Pareto front, and coverage of the Pareto front (Zitzler 1999, Laumanns et al. 2002, Zitzler et al. 2003).

As an example of last generation GA application in surveillance and maintenance optimization, Martorell et al. 2004 propose two approaches SOGA and MOGA respectively to optimize testing and maintenance based on A[U]+C criteria. The first approach is based on a typical Steady State Genetic Algorithm (SSGA) while the second one is based on the SPEA2 algorithm. In Martorell et al. 2006, the performance of the above SPEA2-based approach is compared with a two loops GA based on the NSGA-II algorithm. The optimization of test intervals and strategies of a typical safety system of a NPP based on the mean and maximum time-dependent unavailability and cost criteria is adopted as application example. On the other hand, the MOGA proposed by Fonseca and Fleming 1993 is taken as starting point algorithm in Marseguerra et al. 2004a, Marseguerra et al. 2004b, Podofillini et al. 2006.

3.7 Research Areas

This section discusses current situation and future research in the area of Genetic Algorithms methodology development and application to surveillance and maintenance optimization problems based on RAMS+C.

Martorell et al. 2005 introduce the fundamentals and an example application showing the interest of formulating and solving the multi-objective optimization of testing and maintenance at NPP based on RAMS+C criteria as a whole. However, it is concluded that further research and new cases of application in this area are needed. Thus, introduction of new optimization variables (e.g. maintenance effectiveness, design redundancy, equivalent equipment choices, etc) will extend the number of decision variables but on the other hand the new constraints on the decision criteria may limit the search space (e.g. human resources).

A second topic for future research concerns the development of the methodology and tools to aid the decision-making process itself as concluded in Martorell et al. 2004. It is clear that decision-maker must ultimately make a value judgment; however, the two typical alternatives, i.e. SOP and MOP, which are compared in that work being implemented into a SOGA and a MOGA respectively, represent extreme options for the decision-making process, that is to say, decision-making before search for the SOP alternative (i.e. adopting the SOGA) and decision-making after search for the MOP alternative (i.e. adopting the MOGA). The former may yield to make a decision too early and therefore new trials would be necessary to guide the search towards other regions of the search space. Instead, the later may yield to make a decision too late, once the whole search space has been explored, which may suppose an enormous computational cost to realize finally for example that the Pareto front is not so well defined and that only a portion of the search spacc results of interest for the decision-maker. A third option based on decision-making during search needs further research. A first attempt to bridge the SOGA-MOGA approaches is found in Coit & Baheranwala 2005.

At the present time, the state of the art of MOGA methodologies seems to be in a stagnation stage (Winter et al. 2005). One of the main reasons is the high efficiency reached by the so called "second generation MOGA" where multi-objective optimisation is needed. A wide variety of efficiency test has been used to demonstrate the convergence properties of these MOGA algorithms. As a result of such test efforts it seems to be a consensus about the goodness of the methods.

Despite the great advance represented by these MOGA, the convergence problems when computationally expensive objective functions are optimised remain unresolved. This is in general the case of many real industry optimisation problems and, in particular, the case of testing and maintenance optimization problems based on RAMS+C. As discussed in the preceding sections, the use of time-dependent models, numerical simulation methods and the incorporation of uncertainties in quantifying RAMS+C attributes enlarge drastically complexity and computational effort. For such problems the computational cost to evaluate each candidate solution is very expensive, normally several orders of magnitude (in CPU time) higher than the total cost of the MOGA loops. It is also well known that MOEA algorithms fail converging to the true Pareto front when this situation arises, so the industry decision makers are reluctant to use them and they prefer trusting in the information provided by other, normally classical, methodologies (Winter et al. 2005).

An example of the use of an enhanced evolutionary algorithm for industrial applications is proposed in Naujoks 2005.

However, the need of a deeper change is envisaged in the future development and application of GA to allow testing and maintenance optimization for more complex problems (e.g. increased number of decision variables, objectives and constraints), providing flexibility in the decision-making process and considering uncertain RAMS+C criteria. In addition, the GA-based approaches have to assure convergence reliability and quickness to arrive to a good solution.

New techniques consisting of using single optimization solutions in order to increase the efficiency in the Pareto front (P-front) construction (termed "Pareto Front Reconstruction") are proposed (Winter et al. 2005). Other authors propose alternative methods for generating the entire Pareto frontier (Utyuzhnikov 2005).

Another alternative well consolidated to improve convergence is by means of incorporating hybrid optimization methods, such as for example the ones presented in Neittaanmaki 2005. This approach combines two algorithms coupled to work in a cooperative manner. Typical topologies of hybridizing algorithms consist of a couple heuristic-heuristic (e.g. GA-GA, GA-Multi Directional Search, GA-Simulated Annealing, etc) or heuristic-traditional (GA-interval method, GA-gradient method, etc).

Martorell et al. 2006 propose an approach based on a sort of mixture SOGA-MOGA or MOGA-MOGA (e.g. two loops MOGA) working in a cooperative manner, which needs further research. This approach would increase flexibility for the decision-making process but it will also suppose the need of designing the appropriate man-machine interface to facilitate an interactive MCDM to guide the search step-by-step.

Appropriate selection of genetic operators and tuning of parameters is the usual way of improving GA performance, both SOGA and MOGA. However, this is a very problem dependent task that is normally performed manually. As a consequence of the size and complexity of the RAMS+C problems that increase day-after-day, the analyst has more chance of failing in such a process, then it seems necessary further research on methods to avoid or minimize user errors in optimizing GA performance. This may require a deeper change in the structure and operation of the GA. In Winter et al. 2005, authors introduce the usage of new algorithmic developments like those which use Flexible Self-Adaptable structures (FEA) which consists of a set of different real-coded sampling operators instead of one single crossover or mutation operator. The way the algorithm uses these operators gives the implementation a special flexibility to handle optimization problems and to obtain good solutions. Furthermore, FEA codifies the evolving individuals including information about the variables, the sampling operators and the parameters that control how to sample. This feature enables FEA to tune its parameters on the fly, eliminating the need of choosing the parameters before the optimization process.

The influence of the constraints handling method on the appropriate performance of the GA-based algorithm is well known, as it comes naturally from the fact that GAs are essentially unconstrained search techniques that require adaptation to handled constrained RAMS+C optimization problems. Current research in this area concentrates in proposing new constraint handling methods, with particular attention to provide flexibility in defining the ranges of the decision space (Sasaki 2005, Salazar et al. 2005).

The appropriate incorporation of uncertainty into the analysis of complex system is a topic of importance and widespread interest. Uncertainty is normally associated herein with the "lack of knowledge" or "randomness" in quantifying multiple RAMS+C criteria. Uncertainty normally affects testing and maintenance related parameters, equipment reliability, etc. As a consequence, the general MOP (also SOP) based on such criteria faces multi-objective optimization under uncertain objectives and constraints. An important number of studies have been published in the last decade in the field of RAMS+C informed optimization considering uncertainties, particularly for testing and maintenance optimization problems (Rocco et al. 2000, Bunea and Bedford 2002, Marseguerra et al. 2004c, Marseguerra et al. 2004d).

The earliest published works in the field of testing and maintenance optimization were formulated as single-objective optimization problems, that is, considering one uncertain objective function and a number of constraints. Rocco et al. 2000, Bunea and Bedford 2002 are examples in the field of testing and maintenance optimization. However, recent works formulate testing and maintenance optimization problems (Marseguerra et al. 2004c, Marseguerra et al. 2004d) with multiple objectives subject also to a number of constraints. In general, a Monte Carlo simulation embedded within a Genetic Algorithm based search has proved to be an effective combination to identify optimal testing and maintenance strategies.

However, quantification of RAMS+C models, which have to act as uncertain objective functions and constraints, often require large computational time, even when using deterministic models, in particular for time-dependent RAMS+C models. In addition, to be of value to the GA algorithm using the uncertain results to guide the search in the appropriate direction, there is a general perception that the RAMS+C models have to be quantified for hundreds or thousands of times using for example a Latin Hypercube Sampling or a Simple Random Sampling in a Monte Carlo procedure for propagation of uncertainty, i.e. mapping from input (uncertain model parameters and decision variables) to output results (uncertain RAMS+C quantities). Even more, as the GA must be run for a significant number of generations to reach convergence to the optimal solution or at least to arrive at a good solution, this may suppose a huge number of evaluations of the RAMS+C models and maybe an unaffordable computational effort on even the world's fastest computers. Several authors have proposed methods to solve the problem of reducing the quantification effort, such as the 'drop-by-drop' method proposed in Cantoni et al. 2000, Marseguerra et al. 2000 applied to hybridizing a MC-GA procedure, which is used again in Marseguerra et al. 2002, Marseguerra et al. 2004c. Future research should concentrate in reducing even more the computational effort and assuring at the same time convergence to the real Pareto set in presence of the noise introduced by uncertainty, in order to provide robust, fast and powerful tools for RAMS+C informed multi-objective optimization of testing and maintenance under uncertainty.

References

Bäck, T., Evolutionary Algorithms in Theory and Practice. Evolution Strategies, Evolutionary Programming, Genetic Algorithms (1996), Oxford University Press, NY.

Barata, J., Soares, C.G., Marseguerra, M., Zio, E. (2002). Simulation modelling of repairable multi-component deteriorating systems for 'on condition' maintenance optimisation. Reliability Engineering & System Safety 76, 255-264.

Beasley, D., Bull, D., Martin, R. (1993a). An overview of genetic algorithms: part 1, fundamentals, University Computing 15:2, 58-69.

Beasley, D., Bull, D., Martin, R. (1993b). An overview of genetic algorithms: part 2, research topics, University Computing 15:4, 170-181.

Borgonovo, E., Marseguerra, M., Zio, E. (2000). A Monte Carlo methodological approach to plant availability modeling with maintenance, aging and obsolescence. Reliability Engineering & System Safety 67, 61-73.

Bris, R., Chayelet, E. (2003). New method to minimize the preventive maintenance cost of series–parallel systems. Reliability Engineering & System Safety. 82 247-225.

Bunea, C., Bedford, T. (2002). The effect of model uncertainty on maintenance optimization. IEEE Transactions on Reliability. 51/4: 486-493.

Busacca, P.G., Marseguerra, M., Zio, E. (2001). Multiobjective optimization by genetic algorithms: application to safety systems. Reliability Engineering & System Safety 72, 59-74.

Cantoni, W., Marseguerra, M., Zio, E. (2000). Genetic algorithms and Monte Carlo simulation for optimal plant design. Reliability Engineering & System Safety 2000; 68(1): 29-38.

Carlson, S. (1995). A General Method for Handling Constraints in Genetic Algorithms. University of Virginia.

Carlson, S., Shonkwiler, R., Babar, S., and Aral, M. (1995). Annealing a Genetic Algorithm over Constraints. University of Virginia.

Cepin, M. (2002). Optimization of safety equipment outages improves safety. Rel. Engng & System Safety; 77(1): 71-80.

Coit, D.W., Jin, T.D., and Wattanapongsakorn, N. (2004). System optimization with component reliability estimation uncertainty: A multi-criteria approach. IEEE Transactions on Reliability. 53:3, 369-380.

Coit, D.W., and Baheranwala, F. (2005). Solution of stochastic multi-objective system reliability design problems using genetic algorithms. In Proceedings of ESREL Conference 2005, Ed. Kolowrocki, Tri City, Poland, 391-398.

Davis, L., (Ed.) (1991). Handbook of Genetic Algorithms, Van Nostrand Reinholt, NY.

Deb, K., Pratap, A., Agarwal, S., and Meyarivan, T. (2002). A fast and elitist multiobjective genetic algorithm. NSGA-II. IEEE Transactions on Evolutionary Computation 6(2), 182-197.

De Jong, K. (1975). An analysis of the behaviour of a class of genetic adaptive systems, PhD thesis, University of Michigan, 1975.

Elegbede, C., Adjallah, K. (2003). Availability allocation to repairable systems with genetic algorithms: a multi-objective formulation. Reliability Engineering & System Safety. 82, 319-330.

Fogel, D. (1994). An introduction to simulated evolutionary optimization. IEEE Transactions on Neural Networks 15:1, 3-14.

Fogel, D.B. (1995). Evolutionary Computation, IEEE Press, NY.

Fonseca, C.M., and Fleming, P.J. (1993). Genetic algorithms for multiobjective optimization: Formulation, discussion and generalization. In S. Forrest (Ed.), Proceedings of the Fifth International Conference on Genetic Algorithms, San Mateo, California, pp. 416-423. Morgan Kaufmann.

Goldberg, D.E., (1989). Genetic algorithms in search, optimization and machine learning. Addison-Wesley Pub. Co. Reading MA.

Haruzunnaman, M., Aldemir, T. (1996). Optimization of standby safety system maintenance schedules in nuclear power plants. Nucl. Technol 1996; 113(3): 354-367.

Herrera, F. and Verdegay, J.L., (Eds.) (1996). Genetic Algorithms and Soft Computing, Physica-Verlag Heidelberg.

Holland, J. (1975). Adaptation in Natural and Artificial Systems, Ann Arbor, University of Michigan Press.

Horn, J., Nafpliotis, N., and Goldberg, D.E. (1994). A niched pareto genetic algorithm for multiobjective optimization. In Proceedings of the First IEEE Conference on Evolutionary Computation, IEEE World Congress on Computational Computation, Volume 1, Piscataway, NJ, pp. 82-87. IEEE Press.

Joines, J.A., and Houck, C.R. (1994). On the Use of Non-Stationary Penalty Functions to Solve Nonlinear Constrained Optimization Problems With GAs. In Proceedings of the Evolutionary Computation Conference-Poster Sessions, part of the IEEE World Congress on Computational Intelligence, Orlando, 26-29 June 1994, 579-584.

Joon-Eon Yang, Tae-Yong Sung, Youngho Jin. (2000). Optimization of the Surveillance Test Interval of the Safety Systems at the Plant Level. Nuclear Technology. 132, 352-365.

Knowles, J.D., and Corne D.W. (1999). The pareto archived evolution strategy: A new baseline algorithm for pareto multiobjective optimisation. In Congress on Evolutionary Computation (CEC99), Volume 1, Piscataway, NJ, pp. 98-105. IEEE Press.

Kumar, A., Pathak, R., Grupta, Y. (1995). Genetic-algorithms-based reliability optimization for computer network expansion. IEEE Transactions on Reliability 44:1, 63-68.

Kursawe, F. (1991). A variant of evolution strategies for vector optimization. In H.-P. Schwefel and R. Männer, editors, *Parallel Problem Solving from Nature – Proc. 1st Workshop PPSN*, pages 193-197, Berlin, Springer.

Lapa, C.M.F., Pereira, C.M.N.A., and Melo, P.F.F.E. (2000). Maximization of a nuclear system availability through maintenance scheduling optimization using a genetic algorithm. Nuclear engineering and design, 196, 219-231.

Lapa, C.M.F., Pereira, C.M.N.A., Melo, P.F.F.E. (2003). Surveillance test policy optimization through genetic algorithms using non-periodic intervention frequencies and considering seasonal constraints. Rel. Engng. and System Safety; 81(1): 103-109.

Lapa, C.M.F., Pereira, C.M.N.A. (2006). A model for preventive maintenance planning by genetic algorithms based in cost and reliability. Rel. Engng. and System Safety; 91(2): 233-240.

Laumanns, M., Thiele, L., Deb, K., and Zitzler, E. (2002). Archiving with Guaranteed Convergence And Diversity in Multi-objective Optimization. In GECCO 2002: Proceedings of the Genetic and Evolutionary Computation Conference, Morgan Kaufmann Publishers, New York, NY, USA, pages 439-447, July, 2002.

Leemis, L. (1995). Reliability. Probabilistic Models and Statistical Methods, Prentice-Hall, Englewood Cliffs, New Jersey.

Levitin, G., and Lisnianski, A. (1999). Joint redundancy and maintenance optimization for multistate series–parallel systems. Reliability Engineering & System Safety 64, 33-42.

Levitin, G., and Lisnianski, A. (2000). Optimization of imperfect preventive maintenance for multi-state systems. Reliability Engineering & System Safety 67, 193-203.

Marseguerra, M., Zio, E. (2000). Optimizing maintenance and repair policies via a combination of genetic algorithms and Monte Carlo simulation. Reliability Engineering & System Safety; 68(1): 69-83.

Marseguerra, M., Zio, E., and Podofillini L. (2002). Condition-based maintenance optimization by means of genetic algorithms and Monte Carlo simulation. Reliability Engineering & System Safety 77, 151-166.

Marseguerra, M., Zio, E., and Podofillini L. (2004a). A multiobjective genetic algorithm approach to the optimization of the technical specifications of a nuclear safety system. Reliability Engineering & System Safety 84, 87-99.

Marseguerra, M., Zio, E., and Podofillini, L. (2004b). Optimal reliability/availability of uncertain systems via multi-objective genetic algorithms. Ieee Transactions on Reliability 53, 424-434.

Marseguerra, M., Zio, E., and Podofillini L. (2004c). A multiobjective genetic algorithm approach to the optimization of the technical specifications of a nuclear safety system. Reliability Engineering & System Safety; 84(1):87-99.

Marseguerra, M., Zio, E., and Podofillini, L. (2004d). Optimal reliability/availability of uncertain systems via multi-objective genetic algorithms. Ieee Transactions on Reliability; 53(3): 424-434.

Marseguerra, M., Zio, E., and Martorell, S. (2006). Basics of genetic algorithms optimization for RAMS applications. Reliability Engineering & System Safety. In press.

Martorell, S., Serradell, V., and Samanta, P.K. (1995). Improving allowed outage time and surveillance test interval requirements: a study of their interactions using probabilistic methods. Rel. Engng & System Safety, 47, 119-129.

Martorell, S., Carlos, S., Sanchez, A., and Serradell, V. (2000). Constrained optimization of test intervals using a steady-state genetic algorithm. Rel. Engng. and System Safety 67(3): 215-232.

Martorell, S., Carlos, S., Sanchez, A., and Serradell, V. (2002). Simultaneous and multi-criteria optimization of TS requirements and Maintenance at NPPs. Ann. of Nucl. Energy, 29(2): 147-168.

Martorell, S., Sanchez, A., Carlos, S., and Serradell, V. (2004). Alternatives and challenges in optimizing industrial safety using genetic algorithms. Reliab Eng Syst Safety; 86, 25-38.

Martorell, S., Villanueva, J.F., Carlos, S., Nebot, Y., Sánchez, A., Pitarch, J.L., and Serradell, V. (2005). RAMS+C informed decision-making with application to multi-objective optimization of technical specifications and maintenance using genetic algorithms. Reliab Eng System Safety 87/1: 65-75.

Martorell, S., Carlos, S., Villanueva, J.F., Sanchez, A.I., Galvan, B., Salazar, D., and Cepin M. (2006). Use of multiple objective evolutionary algorithms in optimizing surveillance requirements. Reliability Engineering & System Safety. In press.

McCormick, N.J. (1981). Reliability and Risk Analysis. Methods and Nuclear Power Application. EEUU: Academic Press.

Michalewicz, Z. (1995). A survey of constraint handling techniques in evolutionary computation methods. In Proceedings of the Fourth International Conference on Evolutionary Programming, Ed. McDonnell, J.R. Reynolds, R.G. Fogel, D.B., San Diego, CA., pp. 135-155.

Michalewicz, Z. Genetic Algorithms + Data Structures = Evolution Programs (3 Ed.) (1996), Springer-Verlag, Berlin.

Muñoz, A., Martorell, S., and Serradell, V. (1997). Genetic algorithms in optimizing surveillance and maintenance of components. Rel. Engng. and System Safety; 57(2): 107-120.

Naujoks, B. (2005). Enhanced Evolutionary Algorithms for Industrial Applications. In Proceedings of Evolutionary and Deterministic Methods for Design, Optimization and Control with Applications to Industrial and Societal Problems EUROGEN 2005, FLM, Munich.

Neittaanmaki, P. (2005). Hybrid optimization methods for industrial problems. In Proceedings of Evolutionary and Deterministic Methods for Design, Optimization and Control with Applications to Industrial and Societal Problems EUROGEN 2005, FLM, Munich.

Painton, L., and Campbell, J. (1995). Genetic algorithms in optimization of system reliability. IEEE Transactions on Reliability 44:2, 172-178.

Pereira, C.M.N.A., and Lapa, C.M.F. (2003). Parallel island genetic algorithm applied to a nuclear power plant auxiliary feedwater system surveillance tests policy optimization. Annals of Nuclear Energy; 30, 1665-1675.

Podofillini, L., Zio, E., Vatn, J., and Vatn, J. (2006). Risk-informed optimisation of railway tracks inspection and maintenance procedures. Reliability Engineering & System Safety In press 91(1) 20-35.

Rocco, C.M., Miller, A.J., Moreno, J.A., Carrasquero, N., and Medina, M. (2000). Sensitivity and uncertainty analysis in optimization programs using an evolutionary approach: a maintenance application. Reliability Engineering & System Safety. 67(3) 249-256.

Rocco, C.M. (2002). Maintenance optimization under uncertainties using interval methods & evolutionary strategies. In Annual Reliability and Maintainability Symposium. 254-259.

Salazar, D., Martorell, S., and Galván, B. (2005). Analysis of representation alternatives for a multiple-objective floating bound scheduling problem of a nuclear power plant safety system. In Proceedings of Evolutionary and Deterministic Methods for Design, Optimization and Control with Applications to Industrial and Societal Problems EUROGEN 2005, FLM, Munich.

Sasaki, D. (2005). Adaptive Range Multi-Objective Genetic Algorithms and Self-Organizing Map for Multi-Objective Optimization Problem. In Proceedings of Evolutionary and Deterministic Methods for Design, Optimization and Control with Applications to Industrial and Societal Problems EUROGEN 2005, FLM, Munich.

Schaffer, J.D. (1985). Multiple objective optimization with vector evaluated genetic algorithms. In J.J. Grefenstette (Ed.), Proceedings of an International Conference on Genetic Algorithms and Their Applications, Pittsburgh, PA, pp. 93-100. sponsored by Texas Instruments and U.S. Navy Center for Applied Research in Artificial Intelligence (NCARAI).

Srinivas, N., and Deb, K. (1994). Multiobjective optimization using nondominatedsorting in genetic algorithms. Evolutionary Computation 2(3), 221-248.

Tsai, Y., Wang, K., and Teng, H. (2001). Optimizing preventive maintenance for mechanical components using genetic algorithms. Reliability Engineering & System Safety 74, 89-97.

Tong, J., Mao, D., and Xue, D. (2004). A genetic algorithm solution for a nuclear power plant risk–cost maintenance model. Nuclear engineering and design. 229, 81-89.

Utyuzhnikov, S.V. (2005). Numerical Method for Generating the Entiere Pareto Frontier in Multiobjective Optimization. In Proceedings of Evolutionary and Deterministic Methods for Design, Optimization and Control with Applications to Industrial and Societal Problems EUROGEN 2005, FLM, Munich.

Vesely, W. (1999). Principles of resource-effectiveness and regulatory-effectiveness for risk-informed applications: Reducing burdens by improving effectiveness. Rel. Engng & System Safety; 63, 283-292.

Vinod, G., Kushwaha, A.W. (2004). Optimisation of ISI interval using genetic algorithms for risk informed in-service inspection. Rel. Engng & System Safety; 86, 307-316.

Winter, G., Galván, B., Alonso, S., and Mendez, M. (2005). New Trends in Evolutionary Optimization and its Impact on Dependability Problems. In Proceedings of Evolutionary and Deterministic Methods for Design, Optimization and Control with Applications to Industrial and Societal Problems EUROGEN 2005, FLM, Munich.

Yang, J-E., Hwang, M-J., Sung, T-Y, and Jin, Y. (1999). Application of genetic algorithm for reliability allocation in nuclear power plants. Reliability Engineering & System Safety 1999, (65) 229-238.

Zitzler E. (1999). Evolutionary Algorithms for Multiobjective Optimization: Methods and Applications. PhD Thesis. Swiss Federal Institute of Technology, Zurich.

Zitzler E., Laumanns M. and Thiele L. (2001). SPEA2: Improving the strength pareto evolutionary algorithm. TIK-report 2001. Computer Engineering and Networks Lab (TIK) Swiss Federal Institute of Technology. Zurich. (http://www.tik.ee.ethz.ch/~zitzler/).

Zitzler, E., Laumanns, M., Thiele, L., Fonseca, C.M. And Grunert da Fonseca, V. (2003). Performance Assessment of Multiobjective Optimizers: An Analysis and Review. IEEE Transactions on Evolutionary Computation 7(2), pp 117-132.

Genetic Algorithms and Monte Carlo Simulation for the Optimization of System Design and Operation

Marzio Marseguerra, Enrico Zio

Polytechnic of Milan, Italy

Luca Podofillini

Paul Scherrer Institute, Switzerland

4.1 Introduction

This Chapter presents the effective combination of Genetic Algorithms (GA) and Monte Carlo (MC) simulation for the optimization of system design (e.g. choice of redundancy configuration and components types, amount of spare parts to be allocated) and operation (e.g. choice of surveillance test and/or maintenance intervals). The Chapter addresses both methodological aspects and applications, summarizing the work performed in this area by the authors at the Laboratorio di Analisi di Segnale ed Analisi di Rischio (LASAR, Laboratory of Signal Analysis and Risk Analysis) of the Department of Nuclear Engineering of the Polytechnic of Milan (http://lasar.cesnef.polimi.it).

In particular, the focus of the Chapter is on the application of the combined GA-MC approach to system optimization problems in presence of uncertainty. The issue arises from the fact that in practical industrial applications, the models used for the evaluation of the risk/reliability/ availability/cost performance of a system are based on parameters which are often uncertain, e.g. components' failure rates, repair times, maintenance activity outcomes and the like. Indeed, although in principle several sources are available for the estimation of these parameters, such as historical data, extrapolated life testing, physics-of-failure models, in practice it is often the case that only few, sparse data are available, so that the obtained parameters' values generally lack the necessary accuracy.

M. Marseguerra et al.: *Genetic Algorithms and Monte Carlo Simulation for the Optimization of System Design and Operation*, Computational Intelligence in Reliability Engineering (SCI) **39**, 101–150 (2007)
www.springerlink.com

4.1.1 Motivations for the GA-MC Approach

In general, the optimization of the design and operation of an engineered system implies the development of:

- quantitative models to evaluate the performance of alternative solutions with respect to the relevant risk/reliability/availability/cost measures;
- an efficient searching algorithm to identify the optimal solutions with respect to the relevant system performance measures.

The combined GA-MC approach presented in this Chapter has been propounded to efficiently face the challenges which arise in practice with respect to both of the above development tasks.

4.1.1.1 Use of Monte Carlo Simulation for the System Modeling

Accurate system modeling has become of paramount importance for the operation, control and safety of the complex modern industrial systems. The use of realistic risk models is, for example, mandatory when dealing with hazardous complex systems, such as nuclear or chemical plants and aerospace systems, for their severe safety implications [1, 3-4]. Detailed models are also required to capture the complexity of modern technological network systems, e.g. computer and communication systems [5], power transmission and distribution systems [6, 7], rail and road transportation systems [8], oil/gas systems [8-9], with respect to their dependability, safety and security performance.

The models must capture the realistic aspects of system operation, such as components' ageing and degradation, maintenance strategies, complex physical interconnections and dependencies among the components, operational and load-sharing dependencies, spare parts allocation logics, multiple performance levels of the components, etc. In this respect, analytical modeling approaches encounter severe difficulties.

On the contrary, the MC simulation method offers great modeling flexibility and potential for adherence to reality [1]. For this reason, MC simulation has been chosen for carrying out the modeling task relative to our research work. This has allowed us to capture, in particular, the following relevant features of the management and operation of industrial systems:

- operational dependencies and test/maintenance procedures [10-17]

- multiple performance levels of the components and of the system [18-21]
- spare parts management [22]
- network connections among the systems components [23-25]

4.1.1.2 Use of Multi-Objective Genetic Algorithms for the System Optimization

In practical industrial applications, the choice of the optimal system design (e.g. choice of redundancy configuration and components, amount of spare parts to be allocated) and management (e.g. choice of surveillance test and/or maintenance intervals) involves a search among a combinatorial number of potential alternatives which makes the problem NP-hard, and thus difficult to tackle with classical optimization methods, e.g. of the gradient–descent kind [26].

In our research work, the combinatorial optimization problems of interest have been tackled by means of Genetic Algorithms (GAs). These are numerical search tools operating according to procedures that resemble the principles of natural selection and genetics [27-28]. They are capable of handling multi-variate, nonlinear objective functions and constraints and rely, for their search, simply on information relative to the objective function to be optimized, with no need for the evaluation of any of its derivatives. In a standard single-objective optimization, the GA proposes a set of candidate solutions (e.g. system designs) and the performance of each of them is measured through the value of an objective function, called fitness (e.g. cost, with reliability constraints, or vice versa). The set of candidate solutions is then allowed to evolve according to probabilistic, 'genetic' rules which effectively guide the search towards a near-optimal solution. Because of their flexibility and global perspective, GAs have been successfully used in a wide variety of problems in several areas of engineering and life science. In recent years, an increasing number of GAs applications to single-objective optimization problems has been observed in the field of reliability, maintainability and availability analysis [32-35]. An updated bibliography of such applications can be found in [36].

In practice, when attempting to optimize an engineered system, the analyst is typically faced with the challenge of simultaneously achieving several targets (e.g. low costs, high revenues, high reliability, low accident risks), some of which may very well be in conflict so that the final choice is necessarily a compromised solution. A common approach to this kind of problem is that of focusing the optimization on a single objective constituted by a weighed combination of some of the targets and imposing some constraints to satisfy other targets and requirements [26]. This approach, however, introduces a strong arbitrariness in the a priori definition of the weights and constraints levels, as a consequence of the subjective homogenization of physically different targets, usually all translated in monetary terms.

A more informative approach is one which considers all individual targets separately, aiming at identifying a set of solutions which are equivalent in absence of an assigned preference on the various objectives. Each member of this set is better than or equal to the others of the set with respect to some, but not all, of the targets. Differently from the single objective approach which often implies poor performance of other desired objectives, the set identified by the multiobjective approach provides a spectrum of 'acceptable' solutions among which a compromise may be found a posteriori.

The application of genetic and other evolutionary algorithms to multiobjective optimization problems in the framework of the latter approach just presented is subject to great attention in the technical community, as demonstrated by the flourishing of publications in the field [38-45]. In few anticipatory words, the extension of GAs for the treatment of multiobjective optimization problems entails comparing two solutions (chromosomes, in GA terminology) with respect to the multiple objectives considered [44-45]. In the case of the maximization (minimization) of a single-objective, the comparison is trivial since a vector solution $\boldsymbol{x}$ is better than $\boldsymbol{y}$ if the corresponding objective function (fitness, in GA terminology) value $f(\boldsymbol{x})$ is greater (smaller) than $f(\boldsymbol{y})$. When the situation is extended to N_f objective functions $f_i(\cdot)$, i=1, 2, ..., N_f, the solutions must be compared in terms of *dominance* of one over the other with respect to all N_f objectives. Convergence of the multiobjective search process is achieved on a so-called Pareto-optimal set of nondominated solutions, which represents the information at disposal for the decision maker to identify a posteriori the solution corresponding to his or her preferred Pareto-optimal point in the objective functions space.

4.1.1.3 Combination of GA-MC

In the GA-MC optimization framework, the MC simulation model for the evaluation of the performance of a solution in terms of the risk/reliability/ availability/cost measures is embedded in the GA engine for the search of the optimal solutions. This entails that a Monte Carlo simulation of the system be performed for each candidate solution considered in the successive generations of the search. For a detailed MC model of a realistic system, this would be very time consuming and inefficient. In this respect, the GA evolution strategy itself can be exploited to overcome this obstacle. As we shall see later, during the GA search, the best-performing chromosomes (giving rise to high or low values of the objective functions, depending on whether the optimization problem is a maximization or a minimization) appear a large number of times in the successive generations whereas the bad-performing solutions are readily eliminated by "natural selection" [32, 16-17]. Thus, each chromosome can be evaluated by a limited number of MC trials and whenever it is re-proposed by the natural selection process driving the GA search, its objective functions estimates can be accumulated with the previous ones. The large number of times a 'good' chromosome is proposed by natural selection allows achieving at the end statistically significant results [16-17]. This way of proceeding also avoids wasting time on 'bad' chromosomes which are selected, and MC-simulated, only a small number of times during the search. This approach has been named 'drop-by-drop simulation' for its similarity to this way of filling a glass of water.

4.1.2 Application to System Optimization Under Uncertainty

The focus of the Chapter is on the application of the combined GA-MC approach to system optimization problems in presence of uncertainty. The issue arises from the fact that in practical industrial applications, the models used for the evaluation of the risk/reliability/availability/cost performance of a system are based on parameters which are often uncertain, e.g. components' failure rates, repair times, maintenance activity outcomes and the like. Indeed, although in principle several sources are available for the estimation of these parameters, such as historical data, extrapolated life testing, physics-of-failure models, in practice it is often the case that only few, sparse data are available, so that the obtained paramcters' values generally lack the necessary accuracy. As a cones-quence, the uncertainty on the parameters propagates to the model output, i.e. to the estimate of the system performance.

Particularly when the implications of failures are severe, as for example in the safety systems of hazardous plants such as the nuclear ones, there must be high confidence that the required performance is attained by the system design and operation. Thus, solutions have to be identified which give an optimal performance with a high degree of confidence under the existing uncertainties. This aspect also entails adopting a multiobjective point of view [13, 23] and can be effectively tackled within a genetic algorithm search for the Pareto-optimal solutions with respect to the system objectives of maximum expected reliability/availability (or minimum expected risk/cost) and minimum variance of these measures.

In this respect, the application presented at the end of the Chapter deals with the risk-informed optimization of the technical specifications of a nuclear safety system when uncertainty exists in the parameters of the MC model which evaluates the system availability/safety performance and the costs associated to the occurrence of undesired events, such as system failures, downtimes, etc [13].

4.1.3 Structure of the Chapter

The Chapter is structured as follows. Section 2 summarizes the basics of MC simulation as applied for modeling the risk/reliability/availability behavior of industrial systems. In Section 3, the basic principles behind the GA search procedure are outlined with respect to single objective optimization. In Section 4, the synergic GA-MC approach is detailed. Section 5 shows the potentialities of the method on a single-objective reliability allocation problem [16]. The details for the extension of the GA search procedure to multiobjective optimization problems are presented in Section 6 and a case study regarding the optimization of the technical specifications of a nuclear safety system is given in Section 7 [13]. Conclusions are drawn at closure.

4.2 Fundamentals of Monte Carlo Simulation

The development of computer power has led to a strong increase in the use of MC methods for system engineering calculations.

In the past, restrictive assumptions had to be introduced to fit the system models to the numerical methods available for their solution, at the cost of drifting away from the actual system operation and at the risk of obtaining sometimes dangerously misleading results. Thanks to the inherent modeling flexibility of MC simulation, these assumptions can be relaxed, so that realistic operating rules can be accounted for in the system models for reliability, maintainability and safety applications.

This Section synthesizes the principles underlying the MC simulation method for application to the evaluation of the reliability and availability of complex systems. The majority of the enclosed material is taken from the book *"Basics of the Monte Carlo Method with Application to System Reliability"* by M. Marseguerra and E. Zio, LiLoLe-Verlag GmbH (Publ. Co. Ltd.), 2002 [1].

4.2.1 The System Transport Model

Let us consider a system whose state is defined by the values of a set of variables, i.e. by a point P in the phase space Ω. The evolution of the system, i.e. the succession of the states occupied in time, is a stochastic process, represented by a trajectory of P in Ω. The system dynamics can be studied by calculating the ensemble values of quantities of interest, e.g. probability distributions of state occupancy and expected values.

The MC method allows generating the sample function of the ensemble by simulating a large number of system stochastic evolutions. Every MC trial, or history, simulates one system evolution, i.e. a trajectory of P in Ω. In the course of the simulation, the quantities of interest are accumulated in appropriate counters. At the end of the simulation, after the large number of trials has been generated, the sample averages of the desired quantities represent the MC (ensemble) estimates of the quantities themselves [1-4].

In the case of interest here, the system is made up of N_c physical components (pumps, valves, electronic circuitry, and so on) subject to failures and repairs that occur stochastically in time. Each component can be in a variety of states, e.g. working, failed, standby, etc., and for each component we assign the probabilities for the transitions between different states. Each MC trial represents one possible realization of the ensemble of the stochastic process. It describes what happens to the system during the time interval of interest, called *mission time*, in terms of state transitions, i.e. the sequence of states randomly visited by the N_c physical components, starting from a given initial configuration.

The next Section provides details on the actual MC procedure for generating the system life histories and thereby estimating the system reliability and availability.

4.2.2 Monte Carlo Simulation for Reliability Modeling

The problem of estimating the reliability and availability of a system can be framed in general terms as the problem of evaluating functionals of the kind [1]:

$$G(t) = \sum_{k \in \Gamma} \int_0^t \psi(\tau, k) R_k(\tau, t) d\tau \tag{1}$$

where $\psi(\tau, k)$ is the probability density of entering a state k of the system at time τ and Γ is the set of possible system states which contribute to the function of interest $R_k(\tau, t)$.

In particular, the functionals we are interested in for reliability applications are the system unreliability and unavailability at time t, so that Γ is the subset of all failed states and $R_k(\tau,t)$ is unity, in the former case of unreliability, or the probability of the system not exiting before t from the failed state k entered at $\tau < t$, in the latter case of unavailability. Note that the above expression (1) is quite general, independent of any particular system model for which the $\psi(\tau,k)$'s are computed. In what follows we show the details on how eq (1) can be solved by MC simulation, with reference to the system unavailability estimation problem which provides a more interesting case than unreliability because of the presence of repairs.

Let us consider a single trial and suppose that the system enters a failed state $k \in \Gamma$ at time τ_{in}, exiting from it at the next transition at time τ_{out}. The time is suitably discretized in intervals of length Δt and counters are introduced which accumulate the contributions to $G(t)$ in the time channels: in this case, we accumulate a unitary weight in the counters for all the time channels within $[\tau_{in}, \tau_{out}]$, to count that in this realization the system is unavailable. After performing a large number, M, of MC histories, the content of each counter divided by the time interval Δt and by the number of histories gives an estimate of the unavailability at that counter time (Fig. 1). This procedure corresponds to performing an ensemble average of the realizations of the stochastic process governing the system life.

The system transport formulation of Eq. (1) suggests another analog MC procedure, consistent with the solution of definite integrals by MC sampling [1-4]. In a single MC trial, the contributions to the unavailability at a generic time t are obtained by considering all the preceding entrances, during this trial, into failed states $k \in \Gamma$. Each such entrance at a time τ gives rise to a contribution in the counters of unavailability for all successive times t up to the mission time, represented by the probability $R_k(\tau,t)$ of remaining in that failed state at least up to t. In case of a system made up of components with exponentially distributed failure and repair times, we have

$$R_k(\tau,t) = e^{-\lambda^k (t-\tau)} \tag{2}$$

where λ^k is the sum of the transition rates (repairs or further failures) out of state k.

Again, after performing all the MC histories, the contents of each unavailability counter are divided by the time channel length and by the total number of histories M to provide an estimate of the time-dependent unavailability (Figure 2).

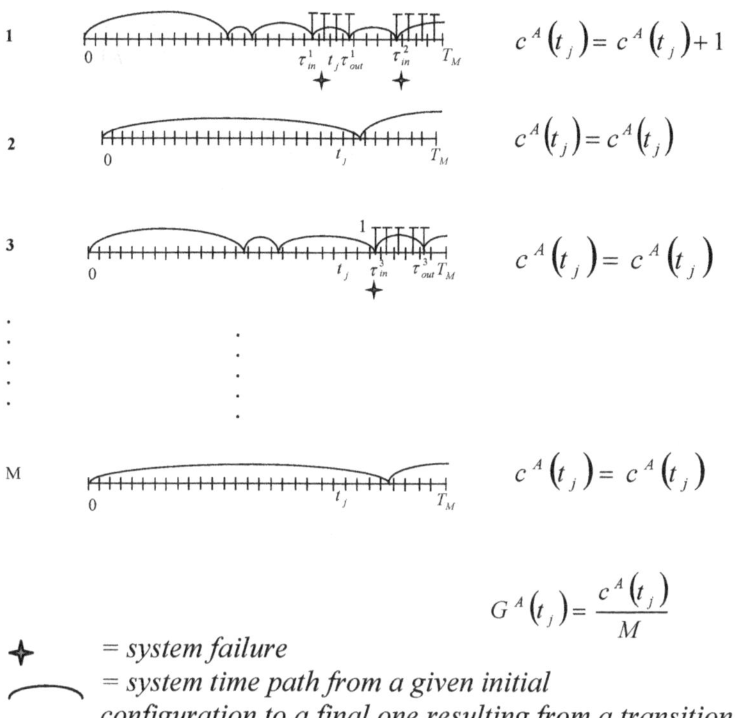

Fig. 1. MC-analog unavailability estimation procedure of collecting unitary weights in the counters $c^A(t_j)$ corresponding to the time channels t_j in which the system is failed. In the first trial the system enters the failed configuration $k_1 \in \Gamma$ at τ^1_{in} and exits at τ^1_{out}; then it enters another, possibly different, failed configuration $k_2 \in \Gamma$ at τ^2_{in} and does not exit before the mission time T_M. In the third trial, the system enters the failed configuration $k_3 \in \Gamma$, possibly different from the previous ones, at τ^3_{in} and exits at τ^3_{out}. The quantity $G^A(t_j)/\Delta t$ is the MC estimate of the instantaneous system unavailability at time t_j.

The two analog MC procedures presented above are equivalent and both lead to satisfactory estimates. Indeed, consider an entrance in state $k \in \Gamma$ at time τ, which occurs with probability density $\psi(\tau,k)$, and a subsequent time t: in the first procedure a one is scored in the counter pertaining to t only if the system has not left the state k before t and this occurs with probability $R_k(\tau,t)$. In this case, the collection of ones in t obeys a Bernoulli process with parameter $\psi(\tau,k)\cdot R_k(\tau,t)$ and after M trials the mean contribution to the unavailability counter at t is given by $M\cdot\psi(\tau,k)\cdot R_k(\tau,t)$.

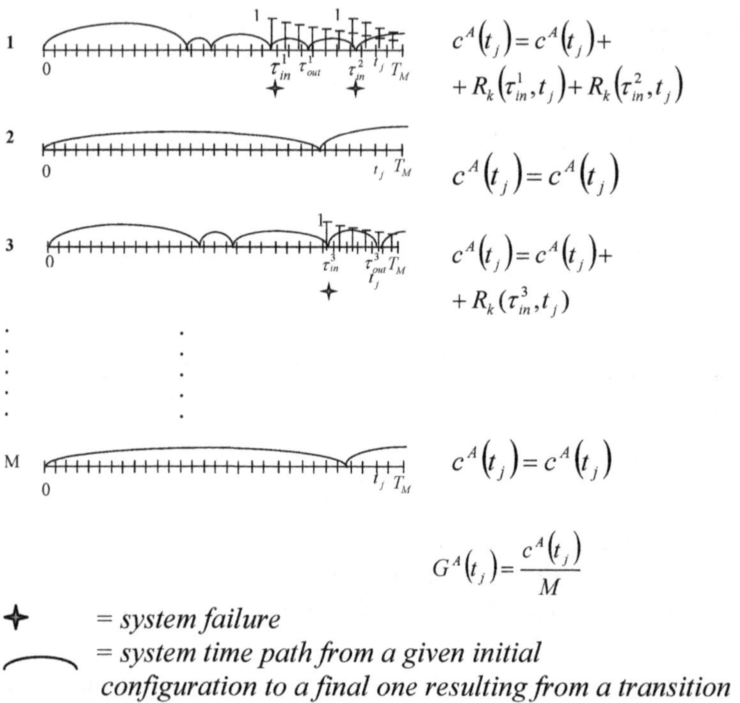

Fig. 2. MC-analog unavailability estimation procedure of collecting $R_K(\tau_{in}, t_j)$ contributions in all counters t_j following a failure to configuration k occurred at τ_{in}. In the first trial the system enters the failed configuration $k_1 \in \Gamma$ at τ^1_{in} and exits at τ^1_{out}; then it enters another, possibly different, failed configuration $k_2 \in \Gamma$ at τ^2_{in} and does not exit before the mission time T_M. In the third trial, the system enters the failed configuration $k_3 \in \Gamma$, possibly different from the previous ones, at τ^3_{in} and exits at τ^3_{out}. The quantity $G^A(t_j)/\Delta t$ is the MC estimate of the instantaneous system unavailability at time t_j.

Thus, the process of collecting ones in correspondence of a given t over M MC trials and then dividing by M and Δt leads to estimating the quantity of interest $G(t)$. On the other hand, the second procedure leads, in correspondence of each entrance in state $k \in \Gamma$ at time τ, which again occurs with probability density $\psi(\tau,k)$, to scoring a contribution $R_k(\tau,t)$ in all the counters corresponding to $t>\tau$ so that the total accumulated contribution in all the M histories is again $M\cdot\psi(\tau,k)\cdot R_k(\tau,t)$.

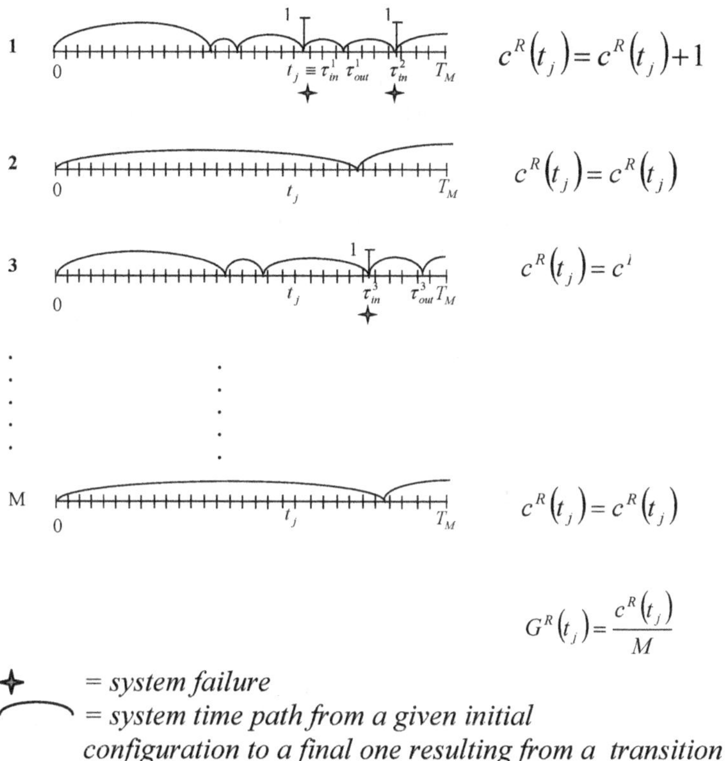

Fig. 3. MC-analog unreliability estimation procedure. In the first trial the system enters the failed configuration $k_1 \in \Gamma$ at τ_{in}^1 and exits at τ_{out}^1; then it enters another, possibly different, failed configuration $k_2 \in \Gamma$ at τ_{in}^2 and does not exit before the mission time T_M. However, if one is interested only in the unreliability estimation, this latter part of the trial does not matter. In the third trial, the system enters the failed configuration $k_3 \in \Gamma$, possibly different from the previous ones, at τ_{in}^3 and exits at τ_{out}^3. The quantity $G^R(t_j)$ is the MC estimate of the system unreliability in $[0,t_j]$.

Dividing the accumulated score by M yields the estimate of $G(t)$. In synthesis, given $\psi(\tau,k)$, with the first procedure for all t's from τ up to the next transition time we collect a one with a Bernoulli probability $R_k(\tau,t)$, while with the second procedure we collect $R_k(\tau,t)$ for all t's from τ up to the mission time: the two procedures lead to equivalent ensemble averages, even if with different variances. We shall not discuss further the subject of the variance, for space limitation.

The MC procedures just described, which rely on sampling realizations of the random transitions from the true probability distributions of the system stochastic process, are called "analog" or "unbiased". Different is the case of a non-analog or biased MC computation in which the probability distributions from which the transitions are sampled are properly varied so as to render the simulation more efficient. The interested reader can consult Refs. [1-4, 10] for further details on this.

Finally, in the case that the quantity of interest $G(t)$, $t \in [0,T_M]$, is the system unreliability, $R_k(\tau,t)$ is set equal to one so that the above MC estimation procedure still applies with the only difference being that in each MC trial a one is scored only once in all time channels following the first system entrance at τ in a failed state $k \in \Gamma$ (Figure 3).

4.3 Genetic Algorithms

4.3.1 Introduction

As a first definition, it may be said that genetic algorithms are numerical search tools aiming at finding the global maxima (or minima) of given real *objective functions* of one or more real variables, possibly subject to various linear or non linear constraints [27]. Genetic algorithms have proven to be very powerful especially when only little about the underlying structure in the model is known. They employ operations similar to those of natural genetics to guide their path through the search space, essentially embedding a survival of the fittest optimization strategy within a structured, yet randomized, information exchange [28].

In synthesis, the relevant features of these algorithms are that the search is conducted *i)* using a population of multiple solution points or candidates, *ii)* using operations inspired by the evolution of species, such as *breeding* and *genetic mutation, iii)* based on probabilistic operations, *iv)* using only information on the objectives or search functions and not on its derivatives. Typical paradigms belonging to the class of evolutionary computing are *genetic algorithms* (*GAs*), *evolution strategies* (*ESs*), *evolutionary programming* (*EP*) and *genetic programming* (*GP*). In the following we shall focus on GAs.

Since the GAs operations are inspired by the rules of natural selection, the corresponding jargon contains many terms borrowed from biology, suitably redefined to fit the algorithmic context. Thus, it is conventional to say that a GA operates on a set of (artificial) *chromosomes*, which are strings of numbers, most often binary. Each chromosome is partitioned in (artificial) *genes*, i.e. each bit-string is partitioned in as many substrings of assigned lengths as the arguments of the objective function to be optimized. Thus, a generic chromosome represents an encoded trial solution to the optimization problem. The binary genes constitute the so called *genotype* of the chromosome; upon decoding, the resulting substrings of real numbers are called *control factors* or *decision variables* and constitute the *phenotype* of the chromosome. The value of the objective function in correspondence of the values of the control factors of a chromosome provides a measure of its 'performance' (*fitness*, in the GA jargon) with respect to the optimization problem at hand.

Figure 4 shows the constituents of a chromosome made up of three genes and the relation between the genotype and the external environment, i.e. the phenotype, constituted by three control factors, x_1, x_2, x_3, one for each gene. The passage from the genotype to the phenotype and vice versa is ruled by the phenotyping parameters of all genes, which perform the coding/decoding actions. Each individual is characterized by a fitness, defined as the value of the objective function f calculated in correspondence of the control factors pertaining to that individual. Thus a population is a collection of points in the solution space, i.e. in the space of f.

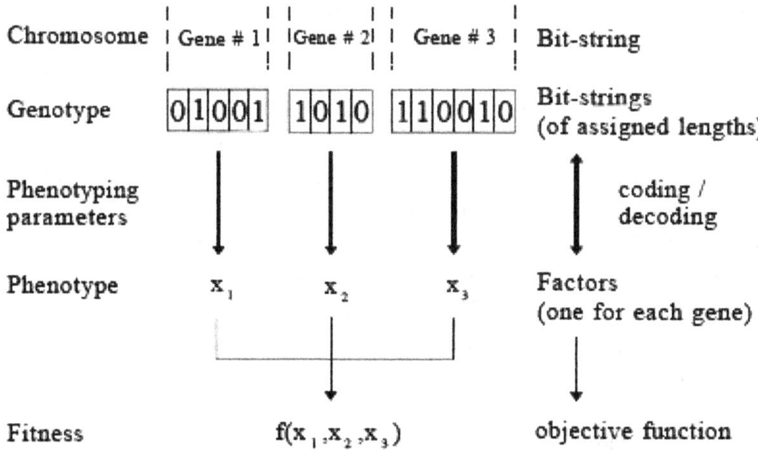

Fig. 4. Components of an individual (a chromosome) and its fitness.

The GA search for the optimal (highest fitness) solution is carried out by generating a succession of *populations* of chromosomes, the individuals of each population being the *children* of those of the previous population and the *parents* of those of the successive population. The initial population is generated by randomly sampling the bits of all the strings. At each step (*generation*) of the search, a new population is constructed by manipulating the strings of the old population in a way devised so as to increase the mean fitness. By analogy with the natural selection of species, the string manipulation consists in selecting and mating pairs of chromosomes of the current population in order to groom chromosomes of the next population. This is done by applying the four fundamental operations of reproduction, crossover, replacement and mutation, all based on random sampling [28]. In the literature, the two operations of crossover and replacement are often unified and the resulting operation is called crossover, so that the fundamental operations are then three and will be illustrated in some details below. The evolution of the chromosomes population is continued until a pre-defined termination criterion is reached.

An important feature of a population, which greatly affects the success of the search, is its genetic diversity, a property which is related to both the size of the population and, as later discussed, to the procedures employed for its manipulation. If the population is too small, the scarcity of genetic diversity may result in a population dominated by almost equal chromosomes and then, after decoding the genes and evaluating the objective function, in the quick convergence of the latter towards an optimum which may well be a local one. At the other extreme, in too large populations, the overabundance of genetic diversity can lead to clustering of individuals around different local optima: then the mating of individuals belonging to different clusters can produce children (newborn strings) lacking the good genetic part of either of the parents. In addition, the manipulation of large populations may be excessively expensive in terms of computer time.

In most computer codes the population size is kept fixed at a value set by the user so as to suit the requirements of the model at hand. The individuals are left unordered, but an index is sorted according to their fitness. During the search, the fitness of the newborn individuals are computed and the fitness index is continuously updated.

Regarding the general properties of GAs, it is acknowledged that they take a more global view of the search space than many other optimization methods and that their main advantages are *i*) fast convergence to near global optimum, *ii*) superior global searching capability in complicated search spaces, *iii*) applicability even when gradient information is not readily achievable. The first two advantages are related to the population-based searching property. Indeed, while the gradient method determines the next searching point using the gradient information at the current searching point, the GA determines the next set of multiple search points using the evaluation of the objective function at the current multiple searching points. When only gradient information is used, the next searching point is strongly influenced by the local information of the current searching point so that the search may remain trapped in a local minimum. On the contrary, the GA determines the next multiple searching points using the fitness values of the current searching points which are spread throughout the searching space, and it can also resort to the additional mutation to escape from local minima.

4.3.2 The Standard GA Procedure

As said in the Introduction, the starting point of a GA search is the uniform random sampling of the bits of the N_p chromosomes constituting the initial population. In general, this procedure corresponds to uniformly sampling each control factor within its range. However, the admissible hypervolume of the control factors values may be only a small portion of that resulting from the Cartesian product of the ranges of the single variables, so that conditional sampling may be applied to reduce the search space by accounting for the possible constraints among the values that the control factors can take [30]. In this case, during the creation of the initial population and the successive phase of *chromosome replacement* below described, a chromosome is accepted only if suitable constraining criteria are satisfied.

Then, the evolution of the chromosomes population from the generic n-th generation to the successive $(n+1)$-th is driven by the so called *breeding algorithm* which develops in four basic steps.

The first step is termed *reproduction* and consists in the construction of a temporary new population. This is obtained by resorting to the following Standard Roulette Selection procedure: *i)* the cumulative sum of the fitnesses of the individuals in the old population is computed and normalized to sum to unity; *ii)* the individuals of the new population are randomly sampled, one at a time with replacement, from this cumulative sum which then plays the role of a cumulative distribution function (cdf) of a discrete random variable (the fitness-based rank of an individual in the old population). By so doing, on the average in the temporary new population individuals are present in proportion to their relative fitness in the old population. Since individuals with large fitness values have more chance to be sampled than those with low fitness values, the mean fitness of the temporary new population is larger than that of the old one.

The second step of the breeding procedure is the *crossover* which is performed as indicated in Figure 5: *i)* $N_p/2$ pairs of individuals are uniformly sampled, without replacement, from the temporary new population to serve as parents; *ii)* the genes of the chromosomes forming a parents pair are divided into two corresponding portions by inserting at random a separator in the same position of both genes (one-site crossover); *iii)* finally, the head portions of the genes are swapped. By so doing, two children chromosomes are produced, which bear a combination of the genetic features of their parents. A variation of this procedure consists in performing the crossover with an assigned probability p_c (generally rather high, say $p_c \geq 0.6$): a random number r is uniformly sampled in (0,1] and the crossover is performed only if $r < p_c$. Viceversa, if $r \geq p_c$, the two children are copies of the parents.

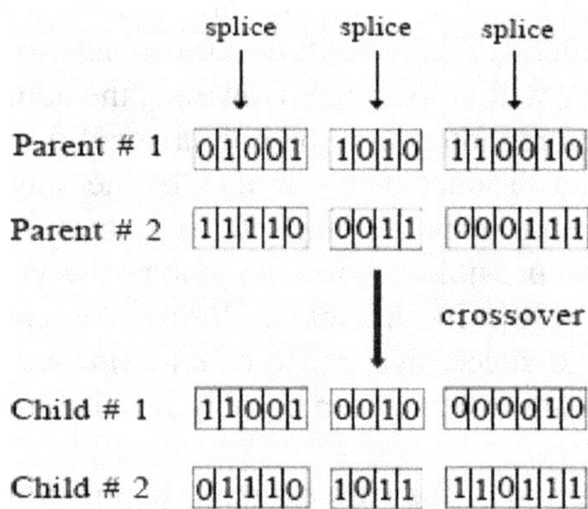

Fig. 5. Example of crossover in a population with chromosomes constituted by three genes.

The third step of the breeding procedure concerns the *replacement* in the new population of two among the four individuals involved in a breeding operation, i.e. the two parents and their two children. The simplest recipe, again inspired by natural selection, simply consists in substituting the parents by their children (children live, parents die). Hence, in this case, each individual breeds only once.

The fourth and last step of the breeding procedure is the random *mutation* of some bits in the population, i.e. the swapping of some bits from their actual values to the opposite ones ($0 \rightarrow 1$ and viceversa). The mutation is performed on the basis of an assigned probability of swapping a single bit (generally quite small, say 10^{-3}). The product of this probability by the total number of bits in the population gives the mean number m of mutations. The bits to be actually mutated are located by uniform random sampling of their positions within the entire bit population. If $m < 1$, a single bit is mutated with probability m.

The evolving sequence of successive population generations is usually stopped according to one of the following criteria of search termination:

1. when the mean fitness of the individuals in the population increases above an assigned convergence value;
2. when the median fitness of the individuals in the population increases above an assigned convergence value;
3. when the fitness of the best individual in the population increases above an assigned convergence value. This criterion guarantees that at least one individual is 'good enough';
4. when the fitness of the weakest individual in the population drops below an assigned convergence value. This criterion guarantees that the whole population is 'good enough';
5. when the assigned number of population generations is reached.

There exist several alternative ways to perform the above described procedures, developed in an attempt to reducing chances for premature convergence of the algorithm to local minima, favouring the genetic diversity in the population, speeding up convergence towards the currently identified best solutions. For a detail discussion on these variants, the reader should refer to [28, 31]

4.4 Embedding the Monte Carlo Simulation in the Genetic Algorithm Search Engine

The need for finding an efficient way to combine the GA search with the MC simulation arises from the fact that a large number of potential alternatives is involved in practical system design and operation optimization problems so that performing a crude search by spanning all possible alternatives and then running a full Monte Carlo simulation (thousands of life histories) for each of them is impractical. The situation is not significantly improved if the search for the optimal solution is performed by means of GAs: still a MC code should be run for each individual of the chromosomes' population considered in the successive generations of the search and this would be very time consuming and inefficient. A possible solution to this problem follows from the consideration that during the GA search, the best-performing (high fitness) chromosomes appear a large number of times in the successive generations whereas the badly-performing chromosomes are readily eliminated [16-17, 32]. Thus, for each proposed chromosome, one can run a limited number of MC trials. More precisely, in our calculations, when a chromosome (coding for example a proposed system configuration or surveillance test and/or maintenance strategy for the system components) is proposed by the GA, the rough estimates of the risk/reliability/availability/cost performances of the system are obtained on the basis of a small number of MC trials. During the GA evolution, an archive of the best chromosome-solutions obtained in previous MC runs and the corresponding MC objective functions estimates, are updated. Whenever a chromosome is re-proposed, the objective functions estimates newly computed can be accumulated with those stored in the archive and the large number of times a 'good' chromosome is proposed by natural selection allows accumulating over and over the results of the few-histories runs, thus achieving at the end statistically significant results. This way of proceeding also avoids wasting time on 'bad' configurations which are simulated only a small number of times. We call this approach 'drop-by-drop simulation' for its similarity to this way of filling a glass of water [16-17]. Obviously, the 'size of the drop', i.e. the number of trials performed each time a chromosome is proposed during the search, is an important parameter and should be calibrated in such a way as to minimizing the possibility that good individuals be lost from the population when they are first proposed because of insufficient accuracy in the early, rough estimates of the objective functions. In other words, one should put care in avoiding the occurrence of situations such that the poorly estimated fitness of a 'good' individual is lower than that of a 'bad' individual.

4.5 Optimization of the Design of a Risky Plant

4.5.1 Problem Statement

In this Section, the combined GA-MC approach is applied for solving a single objective reliability allocation problem [16].

The operation and management of a plant requires proper accounting for the constraints coming from safety and reliability requirements as well as from budget and resource considerations. At the design stage, then, analyses are to be performed in order to guide the design choices in consideration of the many practical aspects which come into play and which typically generate a conflict between safety requirements and economic needs: this renders the design effort an optimization one, aiming at finding the best compromise solution.

In particular, the optimization problem here considered regards a choice among alternative system configurations made up of components which possibly differ for their failure and repair characteristics. The safety vs. economics conflict rises naturally as follows:

Choice of components: choosing the most reliable ones certainly allows the design to be on the safe side and guarantees high system availability but it may be largely non-economic due to excessive component purchase costs; on the other hand, less reliable components provide for lower purchase costs but lose availability and may increase the risk of costly accidents.

Choice of redundancy configuration: choosing highly redundant configurations, with active or standby components, increases the system reliability and availability but also the system purchase costs (and perhaps even the repair costs, if the units composing the redundancy are of low reliability); obviously, for assigned component failure and repair characteristics, low redundancies are economic from the point of view of purchase costs but weaken the system reliability and availability, thus increasing the risk of significant accidents and the system stoppage time.

These very simple, but realistic, aspects of plant design immediately call for compromise choices which optimize plant operation in view of its safety and budget constraints.

Let us assume that technical considerations have suggested that the system at hand be made up of a series of N_n nodes, each one performing a given function, as part, for example, of a manufacturing process. The task of the plant designer is now that of selecting the configuration of each node which may be done in several ways, e.g. by choosing different series/parallel configurations with components of different failure/repair characteristics and therefore of different costs. In order to guide the selection, the designer defines an objective function which accounts for all the relevant aspects of plant operation. In our model we consider as objective function the net profit drawn from the plant during the mission time T_M. This profit is based on the following items: profit from plant operation; purchase and installation costs; repair costs; penalties during downtime, due to missed delivery of agreed service; damage costs, due to damages and consequences to the external environment encountered in case of an accident. Note that the last item allows one to automatically account for the safety aspects of system operation.

Any monetary value S encountered at time t is referred to the initial time t_0 by accounting for an interest rate i_r, as follows:

$$S_0 = S(t)/(1+i_r)^{t-t_0}. \tag{3}$$

The net profit objective function G (gain) can then be written as follows:

$$G = P - (C_A + C_R + C_D + C_{ACC}) \tag{4}$$

where

- $P = P_t \cdot \int_0^{T_M} \frac{A(t)dt}{(1+i_r)^t}$ is the plant profit in which P_t is the amount of money per unit time paid by the customer for the plant service (supposed constant in time) and $A(t)$ is the plant availability at time t.
- $C_A = \sum_{i=1}^{N_C} C_i$ is the acquisition and installation cost of all N_C components of the system. This capital cost is faced at the beginning, at t_0=0, through a bank loan at interest rate i_r. We imagine that the loan be re-paid at annual constant installments of nominal value $C_A \cdot \left(i_r + \frac{i_r}{(1+i_r)^{T_M} - 1} \right)$. If the plant is shut down before T_M (e.g. because of an accident) the residual installments are paid immediately so as to avoid paying the interests in the remaining time up to T_M.
- $C_R = \sum_{i=1}^{N_C} C_{R_i} \cdot \int_0^{T_M} \frac{I_{R_i}(t)dt}{(1+i_r)^t}$ is the repair cost of all N_C components of the system with C_{Ri} being the cost per unit time of repair of component i (supposed constant in time and such that the total repair costs during the component life up to T_M equal a fixed percentage of its acquisition cost C_i) and $I_{Ri}(t)$ being a characteristic function equal to 1 if the component i is under repair at time t, 0 otherwise.
- $C_D = C_U \cdot \int_0^{T_M} \frac{[1 - A(t)]dt}{(1+i_r)^t}$ is the amount of money to be paid to the customer because of missed delivery of agreed service during downtime, with C_U (constant in time) being the monetary penalty per unit of downtime.
- $C_{ACC} = \sum_{h=1}^{N_{ACC}} I_{ACC,h} \cdot \frac{C_{ACC,h}}{(1+i_r)^{t_{ACC,h}}}$ is the amount of money to be paid for damages and consequences to the external environment in case of an accident; N_{ACC} is the number of different types of accidents that

can occur to the plant, $C_{ACC,h}$ is the accident premium to be paid when an accident of type h occurs, $I_{ACC,h}$ is an indicator variable equal to 1 if an accident of type h has happened, 0 otherwise, and $t_{ACC,h}$ is the time of occurrence of the h-accident. We assume that after an accident the plant cannot be repaired and must be shut down.

Following a first qualitative enquiry, the designer ends up considering a pool of a reasonably small number of possible configurations for each node, a priori all offering similar system performances. Now the problem becomes that of choosing a system assembly by selecting one configuration for each node, aiming at maximizing the objective function. Under realistic assumptions, the various terms of the objective function (4) do not lend themselves to an analytical formulation so that the only feasible approach for their evaluation is the MC method. Each system configuration is simulated in a specified number of histories and the mean values and standard deviations of the following quantities are estimated: T_D = system downtime; $T_R(i)$ = total time under repair of component i, i=1,...,N_C; $P_{ACC}(h)$ = probability of accident h; $A(t)$ = instantaneous system availability at time t.

By performing several MC histories of the behaviour of the specified system configuration, one obtains as many independent realizations of the relevant random variables from which estimates of the quantities of interest, T_D, $T_R(i)$, i=1,...,N_C, $P_{ACC}(h) = h = 1, 2, \ldots, N_{ACC}$, $A(t)$, $t \in [0, T_M]$ are obtained. These quantities allow us to compute the various terms constituting the profit in Eq. (4) pertaining to that system configuration and must be estimated anew for each system design examined.

4.5.2 Applying the GA-MC Approach

In order to exploit the genetic algorithm for the search of the optimal design configuration we need to code the available configurations into distinct bit-strings chromosomes. Recalling that our system is made up of N_n nodes, we identify the possible configurations of each node by an integer number so that a system configuration is identified by a sequence of integers, each one indicating a possible node configuration. A chromosome identifying a system configuration then codes N_n integers, one for each node. For the coding, we take a chromosome made up of a single gene containing all the indexes of the node configurations in a string of $n_b = \sum_{n=1}^{N_n} n_{\min}(n)$ bits, where $n_{\min}(n)$ is the minimum number of bits needed to count the alternative configurations pertaining to the n-th node of the system. For example, if for a node there are 6 alternative configurations, a 3-bit string is needed which allows the coding of up to 8 configurations. In this case, the two chromosomes corresponding to nonexistent configurations are discarded automatically before entering the evolving population.

The choice of this coding strategy, as compared to one gene dedicated to each node, is such that the crossover generates children-chromosomes with nodes all equal to the parents except for the one in which the splice occurs. This was experimentally verified to avoid excessive dispersion of the genetic patrimony thus favoring convergence.

As an example of application, let us consider the design of a system with characteristics similar to those of a shale oil plant taken from literature [3]. The system consists of $N_n = 5$ nodes corresponding to the 5 process units indicated in Fig. 6. For each node we have assumed that a design choice is required among several alternatives (Table 1): 7 for nodes A and E, 16 for node B and 14 for nodes C and D. The total number of possible system configurations is 153,664. For nodes C and D although the 4-bits coding would allow 16 possible configurations, only 14 of these are physically significant, as generated by the combination of the various kinds of components in the three different lay-outs considered. The components are assumed to have exponential failure and repair behaviors in time.

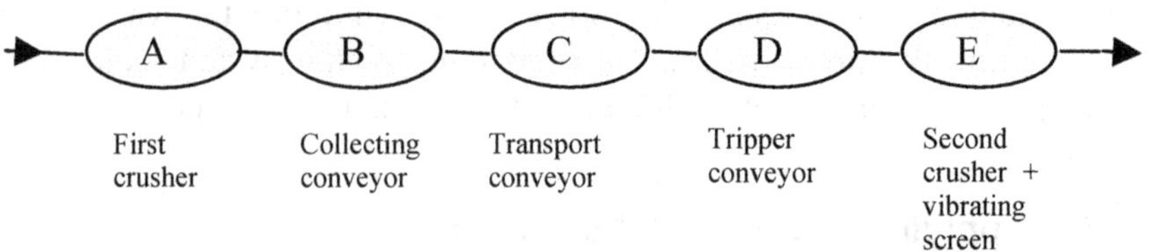

Fig. 6. Sketch of the shale oil plant [16].

The system configuration is coded into a chromosome with one 3+4+4+4+3=18-bit gene. The huge dimension of the search space has made it necessary to consider a large chromosome population of 500 individuals.

All components are assumed to be subject to degrading repairs according to a modified Brown-Proschan model which postulates that a system is repaired to an "as good as before" condition (*minimal repair*) only with a certain probability p and is, otherwise, returned in a "deteriorated" condition (*deteriorating repair*) [46]. Thus, these two conditions obey a Bernoulli distribution. Inclusion of this model within the

MC simulation scheme is straightforward. When a repair action is completed, we sample a uniform random number r in [0,1): if $r<p$, then the repair is minimal and the failure properties of the component are returned to the conditions existing prior to failure; otherwise, repair is deteriorating and the component emerges with a failure rate increased by a given percentage π_λ of its value before the failure. The analyst-defined parameter π_λ specifies the amount of deterioration induced by the failure-repair process. Repairs of deteriorated units are also assumed to become more difficult by reducing the repair rate by a given percentage π_μ of its value before the failure. In our application we take $p = 0.3$, $\pi_\lambda = 1.5$ and $\pi_\mu = 1.3$ so as to enforce the effects of degradation. Moreover, all standby's are cold (no failure when in standby mode) and the failures of nodes A and E are assumed to be accidents with damaging consequences.

Table 1. Potential node configurations [16].

Node	Number of alternative configurations	Type of components	Operational logic
A	7	a	3-out-of-3 G
			3-out-of-4 G
			3-out-of-5 G
			3-out-of-6 G
			3-out-of-7 G
			3-out-of-8 G
			3-out-of-9 G
B	16	b1, b2, b3	2-out-of-2 G
			2-out-of-3 G
C	14	c1, c2	1-out-of-1 G
			1-out-of-1 G + 1 standby
			1-out-of-1 G + 2 standby
D	14	d1, d2	1-out-of-1 G
			1-out-of-1 G + 1 standby
			1-out-of-1 G + 2 standby
E	7	e	3-out-of-3 G
			3-out-of-4 G
			3-out-of-5 G
			3-out-of-6 G
			3-out-of-7 G
			3-out-of-8 G
			3-out-of-9 G

Under these hypotheses, the total plant net profit over the mission time can no longer be evaluated analytically and MC simulation becomes the only feasible approach. Tables 2 and 3 contain the component and system failure and cost data.

Table 2. Component data [16].

Component i	Failure rate λ_i $[y^{-1}]$	Repair rate μ_i $[y^{-1}]$	Purchase cost C_i $[10^6\,\$]$	Repair cost C_{Ri} $[10^6\,\$\cdot y^{-1}]$
a	$1.5\cdot10^{-3}$	$4.0\cdot10^{-2}$	3.0	0.55
b1	$2.0\cdot10^{-4}$	$8.0\cdot10^{-3}$	5.0	10.0
b2	$2.0\cdot10^{-3}$	$8.0\cdot10^{-2}$	3.0	6.2
b3	$2.0\cdot10^{-2}$	$8.0\cdot10^{-1}$	1.0	2.1
c1	$1.0\cdot10^{-4}$	$8.0\cdot10^{-3}$	10.0	41.0
c2	$1.0\cdot10^{-3}$	$8.0\cdot10^{-2}$	5.0	20.0
d1	$1.0\cdot10^{-4}$	$8.0\cdot10^{-3}$	7.0	28.0
d2	$1.0\cdot10^{-3}$	$8.0\cdot10^{-2}$	3.0	12.0
e	$1.7\cdot10^{-3}$	$4.0\cdot10^{-2}$	5.0	0.85

Table 3. System data [16].

Profit per unit time P_t	$[10^6\,\$\cdot y^{-1}]$	20.0
Downtime penalty per unit time C_U	$[10^6\,\$\cdot y^{-1}]$	200.0
Accident 1 (node A) reimbursement cost $C_{ACC,1}$	$[10^6\,\$]$	70.0
Accident 2 (node E) reimbursement cost $C_{ACC,2}$	$[10^6\,\$]$	50.0
Interest rate i		3%
Mission time T_M	[y]	50

The data are adapted from the original data [3] to fit the model. Due to the huge number of potential solutions in the search space, the search procedure was performed in two successive steps. The genetic algorithm parameters and rules utilized in both steps are reported in Table 4. As pointed out in Section 4, such parameters and rules may be critical for the success of the procedure and need to be calibrated carefully, often based on trial and error.

For the replacement of the chromosomes in the population after crossover (Section 3.2), the fittest strategy is used 83% of the times and the weakest 17%. The former entails that out of the four individuals (two parents and two chromosomes) involved in the crossover procedure, the fittest two replace the parents; this procedure should not be used when weak individuals are largely unfavoured as potential parents in the preceding parent selection step or otherwise they have a significant chance to survive forever in the population. The latter entails that the children replace the two weakest individuals in the entire population, parents included, provided that the children fitness is larger; this technique shortens the permanence of weak individuals in the successive populations and it is particularly efficient in large populations.

Table 4. Genetic algorithm parameters and rules.

Number of chromosomes (population size)	500
Number of generations (termination criterion)	1000
Selection	Roulette
Replacement	Fittest 83%; Weakest 17%
Mutation probability	0.001
Crossover probability	1
No elitist selection	

With respect to the embedding of the MC simulation for evaluating the fitness of a GA-proposed chromosome, each MC "drop" is made up of 100 trials. To achieve an initial minimum accuracy, ten drops are run the first time a chromosome is proposed; then, one drop is added to the previous statistics each time the same chromosome is re-proposed in the successive generations. To limit the total computation time, when the fitness evaluation of a chromosome is built on 100 drops already accumulated, no more simulations are run for that chromosome, since reasonable accuracy in the fitness estimate can be considered to have been reached.

The first step has consisted in a usual genetic algorithm search spanning the whole space with a limited number of generations (25) sufficient to achieve convergence to a near optimal solution.

An analysis of the archive of the best chromosomes at convergence has allowed us to conclude that all the potential solutions with highest fitness values where characterized by the same 3-out-of-5 G redundancy configurations at the risky nodes A and E, with the difference in the net profit given by variable combinations of the configurations at the remaining nodes.

In the second step, a fine tuning of the system optimal design configuration was performed by running another GA-search procedure focused only on the 16·14·14=3136 configurations of systems having nodes B, C and D variable and A and E fixed at their optimal configuration (i.e. the 3-out-of-5 G). At convergence, this step has led to the optimal configuration of Figure 7.

Different GA runs starting from different, randomly sampled, initial populations were performed to check the reproducibility of the results.

An a posteriori interpretation of the optimal configuration found can be given on physical grounds. Nodes A and E are the risky ones so that low-redundancy configurations are not acceptable due to the high cost of accident consequences. On the other hand, high-redundancy configurations imply very large components purchase costs so that the optimal choice falls on an averagely-redundant configuration, the 3-out-of-5 G, which is capable of guaranteeing a low probability of risky failures which, in this case, turned out to be equal to $1.15 \cdot 10^{-3}$ and $1.16 \cdot 10^{-3}$ for nodes A and E, respectively (note that the failure of node A leads to worse consequences, as in Table 3).

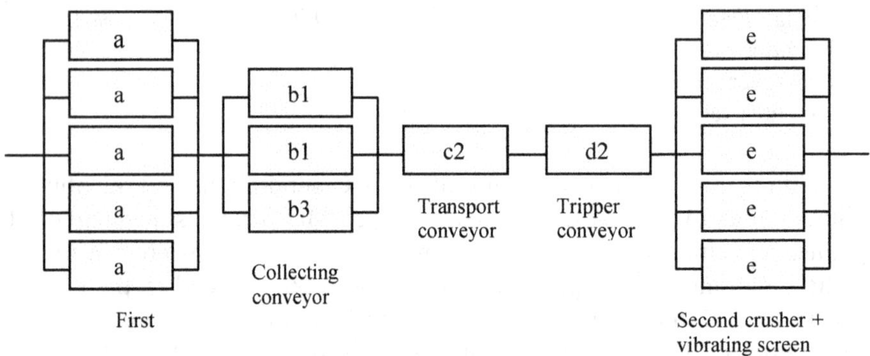

Fig. 7. Sketch of the optimal configuration for the shale oil plant [16].

The types of components which can be used to make up node B, C and D are characterized by a wide range of purchase costs. Economics would then drive the choice towards the least expensive components, but this has to be compatible with the need of minimizing system downtime which bears a strong penalization. These observations explain the choice, for node B, of the 2-out-of-3 G configuration made of a combination of two components of the most expensive and most reliable kind (b_1) and one of the least expensive and least reliable (b_3). For nodes C and D a single component among the cheapest ones is found to suffice to guarantee a high level of reliability, given that the associated mean time to failure is 20 times the mission time T_M. The chosen system configuration has an unreliability of $9.69 \cdot 10^{-3}$ and an average unavailability of $1.27 \cdot 10^{-3}$.

A glance at the archive of the best chromosomes at convergence allows us to see that the selected configuration is better than the second best of an amount of net profit ten times larger than its standard deviation, thus re-assuring on the robustness of the result. For further checking, we have run, for the three best configurations, a standard MC simulation with 10^6 trials. The results are reported in Table 5 and show that the optimal solution is indeed better than the other two, beyond statistical dispersion. The simulation took about 2 minutes of CPU time on an Alpha Station 250 4/266 for each of the three configurations.

Table 5. Monte Carlo results with 10^6 trials for the three best system configurations [16].

Configuration index in decreasing order of optimality	Total net profit at T_M [10^6 \$]
1	471.57 ± 0.08
2	470.20 ± 0.05
3	469.39 ± 0.07

4.6 Multi-objective Genetic Algorithms

The multiobjective optimization problem amounts to considering several objective functions $f_i(\boldsymbol{x})$, i=1, 2, … N_f, in correspondence of each solution $\boldsymbol{x}$ in the search space and then identifying the solution $\boldsymbol{x}^*$ which gives rise to the best compromise among the various objective functions. The comparison of two solutions with respect to several objectives may be achieved through the introduction of the concepts of *Pareto optimality* and *dominance* [44-45] which enable solutions to be compared and ranked without imposing any a priori measure as to the relative importance of individual objectives, neither in the form of subjective weights nor arbitrary constraints.

Solution $\boldsymbol{x}$ is said to (weakly) *dominate* solution $\boldsymbol{y}$ if $\boldsymbol{x}$ is better or equal on all objectives but better on at least one [28, 37, 39], i.e. if

$$f_i(\boldsymbol{x}) \geq f_i(\boldsymbol{y}) \tag{5}$$

with the inequality strictly holding for at least one i. The solutions not dominated by any other are said to be *nondominated* solutions.

Within the genetic approach, in order to treat simultaneously several objective functions, it is necessary to generalize the single-fitness procedure employed in the single-objective GA by assigning N_f fitnesses to each $\boldsymbol{x}$.

Concerning the insertion of an individual (i.e. an $\boldsymbol{x}$ value) in the population, often constraints exist which impose restrictions that the candidate individual has to satisfy and whose introduction speeds up the convergence of the algorithm, due to a reduction in the search space. Such constraints may be handled, just as in the case of single-objective GAs, by testing whether, in the course of the population creation and replacement procedures, the candidate solution fulfills the constraining criteria pertaining to all the N_f fitnesses.

Once a population of chromosomes $\{ \boldsymbol{x} \}$ has been created, they are ranked according to the Pareto dominance criterion by looking at the

N_f-dimensional space of the fitnesses $f_i(\boldsymbol{x})$, i=1, 2, ..., N_f, (see Figure 8 for N_f = 2). All non-dominated individuals in the current population are identified. These solutions are assigned the rank 1. Then, they are virtually removed from the population and the next set of non-dominated individuals are identified and assigned rank 2. This process continues until every solution in the population has been ranked.

The selection and replacement procedures of the multiobjective genetic algorithms are then based on this ranking: every chromosome belonging to the same rank class has to be considered equivalent to any other of the class, i.e. it has the same probability of the others to be selected as a parent and survive the replacement.

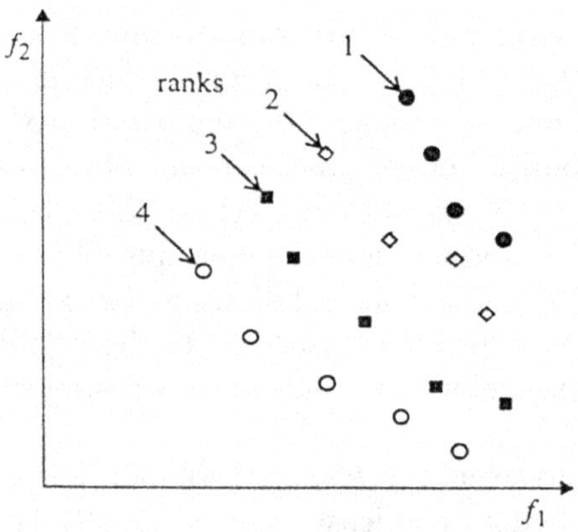

Fig. 8. Example of population ranking for a problem of maximization of two objective functions.

During the optimization search, an archive of vectors, each one constituted by a non-dominated chromosome and by the corresponding N_f fitnesses, representing the dynamic Pareto optimality surface is recorded and updated. At the end of each generation, non-dominated individuals in the current population are compared, in terms of the fitnesses, with those already stored in the archive and the following archival rules are implemented:

1. If the new individual dominates existing members of the archive, these are removed and the new one is added;
2. if the new individual is dominated by any member of the archive, it is not stored;
3. if the new individual neither dominates nor is dominated by any member of the archive then:

   – if the archive is not full, the new individual is stored.

– if the archive is full, the new individual replaces the *most similar* one in the archive (an appropriate concept of distance being introduced to measure the similarity between two individuals: in our research, we have adopted an Euclidean distance based on the values of the fitnesses of the chromosomes normalized to the respective mean values in the archive).

The archive of non-dominated individuals is also exploited by introducing an elitist parents' selection procedure which should in principle be more efficient. Every individual in the archive (or, if the archive's size is too large, a pre-established fraction of the population size N_{ga}, typically $N_{ga}/4$) is chosen once as a parent in each generation. This should guarantee a better propagation of the genetic code of non-dominated solutions, and thus a more efficient evolution of the population towards Pareto optimality.

At the end of the search procedure the result of the optimization is constituted by the archive itself which gives the Pareto optimality region.

4.7 Optimizing the Technical Specifications of the Reactor Protection Instrumentation System of a PWR

The application presented in this Section deals with the risk-informed optimization of the Technical Specifications (TSs) regarding the testing policies of the Reactor Protection Instrumentation System (RPIS) of a Pressurized Water Reactor (PWR) [13]. These TSs define the frequency and procedure with which the RPIS components (instrumentation channels, logic trains, etc.) are to be inspected. The testing frequency is defined in terms of the Surveillance Test Intervals (STIs) of the different components. The test procedure requires the RPIS components to be disconnected from service, thus reducing the safety margin. In this respect, the Allowable Bypass Time (ABT) is defined, for each component, as the time during which the plant is allowed to operate with that component being bypassed and, hence, it is the time during which the plant can operate under the associated reduced safety margin. The values of the STIs and ABTs of each component are specified in the TSs regarding the testing policies of the RPIS system.

The case study considered regards the definition of the TSs of the RPIS, taking into account the uncertainties that affect the components' reliability parameters. In this sense, the TSs need to be set in such a way as to maximize the availability of the RPIS, with a high degree of confidence. To this aim, we frame the problem into an optimization one with two objective functions to be minimized by the genetic algorithms: namely, the mean value and the variance of the system unavailability (actually, the algorithm maximizes the negative of these quantities). The multiobjective search is carried out based on the coupling of GA and MC simulation, which allows to explicitly and transparently account for parameters uncertainties. The modelling flexibility offered by the MC simulation method [1] is exploited as the engine for the evaluation of the objective function, in presence of parameters uncertainty, and Pareto optimality is used as the GA preference criteria.

4.7.1 System Description

The Reactor Protection System keeps the reactor operating within a safe region. During normal control of the reactor, the control rods are raised or lowered into the core by the use of 'magnetic jacks'. If one or more physical monitored parameters (such as neutron flux, temperature, pressurizer pressure, pressurizer water level, etc.) enter an un-acceptable range, a trip signal is produced to de-energize the electromagnetic holding power of the control rod drive mechanisms so that the control rod gravitational drop ensures an orderly shutdown of the nuclear chain reaction.

In this paper, we consider the Reactor Protection Instrumentation System (RPIS) of typical Westinghouse-designed PWRs. This consists of analog channels, logic trains, and trip breakers. The particular hardware configuration that is the subject of this paper is that of a four-loop RESAR-3S PWR-type reactor with solid-state combinatorial logic units [47, 48-49]. Here we only give a short description of the system taken from [48] and [49] where the RPIS is described in greater details.

The analog channels sense the plant parameters and provide binary (on-off) signals to the logic trains. A typical analog channel is composed of a sensor/transmitter, a loop power supply, signal conditioning circuits, and a signal comparator (bistable). The bistable compares the incoming signal to a setpoint and turns its output off if the input voltage exceeds the setpoint. Each bistable feeds two separate input relays—one associated with reactor trip logic train A and the other associated with reactor trip logic train B. Each monitored plant parameter has, in general, its own analog channels with a varying degree of redundancy depending on the parameter (e.g. two out of four for the power range neutron flux and two out of three for the pressurizer water level). However, some plant parameters share the sensors and transmitters. For example, the same sensors and transmitters are used for the pressurizer low-pressure and high-pressure trips. The signal is evaluated by two separate bistables with different setpoints.

There are two logic trains, and each logic train receives signals from the analog channels through input relays. The input signals are then applied to universal boards that are the basic circuits of the protection system. They contain one-out-of-two, two-out-of-three, two-out-of-four coincidence logic circuits depending on the plant parameters and the corresponding analog channels, as mentioned earlier. The trip signals generated in the universal boards are sent to undervoltage output boards or engineered safeguard output boards. The undervoltage board in each logic train has two undervoltage coils—one for the reactor trip breaker and another for the by-pass breaker that is racked out in normal operation of the plant. A trip signal from the undervoltage board will de-energize the undervoltage coils by removing the 48 V output of the undervoltage board. This will open the reactor trip breaker or the bypass breaker (if closed as in the case of test and maintenance of the corresponding trip breaker), removing the power supply holding the control rods.

4.7.2 Testing Procedures and TSs of the RPIS

The RPIS is designed to allow periodic testing during power operation without initiating a protective action unless a trip condition actually exists. An overlapping testing scheme, where only parts of the system are tested at any one time, is used. Typical RPIS testing involves (a) verification of proper channel response to known inputs, (b) proper bistable settings, and (c) proper operation of the coincidence logic and the associated trip breakers. Detailed testing procedures including testing frequency and allowable bypass times are described in [48-49].

Concerning the testing of the generic analog channel, it aims at verifying its proper functioning and that bistable settings are at the desired setpoint. During testing, the test switch disconnects the sensor/transmitter from the channel and the circuit is capable of receiving a test signal through test jacks. The input signal to the test jacks is then adjusted to check operability and setpoints of the bistable. The analog channel under testing is allowed to be bypassed for a duration specified by the TSs and to be put in a trip mode if the allowable bypass time is exceeded.

As for the logic train and trip breaker testing, it encompasses three stages:

1. Testing of input relays places each channel bistable in a trip mode, causing one input relay in logic train A and another in logic train B to de-energize. Each input relay operation will light the status lamp and

annunciator. This stage of the testing provides overlap between the analog channel and logic train positions of the test procedure.

2. Testing of logic trains involves one train at a time. The semiautomatic test device checks through the solid-state logic to the undervoltage coil of the reactor trip breaker; the logic train under testing is also allowed to be bypassed for the specific duration, and the plant must be shut down if the allowable bypass time is exceeded.
3. Testing of the trip breaker requires manual trip and operability verification of the bypass breaker and the manual trip test of the trip breaker through the logic train.

To ensure their proper function, both analog channels and logic trains are tested at predetermined Surveillance Test Intervals (STIs) which are defined in one of the Technical Specifications (TSs) for the RPIS. During testing, both the analog channels and the logic trains are bypassed, and hence, they are unavailable for their intended function in the RPIS. Consequently, during this time, the safety margin built into the RPIS is reduced. The time for which an analog channel or a logic train can remain bypassed, i.e. the Allowable Bypass Time (ABT), is the second TS of the RPIS. If the ABT is exceeded, then the analog channel or the logic train is tripped: that is, it is put in a mode as if that particular channel or logic train were giving a signal for the reactor shutdown. The surveillance frequency and ABTs deeply impact the availability of the RPIS. In principle, the availability of the RPIS is favored by high surveillance frequencies and short ABTs. Actually, excessively high-surveillance frequencies could have the negative counterpart of too much downtime due to test and excessively short ABTs tend to invoke too many inadvertent reactor trips and, as a result, cause unnecessary transients and challenges to other safety systems. It follows that the specification of TSs (STIs and ABTs) requires careful consideration and should be subject to optimization.

4.7.3 Modeling Assumptions

The model developed for this study is adapted from [47] but does not include the mechanical portion (control rod drive mechanisms and control rods) and the operator manual actions to scram the plant by pushing the buttons in the control room or by locally opening trip breakers or output breakers on the rod drive motor-generator sets. A typical four-channel set up was considered to evaluate the effects on system unavailability of changes in the test procedures. A simplified functional block configuration is given in Figure 9 in terms of the analog channels, the logic matrix and the logic trains, including the trip breakers. Each functional block is considered as a supercomponent composed of several basic components in series. The transition times of the components are assumed to follow exponential distributions.

The block representing an analog channel consists of the following components:

1. a sensor/transmitter
2. loop power supply (120-V alternating-current)
3. signal conditioning circuits
4. a bistable
5. an input relay.

Each bistable feeds two input relays—one for each logic train. However, to avoid complexity of the model, it is assumed that each bistable feeds only one input relay.

The operation of the analog channel can be represented in terms of its transitions among five-states:

- *State 1:* the operating state
- *State 2:* the failed state; the failure can be detected in the next test, and the component will be put under repair.
- *State 3:* the tripped state; the channel generates a spurious trip signal, and it may undergo repair.
- *State 4:* the bypass state related to state 1; to perform a test, the channel can be bypassed for a prespecified period of time: the ABT τ_b^{ch}.
- *State 5:* the bypass state related to state 2; if the channel is failed, the testing and repairing can be performed while in a bypass mode provided that the ABT τ_b^{ch} is not exceeded.

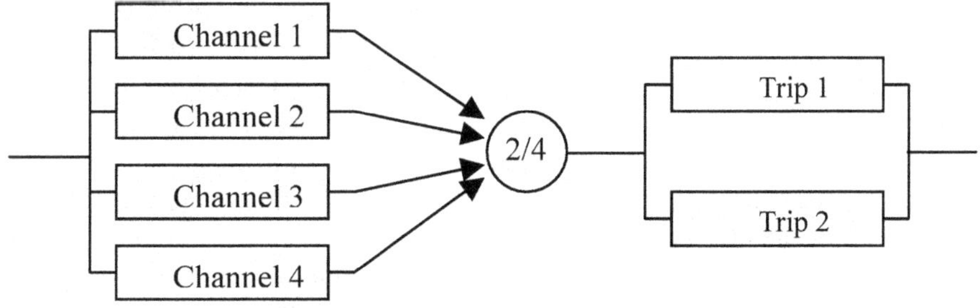

Fig. 9. Block diagram of the RPIS [47].

Given the methodological nature of this work, we have considered a simplified version of the model proposed in [47]. In particular, we do not consider, along with the RPIS unavailability, the other possible risk attributes included in [47], namely, the frequencies of spurious scrams, of anticipate transients without scram and of core damage. Furthermore, in [47] the stochastic behavior of the system is assumed to be dependent on the current system state. For example, the ABTs for an analog channel could depend on whether an analog channel is already tripped; the repair rate of an analog channel might depend on whether another channel is under repair or whether the reactor is shut down or on-line; exceeding the ABT in an analog channel will generate a scram signal depending on whether or not the reactor is on-line. These dependencies have not been considered here.

The state transition diagram for the generic analog channel is given in Fig. 10 [47].

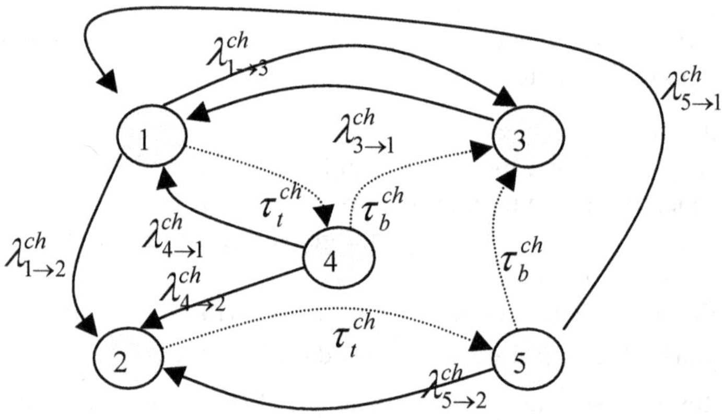

Fig. 10. State transition diagram for the analog channel [13].

From state 1 the analog channel may transit to state 2 with a failure rate $\lambda^{ch}_{1\to2}$, to state 3 when any one of the internal components gives a spurious trip signal with a rate $\lambda^{ch}_{1\to3}$ or to state 4 following a test which takes place every τ^{ch}_{t} hours. Note that this last is not a stochastic transition and, thus, it is represented in the Figure with a dashed line. If the analog channel is in state 2, it transits to state 5 following a test. In other words, every τ^{ch}_{t} hours, if the analog channel is in state 1, it transfers to state 3, if it is in state 5, it transfers to state 4. If the analog channel is in state 3, it transits back to state 1 once the repair is completed, with transition rate $\lambda^{ch}_{3\to1}$. If the analog channel is in state 4, it may transit to state 3 if the testing is not completed within the ABT τ^{ch}_{b} or exit state 4 upon completion of the repair with a transition rate μ_1^{ch}: if there is human error in restoring the channel after the test (occurring with probability P_1^{ch}), the analog channel transfers to state 2 (thus, with actual transition rate $\lambda_{4\to2}^{ch} = \mu_1^{ch} P_1^{ch}$); vice versa, if there is no human error, it transfers to the operating state 1 (with actual transition rate $\lambda_{4\to1}^{ch} = \mu_1^{ch}(1 - P_1^{ch})$). When in state 5, the analog channel behaves similarly: it may transit to state 3 if the test/repair is not completed within the ABT τ_b^{ch}, to state 1 if the test/repair is completed within the ABT and no human error is committed in restoring the channel into its operating state (transition rate $\lambda_{5\to1}^{ch} = \mu_2^{ch}(1 - P_2^{ch})$ or to state 2 if the test is completed within the ABT and there is a human error that leaves the channel failed (transition rate $\lambda_{5\to2}^{ch} = \mu_2^{ch} P_2^{ch}$).

1. However, as mentioned in reference [47], if the ABT is small compared to the mean time to failure of the channel, the components' behavior can be adequately modeled without states 4 and 5 (Figure 11) [47].

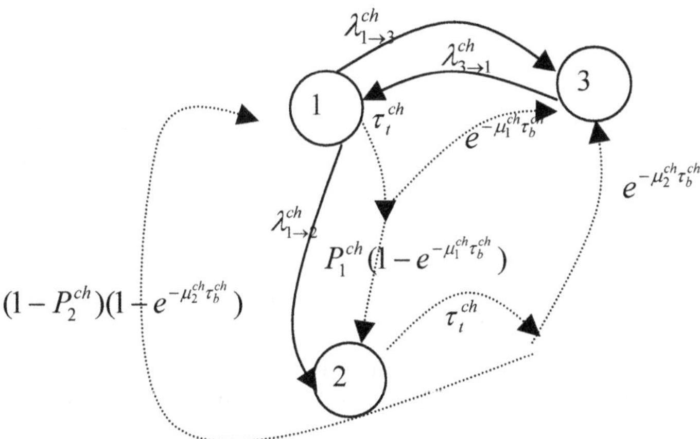

Fig. 11. State transition diagram for the analog channel: equivalent Markov model [13].

This is equivalent to assuming that the transitions in and out of states 4 and 5 occur instantaneously at the time of testing and with the following probabilities :

2. From state 1 to state 3 with probability $e^{-\mu_1^{ch}\tau_b^{ch}}$, i.e. the probability that the test lasts more than τ_b^{ch}.

3. From state 1 to state 2 with probability $P_1^{ch}(1-e^{-\mu_1^{ch}\tau_b^{ch}})$, i.e. the probability of completing the repair in less than τ_b^{ch} and that a human error occurs. Accordingly, the component remains in state 1 with probability $(1-P_1^{ch})(1-e^{-\mu_1^{ch}\tau_b^{ch}})$.

From state 2 to state 3 and from state 2 to state 1 with probabilities $e^{-\mu_2^{ch}\tau_b^{ch}}$ and $(1-P_2^{ch})(1-e^{-\mu_2^{ch}\tau_b^{ch}})$, respectively.

The block for a logic train and trip breaker consists of the following components:

1. solid-state combinational logic circuits
2. direct-current (dc) power for the logic circuits (15-V dc)
3. undervoltage coils
4. dc power for the undervoltage coils (48-V dc)
5. a trip breaker.

The state transition diagram for this block is equivalent to that of the analog channel.

As shown in Figure 9, the four analog channels and the two logic trains operate in a two-out-of-four logic and in a one-out-of-two logic, respectively. This means that the signal to de-energize magnetic jacks and, thus, to operate the reactor scram requires at least that two analog channels signal the trip and one logic train succeeds in switching to the trip state. Thus, the RPIS is available whenever at least two analog channels and one logic train are operative (i.e. in their state 1). Also, the RPIS functions correctly if there is one analog channel in the trip state and there is at least one more analog channel and logic train operative. If at least two analog channels or one logic train are in the trip state, a spurious signal to reactor shutdown is generated. If the shutdown is correctly performed (the control roads do not fail to enter the core and the decay heat removal system works properly), the RPIS system is renewed after a mean time $1/\mu_s$. The conditional probability of unsafe shutdown is P_s. If unsafe shutdown occurs, it is assumed that the system cannot be brought back to the operative condition. Under these hypotheses, we can identify the states in which the system is unavailable: if three or more analog channels or two logic trains are failed or if an unsafe reactor shutdown has occurred.

The model parameters, i.e. the failure and spurious signals rates, the mean times to repair, the probabilities of human error are affected by uncertainties and must be considered as random variables. As in [47], the parameters are here assumed to have lognormal probability density functions. The means and medians of the lognormal distributions considered in [47] are given in Table 6.

Table 6. Reliability data for the components of the RPIS: mean and median of the lognormal distributions [47].

Parameter	Mean	Median
$\lambda_{1\to2}^{ch}$	3.51 10^{-6} [h^{-1}]	2.1749 10^{-6} [h^{-1}]
$\lambda_{1\to3}^{ch}$	6.16 10^{-6} [h^{-1}]	3.817 10^{-6} [h^{-1}]
$\lambda_{3\to1}^{ch}$	16 [h]	12.8017 [h]
$1/\mu_1^{ch}$	2 [h]	1.2393 [h]
P_1^{ch}	7.99 10^{-3} [h^{-1}]	3.00 10^{-3} [h^{-1}]
$1/\mu_2^{ch}$	8 [h]	4.9571 [h]
P_2^{ch}	1.82 10^{-3} [h^{-1}]	6.84 10^{-3} [h^{-1}]
$\lambda_{1\to2}^{tr}$	2.52 10^{-6} [h^{-1}]	1.5615 10^{-6} [h^{-1}]
$\lambda_{1\to3}^{tr}$	3.28 10^{-6} [h^{-1}]	2.0324 10^{-6} [h^{-1}]
$\lambda_{3\to1}^{tr}$	16 [h^{-1}]	12.8017 [h]
$1/\mu_1^{tr}$	2 [h]	1.2393 [h]
P_1^{tr}	7.99 10^{-3} [h^{-1}]	3.00 10^{-3} [h^{-1}]
$1/\mu_2^{tr}$	8 [h]	4.9571 [h]
P_2^{tr}	7.99 10^{-3} [h^{-1}]	3.00 10^{-3} [h^{-1}]
$1/\mu_s$	13.7 [h]	9.6051 [h]
P_s	5.21 10^{-7}	1.9561 10^{-7}

In addition, we assume that the uncertainties on the parameters governing the transitions of failure, repair or test actions and human error during repair or test (namely, $1/\mu_1^{ch}$, $1/\mu_2^{ch}$, $1/\mu_1^{tr}$, $1/\mu_2^{tr}$, $1/\mu_s$, $\lambda_{3\to1}^{ch}$, $\lambda_{3\to1}^{tr}$, P_1^{ch}, P_2^{ch}, P_1^{tr}, P_2^{tr} and P_s) depend on the frequency with which the components are tested. In particular, we want to reflect the fact that the more frequently the test is performed, the less uncertain the timing and outcome of the transition is. For example, it seems reasonable to expect that when a test action is performed frequently, there is more confidence in the estimate of its duration. If the test action is performed rarely, there is a higher probability of occurrence of unexpected events which can cause the test duration to be more uncertain. The uncertainty on the failure parameters (namely, $\lambda^{ch}_{1\to2}$, $\lambda^{ch}_{1\to3}$, $\lambda^{tr}_{1\to2}$ and $\lambda^{tr}_{1\to3}$) is assumed dependent on the frequency of test action as well, so as to reflect that the amount of physical information available on the component increases (and thus, the amount of uncertainty on its failure behaviour decreases) with the increase of the number of inspections. As a 'global' indicator of the frequency of inspections performed on the system, we consider the average surveillance time interval of the six components, $\tau_{avg} = (4\tau_t^{ch} + 2\tau_t^{tr})/6$: low values of τ_{avg} indicate frequent inspections whereas rare inspections correspond to high values of τ_{avg}. Hence, the amount of uncertainty is assumed to be a function of the average surveillance time interval τ_{avg}. More precisely, to model the growth of uncertainty with the RPIS components surveillance time intervals, the value of the median p^{50} of the lognormal distribution of the generic parameter p is here arbitrarily assumed to vary with τ_{avg} as follows:

$$p^{50} = \overline{p} - (\overline{p} - p^{50}_{ref})\, a_{unc}(\tau_{avg}) \tag{6}$$

where $\overline{p}$ is the mean of the parameter (independent on τ_{avg}) and p^{50}_{ref} is the value considered in [47] and reported in Table 6.

The function $a_{unc}(\tau_{avg})$ is arbitrarily chosen as the non-decreasing stepwise function reported numerically in Table 7.

Table 7. Numerical values of the arbitrary function $a_{unc}(\tau_{avg})$ [13].

τ_{avg}	$a_{unc}(\tau_{avg})$
0 d < τ_{avg} ≤ 70 d	0.001
70 d < τ_{avg} ≤ 140 d	0.04
140 d < τ_{avg} ≤ 280 d	0.05
280 d < τ_{avg} ≤ 350 d	0.07
350 d < τ_{avg} ≤ 420 d	0.1
τ_{avg} > 420 d	1

Note that if $a_{unc}(\tau_{avg}) = 0$ the lognormal distribution reduces to a Dirac delta function and for $a_{unc}(\tau_{avg}) = 1$, $p^{50} = p^{50}_{ref}$, i.e. the reference value of Table 6. The behavior of such function is intended to account for the different levels of uncertainty that possibly affect the parameters' estimates as a function of the system average surveillance time interval. An example of these effects are shown in Fig. 12 with reference to the failure rate of the analog channel, $\lambda_{1\to2}^{ch}$. In the Figure, the lognormal probability distribution function of $\lambda_{1\to2}^{ch}$, $f(\lambda_{1\to2}^{ch})$, is reported for the three different levels of uncertainty identified by $a_{unc}(\tau_{avg})= 1$, 0.1, 0.05, respectively. Note that the value of the median of the distribution $\lambda_{1\to2}^{ch,50}$ assumes the reference value of Table 6 for a_{unc}=1, whereas it approaches that of the mean as the value of a_{unc} decreases. Accordingly, the parameter distribution gets visibly narrower, maintaining the same value of the mean.

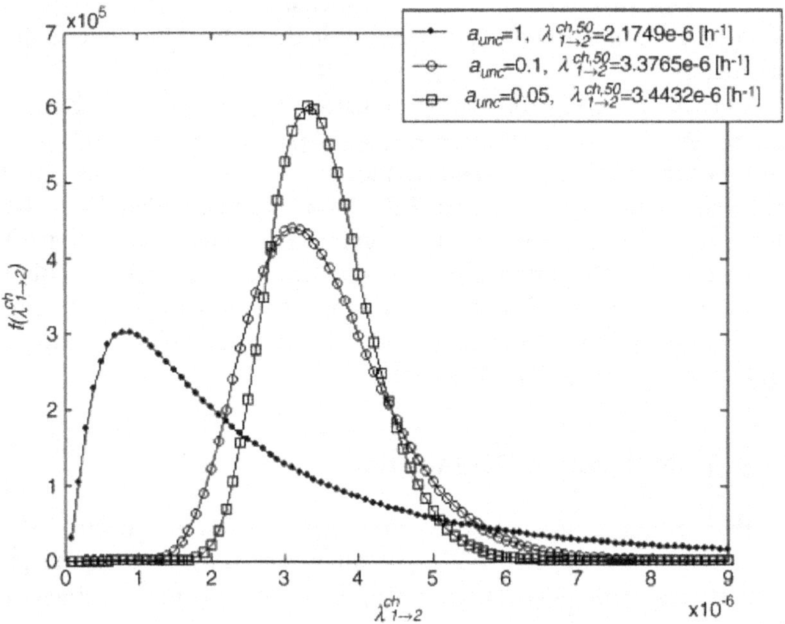

Fig. 12. Probability density function of the failure rate parameter $\lambda_{1\to2}^{ch}$ for three different values of a_{unc} [13].

A Monte Carlo model has been built to catch the complex dynamics of the multi-state components of the RPIS system in presence of uncertainty. It allows to quantitatively evaluate the alternative TSs in terms of different values of the STIs and ABTs, within the search for an optimal solution giving proper account to the uncertainties in the parameters values. To this aim, we consider two objective functions: the negative of the expected value of the mean system unavailability over the mission time of one year, $-E[U(\tau_t^{ch}, \tau_t^{tr}, \tau_b^{ch}, \tau_b^{tr})]$ and the negative of its variance, $-Var[U(\tau_t^{ch}, \tau_t^{tr}, \tau_b^{ch}, \tau_b^{tr})]$. The multiobjective maximization of these two functions allows identifying those solutions which are optimal from the point of view of the expected availability behavior of the system, and, at the same time, of its intrinsic variability, due to the uncertainties in the system input parameters.

In the presence of parameter uncertainty, the Monte Carlo simulation for the assessment of the unavailability of a system proceeds as follows. First, N_R realizations of the N_P-dimensional vector of parameters are sampled from the multivariate distribution describing the uncertainty in the parameters.

For the generic *r*-th vector of parameter values, *r*=1, 2,..., N_R, a number N_S of system life histories are simulated by Monte Carlo, each of these histories being governed by the same parameters values. At the end of the simulation, the collected realizations are manipulated statistically to obtain ensemble estimates of the moments (e.g. mean and variance) of the output quantities of interest (e.g. availability and reliability). This way of proceeding is often referred to as 'double randomization' in the literature, the randomization of the parameters being superposed to the inherent randomness of the stochastic system.

4.7.4 Multiobjective Optimization

In this Section we present the results of the genetic algorithm maximization of the two objective functions introduced in the previous Section. The three decision variables of the optimization problem are: the STI of the analog channels τ_t^{ch} (the same for all four channels), the STI of the logic trains τ_t^{tr} (the same for both trains) and the ABT τ_b, which is assumed to be equal for all the six components (the four analog channels and two logic trains). The STIs and the ABT are to be selected within the range (30, 365) days and (1, 50) h, respectively. The choice of such search intervals has been suggested by those considered in [47]. Note that the simplifications introduced in the model with respect to [47] lead to three decision variables against the five considered in the original work. Although currently not an issue in the requirements of TS for nuclear power plants, but being of utmost importance for the plants effective operation, we introduce also the economic aspect related to STI activities to embrace a risk-informed approach. In particular, as a means to control the costs of plant operation, we consider constraints on the total cost of surveillance C_s (in arbitrary units) normalized to the month. Given the STI τ_t^i (in days) of the generic component *i=ch, tr*, the number of surveillance actions within a month is $30/\tau_t^i$, and the cost of surveillance actions in a month is $(30/\tau_t^i)\,c_s^i$, where c_s^i is the cost per surveillance of component *i*. Thus, we consider the following arbitrary constraint on the total cost of surveillance actions for the six components, over a month:

$$C_s = 4\,(30/\tau_t^{ch})\,c_s^{ch} + 2\,(30/\tau_t^{ch})\,c_s^{tr} \le 1 \tag{7}$$

The values considered for the costs c_s^{ch} and c_s^{tr} per surveillance action on the analog channels and on the logic trains are 0.25 and 0.5, in arbitrary units, respectively.

The resolution level of the search is defined by the coding of the chromosomes: two 7-bit genes and one 5-bit gene code the values of the STIs and ABT respectively. The search space is then made up of 524,288 candidate solutions. The internal rules and parameters for the genetic algorithm search are reported in Table 8. The Fit-Fit selection procedure adopted entails that the population, rank-ordered on the basis of the individual fitnesses, is scanned and each individual is parent-paired with the next fittest one. Note that the crossover probability is set to 1, meaning that it is performed for every couple of individuals selected as parents. In principle, adopting a lower probability value would favour the persistence of the same chromosomes in the population and the chances for accumulating statistical accuracy in the fitnesses estimates. However, this would be at the cost of a lower genetic diversity in the population. The value of 1 was found to be adequate for this application.

Table 8. Genetic algorithm parameters and rules [13].

Number of chromosomes (population size)	50
Number of generations (termination criterion)	100
Selection	Fit-Fit
Replacement	Children-parents
Mutation probability	0.001
Crossover probability	1
Number of generations without elitist selection	5
Fraction of parents chosen with elitist selection	0.25

Each drop of MC simulation is made up of 10 realizations of the dimensional vector of parameters, sampled from the multivariate distribution describing the uncertainty in the parameters, i.e. the outer loop of the double randomization process described at the end of the previous Section 7.3. The internal loop consists of a full MC simulation. Ten drops are run the first time a chromosome is proposed, then one drop is added each time the same chromosome is re-proposed in the successive GA evolution.

The CPU time required to complete the 100 generations of the GA search was approximately 20 hours in an Athlon 1400 MHz computer. However, the algorithm had converged after only 18 generations. An estimated 10^4 hours would be necessary for a crude, enumerative search of the entire decision variables space.

As a result of the search, the multiobjective GA identified only two non-dominated solutions. Different GA runs starting from different, randomly sampled, initial populations were performed to check the reproducibility of the results. Table 9 reports their corresponding MC estimates of $E(U)$ and $Var(U)$ and Figure 13 presents graphically the estimated values of $E(U)$ with error bars of one standard deviation.

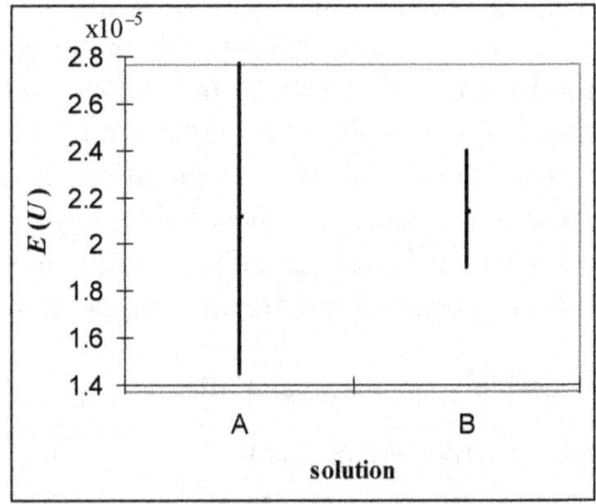

Fig. 13. RPIS expected mean unavailability and corresponding standard deviation (error bar) of the non-dominated solutions A and B found by the algorithm [13].

The two solutions are both characterized by an optimal value of the allowable bypass time $\tau_b = 1\,h$. This value corresponds to the lower extreme in the search interval for this parameter. This result was actually foreseeable a priori but we proceeded anyway to perform the GA search on this parameter, for internal validation of the approach. Indeed, as illustrated in the previous Section 5.3, the ABT is the allowable time in which a reduced safety margin of the protection system is acceptable. Thus, intuitively, short ABTs favor lower unavailabilities. For the system under analysis, it is possible to verify by direct calculation that the values given in [47] for the model parameters, are such that the system unavailability monotonically increases with τ_b : hence, the lower τ_b is the lower the unavailability is. Obviously, when other risk attributes are considered along with the RPIS unavailability, e.g. the frequency of spurious scrams or the core damage frequency as in [47], different results may be obtained. For example, as shown in [47], too short ABTs can cause too many spurious signals of reactor shutdown. Spurious scrams, indeed, represent a challenge to the decay heat removal and other safety systems of the plant and have the potential of leading to core damage. For a detailed discussion on the effects of short ABTs on the number of spurious scrams the reader should consult [47].

Solution A is the solution with lower $E(U)$ whereas solution B has the lower variance $Var(U)$. Both solutions are on the frontier of acceptable values identified by the introduced constraint on the total cost of surveillance actions, the corresponding values of C_s in the last column of Table 9 being very close to unity. Hence, the best solutions are characterized by the smallest values of STIs τ_t^{ch} and τ_t^{tr} which meet the cost constraint of Eq. (7). Based on this information, we checked for possible misses of non-dominated solutions by calculating the values of the two objective functions $-E(U)$ and $-Var(U)$ in correspondence of all the possible τ_t^{ch} and τ_t^{tr} on the frontier defined by the constraining relation of Eq. (7). The two solutions A and B identified by the GA were indeed confirmed to dominate all the others, thus partially validating the results of the search. Further, an analysis of the search intermediate results has shown that the parameter that drives the search is the STI τ_t^{tr} of the logic train: this is physically reasonable since the redundancy level of this component (two) is lower than that of the analog channel (four) so that the overall system availability performance is likely to be more sensitive to its characteristic parameter. Solution A is characterized by a shorter τ_t^{tr} than solution B, and its numerical value seems to be a good compromise between a too small value (causing unavailability due to too much test downtime) and a too large value (causing unavailability due to too many undetected failures). The level of uncertainty corresponding to solution A is driven by the average surveillance interval τ_{avg}=86.67 h, for which the function $a_{unc}(\tau_{avg})$ assumes the value 0.04.

Solution B is instead characterized by a larger value of the logic train STI and a lower average surveillance interval τ_{avg} than A. The higher value of τ_t^{tr} is responsible for the slightly higher unavailability of solution B with respect to that of solution A; as a counterpart, solution B is characterized by a lower uncertainty in the estimate of U, due to the lower value of τ_{avg} for which $a_{unc}(\tau_{avg})$=0.001.

The number of non-dominated solutions found in the application is driven by the particular cost constraint considered and the stepwise behavior of the function $a_{unc}(\tau_{avg})$ characterizing the uncertainty in the parameters distribution. Intuitively, in the region of the acceptable proposals of TSs with 'good' values of unavailability, the function $a_{unc}(\tau_{avg})$ presents one step in correspondence of the value τ_{avg}=70. Solutions A and B are located on opposite sides of the step, the former being on the right-hand side and the latter being on the left-hand side. The potential solutions on the right-hand side are characterized by high unavailabilities and low variances whereas the potential solutions on the left-hand side are characterized by low unavailabilities and high variances. Then, solution B can be regarded as the best solution in the left-hand region, whereas solution A as the best in the right-hand region.

Given the non-dominated solutions, characterized by different levels of system unavailability and associated uncertainty, the decision makers can select, according to their respective attitudes towards risk, a preferred TS alternative. If, as often done in practice, the decision were to consider only the expected value of the system unavailability, the best solution would be A in Table 9.

Table 9. Relevant data of the two solutions identified by the GA [13].

	$E(U)$	$Var(U)$	τ_t^{ch} [d]	τ_t^{tr} [d]	τ_b [h]	C_s	τ_{avg} [h]
A	$2.13\ 10^{-5}$	$4.37\ 10^{-11}$	109.28	41.43	1	0.998	86.67
B	$2.16\ 10^{-5}$	$5.62\ 10^{-12}$	77.96	49.26	1	0.993	68.40

In reality, when considering safety systems, decision makers are often risk-averse, in the sense that they are not satisfied with expected predictions, but request also high confidence that the actual performance is the one predicted. In this respect, the multiobjective approach embraced would allow the decision-maker to sustain a shift towards solution B, which presents a lower uncertainty on an only slightly higher system unavailability.

Obviously, from a practical viewpoint the choice of the proper STI would need to take into account also the logistics of the actual test scheduling with respect to the overall plant operation.

4.8 Conclusions

This Chapter has illustrated the combined use of Genetic Algorithms and Monte Carlo simulation as an approach for the optimization of the reliability design (e.g. choice of redundancy configuration and of components types, amount of spare parts to be allocated) and operation (e.g. choice of surveillance test and/or maintenance intervals) of engineered systems.

The approach has been developed to satisfy two basic requirements in practical system risk/reliability/availability/cost optimization problems:

1. the need of realistic quantitative models to evaluate the risk/reliability/availability/cost performance of alternative design solutions and operation strategies;
2. the need of an efficient search algorithm for identifying the optimal solution with respect to the system risk/reliability/availability/cost performance.

Regarding the first requirement, realistic models are necessary for a proper description of hazardous complex systems with significant safety implications, such as the nuclear plants (but also the chemical ones and the aerospace systems, for example). In this respect, Monte Carlo simulation is a suitable tool for carrying out the modeling task as it allows to give account to many realistic aspects of system operation such as components' ageing and degradation, maintenance and renovation strategies, spare parts allocation, multiple performance levels of the components, operational and load-sharing dependencies, etc.

With respect to the second requirement regarding the effective search for the optimal design solutions and operation strategies, a combinatorial number of potential alternatives needs to be evaluated. For real systems, the problem becomes intractable by classical optimization methods, e.g. of the gradient–descent kind, because of the size of the search space and because the objective functions to be optimized are typically embedded in complicated computer codes from which the needed differential information which guides the search is not easily retrieved. In this respect, genetic algorithms offer a powerful alternative as search engines as they allow handling multi-variate, nonlinear objective functions and constraints and base their search only on information relative to the objective function to be optimized and not on its derivatives.

In practice, system design and operation optimization entails considering various conflicting targets, e.g. low costs and high revenues on one side, and high reliability and low accident risks on the other side. To handle this situation, a multiobjective approach can be embraced in which all targets are treated separately and the search is conducted so as to identify the set of solutions, among all possible ones, which are equivalent in that each member of the set is better or equal to the others with respect to some, but not all, of the targets. This set is called the Pareto set and constitutes the full spectrum of qualified candidate solutions at disposal of the decision maker for informing the selection of the one which best satisfies his or her preference priorities. The multiobjective approach thus frees, in general, the decision maker from the difficult and controversial task of introducing 'a priori', arbitrary weights and/or constraints on the objective functions which, instead, can be brought into play 'a posteriori' in a fully informed way, at the decision level.

The multi-objective search framework can effectively be implemented within a genetic algorithm search approach in which the evaluation of the objective functions in correspondence of each solution proposed by the genetic algorithm is performed by means of realistic Monte Carlo simulation models. To limit the computational burden associated to the Monte Carlo estimate of system performance for each potential solution, an effective coupling of the genetic algorithm search and the Monte Carlo estimation has been propounded. This has been shown effective in allowing exploiting the efficiency of genetic algorithms in the search among large numbers of possible alternatives, while being able to realistically model, by Monte Carlo simulation, the dynamics of complex systems.

Two applications of the approach have been presented in support of the methodological developments. The first application demonstrates the potentiality of the approach on a single-objective redundancy allocation problem of system reliability design.

The second application concerns the multi-objective risk-informed optimization of the technical specifications (e.g. test/maintenance time intervals and/or procedures, etc.) of a safety system of a nuclear power plant. The identification of the optimal technical specifications entails resorting to probabilistic models of the system reliability and availability behavior. Inevitably, the model predictions thereby obtained are affected by uncertainties due to the simplifying assumptions introduced in the models and to the estimates of the unknown parameters values. In the application presented, only parameter uncertainty has been considered and a multiobjective point of view has been embraced to allow for the explicit consideration of such uncertainties in the model predictions. This is accomplished by simultaneously optimizing, in a Pareto dominance sense, two objective functions: the negative of the expectation of the system unavailability and the negative of the variance of its estimate.

The combined genetic algorithms and Monte Carlo simulation approach has been successfully applied to the Reactor Protection Instrumentation System of a Pressurized Water Reactor. The results show the value of the method in that it provides the decision maker with a useful tool for distinguishing those solutions that, besides being optimal with respect to the expected availability behavior, give a high degree of confidence on the actual system performance.

Alternative strategies of efficiently embedding the Monte Carlo simulation approach into the genetic algorithm search engine are possible. For example, the number of Monte Carlo trials can be increased in a progressive manner as the selection process proceeds over the best performing individuals. Further research can be undertaken to investigate the effectiveness of these alternatives.

Finally, although more informative, the multi-objective Pareto optimality approach does not solve the decision problem: the decision maker, provided with the whole spectrum of Pareto solutions, equivalently optimal with respect to the objectives, still needs to select the preferred one according to his or her subjective preference values. The closure of the problem then still relies on techniques of decision analysis such as utility theory, multi-attribute value theory or fuzzy decision making, to name a few.

All the results reported in this Chapter have been obtained by combining the user-friendly, multipurpose MOGA (Multi-Objective Genetic Algorithm) and MARA (Montecarlo Availability and Reliability Analysis) codes developed at the Department of Nuclear Engineering of the Polytechnic of Milan, Italy (website http://lasar.cesnef.polimi.it).

References

1. Marseguerra M, Zio E (2002) Basics of the Monte Carlo Method with Application to System Reliability. LiLoLe- Verlag GmbH (Publ. Co. Ltd.)
2. Lux I, Koblinger L (1991) Monte Carlo particle transport methods: neutron and photon calculations, CRC Press.

3. Henley EJ, Kumamoto H (1991) Probabilistic Risk Assessment, IEEE Press.
4. Dubi A (1999), Monte Carlo Applications in Systems Engineering, Wiley.
5. Kubat P (1989) Estimation of reliability for communication/ computer networks simulation/analytical approach. IEEE Trans Commun 1989;37:927-33.
6. Jane CC, Lin JS, Yuan J (1993) Reliability evaluation of a limited-flow network in terms of MC sets. IEEE Trans Reliab;R-42:354-61.
7. Yeh WC (1998) Layered-network algorithm to search for all d-minpaths of a limited-flow acyclic network. IEEE Trans Reliab 1998;R-46:436-42.
8. Aven T (1987) Availability evaluation of oil/gas production and transportation systems. Reliab Eng Syst Safety;18:35-44.
9. Aven T (1988) Some considerations on reliability theory and its applications. Reliab Eng Syst Safety;21:215-23.
10. Marseguerra M, Zio E (2000), System Unavailability Calculations in Biased Monte Carlo Simulation: a Possible Pitfall, Annals of Nuclear Energy, 27:1589-1605.
11. Marseguerra M and Zio E (1993) Nonlinear Monte Carlo reliability analysis with biasing towards top event. Reliability Engineering & System Safety 40;1:31-42
12. Borgonovo E, Marseguerra M and Zio E (2000) A Monte Carlo methodological approach to plant availability modeling with maintenance, aging and obsolescence. Reliability Engineering & System Safety, 67;1:Pages 61-73
13. Marseguerra M, Zio E, Podofillini L (2004) A multiobjective genetic algorithm approach to the optimization of the technical specifications of a nuclear safety system. Reliability Engineering and System Safety 84:87-99.
14. Marseguerra M, Zio E, Podofillini L (2002) Condition-based maintenance optimization by means of genetic algorithms and Monte Carlo simulation. Reliability Engineering and System Safety 77:151-165.
15. Barata J, Guedes Soares C, Marseguerra M and Zio E (2002) Simulation modeling of repairable multi-component deteriorating systems for 'on condition' maintenance optimization. Reliability Engineering & System Safety 76;3:255-264
16. Cantoni M, Marseguerra M and Zio E (2000) Genetic algorithms and Monte Carlo simulation for optimal plant design. Reliability Engineering & System Safety 68;1:29-38
17. Marseguerra M and Zio E (2000) Optimizing maintenance and repair policies via a combination of genetic algorithms and Monte Carlo simulation. Reliability Engineering & System Safety 68:69-83
18. Zio E, Podofillini L (2003) Importance measures of multi-state components in multi-state systems. International Journal of Reliability, Quality and Safety Engineering 10;3:289-310
19. Levitin G, Podofillini L, Zio E (2003) Generalized importance measures for multi-state systems based on performance level restrictions. Reliability Engineering and System Safety 82:235-349

20. Podofillini L, Zio E, Levitin G, (2004) Estimation of importance measures for multi-state elements by Monte Carlo simulation. Reliability Engineering and System Safety 86;3:191-204
21. Zio E, Podofillini L (2003) Monte Carlo simulation analysis of the effects of different system performance levels on the importance of multi-state components. Reliability Engineering and System Safety 82:63-73
22. Marseguerra M, Zio E and Podofillini L (2005) Multiobjective spare part allocation by means of genetic algorithms and Monte Carlo simulation. Reliability Engineering & System Safety 87:325-335
23. Marseguerra M, Zio E, Podofillini L, Coit DW (2005) Optimal Design of Reliable Network Systems in Presence of Uncertainty. Accepted for publication to IEEE Transactions on Reliability
24. Rocco CM and Zio E (2005) Solving advanced network reliability problems by means of cellular automata and Monte Carlo sampling Reliability Engineering & System Safety 89;2:219-226
25. Zio E, Podofillini L and Zille V (2005) A combination of Monte Carlo simulation and cellular automata for computing the availability of complex network systems Reliability Engineering & System Safety, In Press, Corrected Proof, Available online 17 March 2005
26. Belegundu AD, Chandrupatla TR (1999) Optimization Concepts and Applications in Engineering, Prentice Hall Editions, Chapter 11 pp. 373-381
27. Holland JH (1975) Adaptation in natural and artificial system. Ann Arbor, MI: University of Michigan Press
28. Goldberg DE (1989) Genetic Algorithms in Search, Optimization, and Machine Learning. Addison-Wesley Publishing Company
29. Herrera F, Vergeday JL (1996) Genetic algorithm and soft computing, Heidelberg, Ohysica-Verlag
30. Marseguerra M and Zio E (2001) Genetic Algorithms: Theory and Applications in the Safety Domain, In: Paver N, Herman N and Gandini A The Abdus Salam International Centre for Theoretical Physics: Nuclear Reaction Data and Nuclear Reactors, Eds., World Scientific Publisher pp. 655-695.
31. Marseguerra M, Zio E, Martorell S (2006) Basics of genetic algorithms optimization for RAMS applications. To be published in Reliability Engineering and System Safety
32. Joyce PA, Withers TA and Hickling PJ (1998), Application of Genetic Algorithms to Optimum Offshore Plant Design Proceedings of ESREL 98, Trondheim (Norway), June 16-19, 1998, pp. 665-671
33. Martorell S, Carlos S, Sanchez A, Serradell V (2000) Constrained optimization of test intervals using a steady-state genetic algorithm. Reliab Engng Sys Safety 2000; 67:215-232
34. Munoz A, Martorell S, Serradell V (1997) Genetic algorithms in optimizing surveillance and maintenance of components. Reliab Engng Sys Safety 57: 107-20

35. Coit DW, Smith AE (1994) Use of genetic algorithm to optimize a combinatorial reliability design problem, Proc. Third IIE Research Conf. pp. 467-472.
36. Levitin G Genetic Algorithms in Reliability Engineering, bibliography: http://iew3.technion.ac.il/~levitin/GA+Rel.html
37. Sawaragi Y, Nakayama H, Tanino T (1985) Theory of multiobjective optimization. Academic Press, Orlando, Florida
38. Giuggioli Busacca P, Marseguerra M, Zio E (2001) Multiobjective optimization by genetic algorithm: application to safety systems. Reliability Engineering and System Safety 72:59-74.
39. Zitzler E and Thiele L (1999). Multiobjective Evolutionary Algorithms: A Comparative Case Study and the Strength Pareto Approach. IEEE Transactions on Evolutionary Computation, 3;4:257-271.
40. Zitzler E (1999) Evolutionary Algorithms for Multiobjective Optimization: Methods and Applications. PhD thesis: Swiss Federal Institute of Technology (ETH) Zurich. TIK-Schriftenreihe Nr. 30, Diss ETH No. 13398, Shaker Verlag, Germany, ISBN 3-8265-6831-1.
41. Fonseca CM and Fleming PJ (1995) An Overview of Evolutionary Algorithms in Multiobjective Optimization. Evolutionary Computation, 3;1:1-16.
42. Fonseca CM and Fleming PJ (1993) Genetic Algorithms for Multiobjective Optimization: Formulation, Discussion and Generalization, In S. Forrest (Ed.), Genetic Algorithms: Proceedings of the Fifth International Conference, San Mateo, CA: Morgan Kaufmann.
43. Srinivas, N and Deb K (1995) Multiobjective function optimization using nondominated sorting genetic algorithms. Evolutionary Computation, 2;3:221-248
44. Parks GT (1997) Multiobjective pressurized water reactor reload core design using genetic algorithm search. Nuclear Science and Engineering 124:178-187
45. Toshinsky VG, Sekimoto H, and Toshinsky GI (2000) A method to improve multiobjective genetic algorithm optimization of a self-fuel-providing LMFBR by niche induction among nondominated solutions. Annals of Nuclear Energy 27:397-410
46. Brown M and Proschan F (1983) Imperfect repair. Journal of Applied Probabilty 20:851-859
47. Papazoglou IA (2000) Risk informed assessment of the technical specifications of PWR RPS instrumentation. Nuclear Technology 130:329-350.
48. Reference Safety Analysis Report RESAR-3S, Westinghouse Electric Corporation (1975)
49. Jansen RL, Lijewski LM, Masarik RJ (1983) Evaluation of the Surveillance Frequencies and Out of Service Times for the Reactor Protection Instrumentation System WCA.P-10271 Westinghouse Electric Corporation

New Evolutionary Methodologies for Integrated Safety System Design and Maintenance Optimization

B. Galván, G. Winter, D. Greiner, D. Salazar

Evolutionary computation Division (CEANI), Institute of Intelligent Systems and Numerical Applications in Engineering (IUSIANI), University of Las Palmas de Gran Canaria (ULPGC)

M. Méndez

Computer Science Department, University of Las Palmas de Gran Canaria (ULPGC)

5.1 Introduction

The concepts considered in the Engineering domain known as Systems Design have experienced considerable evolution in the last decade, not only because the growing complexity of the modern systems, but by the change of criteria which implies the necessity to obtain optimal designs instead of merely adequate ones. Design requirements are especially strict for systems in which failure implies damage to people, environment or facilities with social-technical importance. For such systems, modern safety requirements force to consider complex scenarios with many variables normally developed under Probabilistic Risk Assessment frameworks. In such systems, special attention must be paid to the Safety Systems Design, considering design alternatives (at the design stage) and maintenance strategies (during the system operation), in order to perform a global optimization.

Although big advances have been introduced in system design, maintenance modeling and optimization during last years, some key issues require more effort that allows us to face up a broad variety of complex technical systems. In this context, some research targets are: To select/develop

B. Galván et al.: *New Evolutionary Methodologies for Integrated Safety System Design and Maintenance Optimization*, Computational Intelligence in Reliability Engineering (SCI) **39**, 151–190 (2007)
www.springerlink.com

adequate optimization methods, to simplify/improve the systems modeling methodologies and to evaluate the system model with appropriate quantitative methods.

In this work a new and efficient approach which includes the features mentioned before is introduced. The novel approach is named Integrated Safety Systems Design and Maintenance Optimization (ISSDMO) and integrates Fault Tree Analysis (FTA) which is a widely used modeling methodology with Evolutionary Algorithms (EA) as a successful tool used for optimization. The incorporation of all these issues allows considering the influence of each aspect upon the others, while preserving the global efficiency of the optimization process.

The research in ISSDMO is not new, but so far none of the previous attempts have succeeded in solving complex problems efficiently. There is a long upheld tradition of optimization of systems in dependability (e.g. see [41]) as well as the use of evolutionary methods as optimization tool. As a matter of fact, Genetic Algorithms (GA) [34,40] have been the heuristic preferred by many authors [31], mainly because their ability to search for solutions in complex and big search spaces. Nevertheless, there are only a few studies –and relatively recent- devoted to ISSDMO. They commenced a little bit more than a decade with Andrews, who passed from the classical approach [1] to the heuristic-based approach [2-4] and continued in the last few years [54] towards more complex systems. However the complexity in modeling and evaluating the fault trees as has been done so far has led the ISSDMO research to come to a standstill. The preceding experiences in ISSDMO [2-4][54] were based on single objective formulation solved with binary coded GA and a Binary Decision Diagram (BDD) for Fault Tree assessment. The authors reported convergence problems [54] due to the different nature of the variables involved: on the one hand strict binary variables for the design alternatives, and on the other hand real variables for maintenance strategies.

The lessons learned up to now point to a last generation ISSDMO characterized by: a multiple objective formulation of the problem; the use of evolutionary algorithms tailored to deal with multiple objectives and a heterogeneous representation of the variables, and capable to incorporate the more recent advances in maintenance modeling and optimization. The reasons are straightforward:

On one hand, there is a growing trend to develop modern and versatile maintenance models. The optimization of the maintenance strategy is top issue now [48][50] with recent and exciting advances [53][60]. Unavailability and Cost have been considered as the main functions to optimize in both single objective (minimization of Unavailability subject to some Cost constraints) and multiple objective (simultaneously minimization of Unavailability and Cost) [13,37,38] formulations.

On the other hand the multiple objective approach which was introduced earlier in reliability-related design problems [31] has been extended successfully to safety systems during the last few years, yielding significant advances in both surveillance requirements [47,49] and systems design [37,38] optimization.

Finally, there are relevant contributions to system modeling and evaluation that must be incorporated to ISSDMO. Since the introduction the first attempts in System Safety modeling with classic logic gates (that permit component and redundancy-diversity levels selection [1]) until now, new logic gates have been developed, allowing to model complex, Dynamic [25] or Fault Tolerant [23], systems.

The main task concerned to Fault Tree models is the quantitative evaluation, which have been performed using the classical Minimal Cut Sets (MCS) approach [63], or the BDD [12,19]. Both approaches have well known drawbacks: The former have exponential growth with the number of basic events and/or logic gates of the tree, while the later depends of a basic events ordering without mathematical foundations, supported only in a few existing heuristics [2,8,9]. These approaches, when embedded in optimization problems with real systems, demonstrate poor results when using MCS, and prohibitive CPU cost of fault tree assessment when using BDD. The main reason of that problems is the necessity to run the fault tree evaluation procedure thousands of times per each solution studied. So from this point of view becomes important to develop fast and/or approximate fault tree evaluation methods. Consequently, new approaches are now under develop like the use of Boolean Expression Diagrams (BED) [38], Direct Methods based on an Intrinsic Order Criterion (IOC) of basic events [35][36], and Monte Carlo methods based on powerful Variance Reduction Techniques [30].

In this work we introduce new methodologies for ISSDMO, the main characteristics of the approaches are:

1. Fault trees with some new logic arrangements are used to model the system in such a way that, it is possible to add/deduct components and redundancy levels without changes on its equations. This feature permits that only one fault tree represents many different layouts designs of a system.
2. The fault tree is evaluated by means of an efficient and direct Monte Carlo method called Restricted Sampling. The method has two stages: during the first one the utilization of a deterministic procedure binds the real value into a proper interval; then a restricted sampling scheme

makes use of the proper interval during the second stage to obtain accurate estimates. In many cases the efficiency of the first stage permits to find the exact system unavailability being unnecessary the second.

3. The model selection for each redundant component is performed directly by the evolutionary process, simplifying the system fault tree. This is an important difference with the previous works on the matter which included the model selection into the fault tree logic. These works included one house event, one basic event and one OR gate to represent each available model on the market for all the system components. In the approach introduced in this work these elements are not included into the fault tree logic, which means in practice that the fault tree size used are significantly lesser, being an integer variable, included in the algorithm, in charge to select the appropriate model for each possible allocation.
4. Binary and Integer coding is used to model both design and maintenance variables.
5. A new version for integer variables of the so called the Flexible Evolutionary Agent (FEA) is introduced, being the search engine for the two loops of a double loop evolutionary Algorithm which is introduced as well for ISSDMO.
6. Results from both Single and Double Loop Multiple objective Evolutionary Algorithms are compared. The so called NSGA-II Multiple objective method is used for all algorithms.

The combined effect of these characteristics overcomes difficulties reported when using classical approaches, based on single objective Genetic Algorithms and Binary Decision Diagrams, for ISSDMO problems. In the following sections the main properties of these new methodologies are described. In Section 2, the main characteristics of the Evolutionary Algorithms used in this work are described. Special attention is paid to Coding Schemes, Algorithm Structures, Multiple Objective Approaches, Flexible Evolutionary Algorithms and Double Loop Algorithms. Section 3 describes the Automated Safety System Design Optimization problem. New features are introduced when the system is modeled with Fault Trees and the Restricted Sampling Technique is described. Coding schemes adopted for the design and maintenance variables and the multiple objective approaches are covered too. In Section 4 a practical example is developed for a Safety System of a Nuclear Power Plant, analyzing and comparing the solutions found by the single and double loop multiple objective algorithms.

5.2 Evolutionary Algorithms (EA)

Under this term are grouped a wide set of learning techniques inspired in the behavior of some natural systems. Experience has shown that when the subject of the learning process is an optimization problem, evolutionary algorithms become a useful tool for optimizing hard and complex problems, even when the domain is non continuous, non linear and/or discrete. On the other hand, the main drawback of this type of algorithms is that, due to its stochastic nature, the different instances cannot assure the final outcome to be a true global optimal solution.

5.2.1 Multiple Objective Evolutionary Algorithms

In general words, evolutionary algorithms evolve an initial sample of alternatives or decision vectors (often called parents) by applying some rules or recombination operators to produce a new sample (often called offspring). Then, the initial sample is updated with the best new values and the process is repeated until a stop criterion is reached (see Fig. 1). Both the selection of those solutions which serve as basis for generating the new sample as well as the updating are related to the performance or fitness scheme. In single objective optimization problems the fitness is commonly a function of the value of the objective under optimization. Likewise in multiple-objective problems, the fitness is usually assigned regarding to the objectives values in terms of Pareto dominance and density.

1st step:	Generate an initial population of decision vectors or individuals according to some scheme
2nd step:	Assign a fitness value to each individual into the population according to some performance function
3rd step:	Select those individuals which will serve as parents according to their fitness
4th step:	Apply the crossover operators over the selected individuals
5th step:	Apply the mutation operators over the selected individuals
6th step:	Integrate into the population the new individuals according to some scheme
7th step:	If the finalization criterion is not met go to step 2, otherwise stop and report the outcome

Fig. 1. General design for an Evolutionary Algorithm.

On the other hand, the recombination is realized by applying different sorts of operators. Some typical examples are the crossover and the mutation operator. The former is applied with a given probability to combine the information of pre-existing parents to generate new decision vectors by setting each component of the new vector to the component value of any of the parents. On the other hand, the mutation operator is applied not over the whole vector at the same time but over each component to shift its value with given probability.

During the last years a great effort has been devoted to develop optimization procedures for problems with conflicting objectives. These problems have many solutions instead of only one, which are termed as Non-Dominated solutions. The set of non-dominated solutions is called the "Pareto Set" while the set of the corresponding objective function values is termed as the "Pareto Front". The **M**ultiple **O**bjective **E**volutionary **A**lgorithms (MOEA) were developed with the aim to conduct the search towards the Pareto Front and to maintain the diversity of the population in the front. Among the first developed algorithms, the following are cited:

- *Vector Evaluated Genetic Algorithm* (VEGA): Schaffer (1984); It divides the population in a number of subpopulations equal to the number of different objective functions. The offspring population is created mixing the result of selected individuals of each subpopulation.
- *Multi Objective Genetic Algorithm* (MOGA): Fonseca & Fleming (1993): The ranking of each individual is based in the number of individuals it is dominated by.
- *Non-dominated sorting genetic algorithm* (NSGA): Srinivas & Deb (1994): The ranking of each individual is based in the ranking of the front by the Pareto non-domination criteria.
- *Niched Pareto Genetic Algorithm* (NPGA): Horn, Nafploitis and Goldberg (1994); It uses a tournament selection procedure in the front. The final result is very sensitive to the tournament population size.

An important feature that has demonstrated to improve significantly the performance of evolutionary multiple objective algorithms is elitism, which maintains the knowledge acquired during the algorithm execution. It is materialized by preserving the individuals with best fitness in the population or in an auxiliary population. Also, the parameter independence in the process of diversity maintenance along the frontier is a beneficial algorithm characteristic. Among algorithms with lack of parameter dependence and including elitism, the following were developed:

- *SPEA* (*Strength Pareto Evolutionary Algorithm*), Zitzler and Thiele, 1999: It stores the solutions of the best obtained front in an external auxiliary population (elitism). The rank of each individual is based on its strength factor.
- *PAES* (*Pareto Archived Evolution Strategy*), Knowles and Corne, 1999: It stores the solutions of the best located front in an external auxiliary

population (elitism). A new crowding method is introduced to promote diversity in the population. The objective function domain is subdivided into hypercubes by a grid that determines the individual density.

- *NSGA-II*; Deb, Agraval, Pratap and Meyarivan, 2000: It maintains the solutions of the best front found including them into the next generation (elitism). A crowding distance is evaluated, avoiding the use of sharing factor.
- *SPEA2*; Zitzler, Laumanns and Thiele, 2001: Based on SPEA, the SPEA2 algorithm has some differences oriented to eliminate the possible weaknesses of its ancestor. A truncation operator is introduced instead of the clustering technique of SPEA to keep diversity in the population; it avoids the possible loss of outer solutions, preserving during the whole algorithm execution, the range of Pareto solutions achieved. In a comparative with five different test functions among some evolutionary multi-objective algorithms, NSGA-II and SPEA2 are the ones which appear to achieve the better whole results [70].

One of the key issues in evolutionary computation is the representation or coding scheme to adopt. Normally the same problem is susceptible to be represented in different ways, but since the topology of the search space is related to the code scheme, the ease or efficiency in solving the problem could be sensibly affected by the codification.

Among the most extendedly used schemes we have the binary, the integer and the real representation. The first one is applied to represent both integer and real variables by means of binary numbers, whereas the others represent the variables in a natural way, i.e. by using integer numbers for integers variables and real numbers for real components.

5.2.2 Evolutionary Algorithms with Mixed Coding Schemes: A Double Loop Approach

In complex problems, like the one studied here, premature convergence can appear due to different reasons. Those cases related to the EA have been studied in depth during the last years, allowing to state that the second-generation MOEA (e.g. NSGA-II, SPEA2) are remarkably efficient, as is recognized by many authors [21].

However, there remain other sources of premature convergence directly related with the coding scheme adopted. Normally, it is verified when the vector of coded variables presents at least one of the following characteristics:

- Variables of different nature, e.g. binary and real.

- Variables of different ranges.
- Dependent and independent variables: the values of some variables are dependent of some others or some trigger variables are the cause of the values of some others. This is the typical case in which new fault tree gates (e.g. Functional Dependency Gates) are needed to model the system [24].
- Temporal and non-temporal variables: e.g. in some systems there are variable which are delayed respect to the rest of the input, or the ranges for some variables change with time (e.g. Aircraft Landing Systems) [23][25].
- The sensitivity of the objective function differs much from some variables to others.

A glance to the state of the art of EA applications in Reliability and Dependability problems reveals an important occurrence of the abovementioned problem. Among others, we have: Reliability Optimization, Allocation or Redundancy [22,26,27,46,55,57], System Optimization and synthesis integrating Reliability [17,33], Maintenance policies [10,20,27,42,43], Redundancy and Maintenance optimization [52,58].

In general the convergence problem takes place because the whole decision vectors are evolved in the same process, neglecting the different speeds of convergence of each group of variables. In order to tackle this difficulty, we can imagine diverse strategies like:

1. To induce delay criteria for the group of variables with faster convergence. A typical example of this procedure is the use of a specific penalty for the studied problem [54].
2. To alternate the group of variables susceptible to be modified during the evolutionary process. It entails to restrict the application of the EA to each group of variables during a given number of generations, taking care of selecting suitable recombination operators and parameters in each stage. This alternative corresponds to the concept of Nash equilibrium [51,61] during the optimization and has the advantage of keeping the EA virtually unaltered.
3. To enumerate exhaustively some of the variables with smaller range, exploring the rest of the vector by means of EA. It is clear that the exhaustive search is an evident drawback; however this strategy is useful when there are dependencies between variables. In [47] this technique is employed to solve the simultaneous optimization of surveillance test intervals and test planning for a nuclear power plant safety system.
4. To employ evolutionary nested loops in such way that each loop correspond to a heuristic search over a specific group of variables. This method has the advantage of limiting the exponential growth of an enumerative

approach whereas explores the search space in depth with suitable operators for each group of variables. Nevertheless, the total number of evaluations remains high compared with other alternatives if there are several groups of variables that should be associated with a nested loop.

To the present, the advantages and drawbacks of each approach have not been studied and compared exhaustively, but there exist some previous works like the Integrated Design of Safety Systems [54] where the authors employed penalization to delay the convergence.

We call a double loop algorithm an algorithm comprising two nested optimizers, one inside the other. Each algorithm acts over a different group of variables of the same chromosome, attempting to find the optimal individuals by means of a heuristic evolution. It has been implemented successfully in some engineering design problems, as for example shown in [64]. The double loop algorithm might be implemented by means of different heuristic of local and global search; however in this chapter we restrict our exposition to genetic algorithms for single and multiple-objective optimization. Fig. 2 shows the structure of a single objective optimization double loop algorithm.

// $NILV = Number\_of\_Inner\_Loop\_Variables$

// $NOLV = Number\_of\_Outer\_Loop\_Variables$

// $NumVar = NOLV + NILV$

// $NGOL$ =Number of Generations for the Outer Loop

// $NGIL$ =Number of Generations for the Inner Loop

// $R_{NOLV}^{Pop\_P\_size}$ = Set of variables managed by the outer loop

// R_{NOLV}^{k} = kth solution of $R_{NOLV}^{Pop\_P\_size}$ $(1 \le k \le Pop\_P\_size)$

// $S_{NILV}^{Pop\_P\_size}$ = Set of variables managed by the inner loop

// $P_{NumVar}^{Pop\_P\_size} = R_{NVOL}^{Pop\_P\_size} \bigcup S_{NVIL}^{Pop\_P\_size}$ = Main Population (outer loop)

// $Q_{NumVar}^{Pop\_Q\_size} = R_{NVOL}^{k} \bigcup S_{NVIL}^{Pop\_Q\_size}$ = Auxiliary Population (inner loop)

Create Initial population $(P_{NumVar}^{Pop\_P\_size})_0$

Evaluate & Sort $\{(P_{NumVar}^{Pop\_P\_size})_0\}$

```
i=1 // Initialize the number of generations of the outer loop
While (i<=NGOL)  //Outer loop
```

Use GA to generate "n_size" Child. for outer variables $R_{NOLV}^{n\_size}$

k=1 //Initialize the pointer over the set $R_{NOLV}^{n\_size}$

While (k<= n_size)

$(Q_{NumVar}^{Pop\_Q\_size})_0$ `// Initialize population for inner loop`
```
j=1 // Initialize the number of gen. of the inner loop
While (j<=NGIL)  // Inner loop
```
Use GA to gen. "m_size" Child. for inner var $S_{NILV}^{m\_size}$

$$\left(Q_{NumVar}^{Pop\_Q\_size+m\_size}\right)_j = \left(Q_{NumVar}^{Pop\_Q\_size}\right)_j \cup \left\{R_{NVOL}^{k} \cup S_{NVIL}^{m\_size}\right\}$$

Evaluate & Sort $\left\{\left(Q_{NumVar}^{Pop\_Q\_size+m\_size}\right)_j\right\}$

$$\left(Q_{NumVar}^{Pop\_Q\_size}\right)_{j+1} \leftarrow \left(Q_{NumVar}^{Pop\_Q\_size+m\_size}\right)_j$$ `//Del.worst sol.`

```
j++
End While
```
$$(P_{NumVar}^{NumSol})_i = (P_{NumVar}^{NumSol-1})_i \cup \left(Q_{NumVar}^{1}\right)_{NGIL}$$ `//Actualize ext. pop.`

```
k=k+1
End While
```
Sort $\left\{(P_{NumVar}^{NumSol})_i\right\}$ `//Sort from best to worst`

$$\left(P_{NumVar}^{Pop\_P\_size}\right)_{i+1} \leftarrow \left(P_{NumVar}^{NumSol}\right)_i$$ `//Delete the worst solutions`

```
i++
End While
```

Fig. 2. Double Loop Genetic Algorithm

In this work we have developed a multiple-objective double loop algorithm based on the well known Non-dominated Sorting Genetic Algorithm NSGAII [21]. The core of the approach considers the following aspects (Fig. 3):

- Both the inner and the outer loop follows the definition formulation of the NSGA-II
- The external loop search the Pareto front of the global problem by proposing to the inner loop partial problems extracted from the global search.
- The inner loop looks for the optimal solution of the partial problems and gives the non-dominated solutions back to the external loop.
- The algorithm has two different non-dominated solutions archives.

// *NGOL =Number of Generations for the Outer Loop*

// *NGIL =Number of Generations for the Inner Loop*

```
Initialize population P for the outer loop
Compute Objective Functions for P
Order P using Non Dominated Criterion
While (i<=NGOL)   //Outer loop
    Use Evo. Alg. to generate "n_size" Children for outer variables
       For each "n_size" children
          Initialize population Q for inner loop
```

```
        While (j<=NGIL)  // Inner loop
            Use Evo. Alg. to gen. "m_size" Children for inner vars
            Compose solutions (outer loop+inner loop) into pop. Q
            Compute Objetive Functions for Q
            Classify Q using Non Dominated Criterion.
            Obtain the best(s) front(s)
        End While
        Actualize external Population P
        Order P using Non Dominated Criterion
        Obtain the best(s) front(s)
    End For
End While
```

Fig. 3. Double loop NSGA-II-based algorithm design

Fig. 3 describes in detail the structure of the double loop NSGA-II. Notice that the **Sort** procedure sorts the populations classifying them in fronts and then applying a density measure defined by the "crowding operator" according to [21].

As "a priori" insight, we expect the algorithm to show the following behavior:

- The method has the potential of outperforming single loop heuristics due to the partition of the search space which allows it to explore in depth the problem's domain.
- The algorithm is highly sensitive to the number of iterations set for each loop, which constitutes the key issue in assessing its efficiency. With just a few iterations of the inner loop the algorithm will bring poor results whereas with a large number of iterations the results will be good but at the cost of a slow convergence.
- The efficiency of the single NSGA-II faced with a double-loop NSGA-II depends on the complexity and the structure of the search space. Since there are no previous results for this kind of approach, it is necessary to perform several experiments that lead to upheld conclusions.

Until now the applications of the double loop structure were restricted to those cases where an external loop -evolutionary or not- fixes the parameters and/or the operators of the global search heuristic. Only recently the double loop structure has started to be applied in optimization of the maintenance schedule [47,49]. In this work we test the algorithm described in Fig. 4, improved with the introduction of Flexible Evolution strategies (Figs. 5-73) as well as the original formulation of the NSGA-II [21] over the simultaneous design of the structure and the maintenance strategy of Safety Systems.

$$// \; NILV = Number\_of\_Inner\_Loop\_Variables$$
$$// \; NOLV = Number\_of\_Outer\_Loop\_Variables$$

// $NumVar = NOLV + NILV$

// $NGOL$ =*Number of Generations for the Outer Loop*

// $NGIL$ =*Number of Generations for the Inner Loop*

// $R_{NOLV}^{Pop\_P\_size}$ = *Set of variables managed by the outer loop*

// R_{NOLV}^{k} = k*th solution of* $R_{NOLV}^{Pop\_P\_size}$ $(1 \le k \le Pop\_P\_size)$

// $S_{NILV}^{Pop\_P\_size}$ = *Set of variables managed by the inner loop*

// $P_{NumVar}^{Pop\_P\_size} = R_{NVOL}^{Pop\_P\_size} \cup S_{NVIL}^{Pop\_P\_size}$ = *Main Population (outer loop)*

// $Q_{NumVar}^{Pop\_Q\_size} = R_{NVOL}^{k} \cup S_{NVIL}^{Pop\_Q\_size}$ = *Auxiliary Population (inner loop)*

Create Initial population $(P_{NumVar}^{Pop\_P\_size})_0$

Evaluate & Sort $\left\{(P_{NumVar}^{Pop\_P\_size})_0\right\}$ //Sorting based in non-dom. Crit.

i=1 // Initialize the number of generations of the outer loop
While (i<=NGOL) //Outer loop

    Use Evo. Alg. to gen. "n_size" Child. for outer variables $R_{NOLV}^{n\_size}$

    k=1 //Initialize the pointer over the set $R_{NOLV}^{n\_size}$
    While (k<= n_size)

        $(Q_{NumVar}^{Pop\_Q\_size})_0$ // Initialize population for inner loop
        j=1 // Initialize number of gen. (inner loop)
        While (j<=NGIL) // Inner loop

            Use EA to gen. "m_size" Child. for inner vars. $S_{NILV}^{m\_size}$

$$\left(Q_{NumVar}^{Pop\_Q\_size+m\_size}\right)_j = \left(Q_{NumVar}^{Pop\_Q\_size}\right)_j \cup \left\{R_{NVOL}^{k} \cup S_{NVIL}^{m\_size}\right\}$$

            //Evaluate Obj. Func. and sort based in non-dom. Crit.

            Evaluate & Sort $\left\{\left(Q_{NumVar}^{Pop\_Q\_size+m\_size}\right)_j\right\}$

            //Actualize population with deletion criterion

$$\left(Q_{NumVar}^{Pop\_Q\_size}\right)_{j+1} \leftarrow \left(Q_{NumVar}^{Pop\_Q\_size+m\_size}\right)_j$$

            j++
        End While

$$(P_{NumVar}^{Pop\_P\_size+Pop\_Q\_size})_i = (P_{NumVar}^{Pop\_P\_size})_i \cup \left(Q_{NumVar}^{Pop\_Q\_size}\right)_{NGIL}$$

        // Sort based in non-dominated criterion

          Sort $\left\{(P_{NumVar}^{Pop\_P\_size+Pop\_Q\_size})_i\right\}$

          //Actualize population with deletion criterion

          $(P_{NumVar}^{Pop\_P\_size})_i \leftarrow (P_{NumVar}^{Pop\_P\_size+Pop\_Q\_size})_i$ **k=k+1**

    End While

```
    (P_NumVar^Pop_P_size)_{i+1} = (P_NumVar^Pop_P_size)_i
    i++
End While
```

$$(P_{NumVar}^{Pop\_P\_size})_{i+1} = (P_{NumVar}^{Pop\_P\_size})_i$$

Fig. 4. NSGA-II-based double loop evolutionary algorithm

5.2.3 Flexible Evolutionary Algorithms

The Flexible Evolutionary Algorithms (FEA) have been developed during the last five years [29, 66, 67, 68], trying to take advantage of lessons learned by many researchers developing Evolutionary Algorithms. The main feature of FEA is the ability of self-adapting its internal structure, parameters and operators on the fly, during the search for optimum solutions. Fig. 5 depicts the FEA basic pseudo-code.

The FEA algorithm is divided into functions, called *Engines*, designed to group the different actions occurring during the optimization:

- *Learning Engine* will calculate statistical measures (e.g.: convergence rate) that will be stored to be used afterwards. With the aim of clarify the basic principles of FEA, the release of FEA used in this work do not use any learning function. For more advanced algorithms the reader is referred to more specialized works [68].
- In order to generate more diversity in the samples a one-point crossover acts before launching the selection and sampling processes.
- The *Selection Engine* may contain different selection operators. Actual releases of the FEA algorithms contain only one Tournament Selection (2:1) but the authors announced recently [68] engines with several selection methods.
- The *Sampling Engine* contains a set of different mutation and crossover operators [66]. Among the crossover operators considered we can cite the Geometrical and the Arithmetical ones, while among the mutation operators used are included the Gaussian Mutation, the Non-Uniform Mutation or the Mutation Operator of the Breeder Genetic Algorithm. None of these sampling methods have extra parameters that may be adjusted previously. All the existing parameters are calculated randomly.
- There exists a central function called *Decision Engine* which rules all the processes. The way the *Sampling Engine* is managed by the *Decision Engine* is the main responsible for the FEA flexibility and capability to explore/exploit the search space.

// **Algorithm parameters**
// *NG =Number of Generations*
// Pop_P_size = *Size of Population P*

```
// NumVar = Number of Variables
// P_{NumVar}^{Pop_P_size} = Main Population
Create Initial (P_{NumVar}^{Pop_P_size})_0
Evaluate & Sort {(P_{NumVar}^{Pop_P_size})_0} ////Evaluate Objective Function and
sort from best to worst
While (i<=NG)  //Main loop
        Learning Engine
        One Point Crossover
        // Use Evo. Alg. to generate "n_size" Children R_{NumVar}^{n_size}
        Selection Engine
        Sampling Engine & Decision Engine
        (P_{NumVar}^{Pop_P_size+n_size})_i = (P_{NumVar}^{Pop_P_size})_i ∪ (R_{NumVar}^{n_size})_i
        Evaluate & Sort (P_{NumVar}^{Pop_P_size+n_size})_i  // Eval. Obj. Func. & Sort
        (P_{NumVar}^{Pop_P_size})_i ← (P_{NumVar}^{Pop_P_size+n_size})_i
        i++
End While
```

Fig. 5. FEA-based algorithm design

FEA works with only few external parameters, since most of them have been incorporated into the internal structure of the algorithm; it evolves the internal parameter setting during the optimization and does not need to be tuned for every specific problem. The remaining parameters are the population size and the number of generations. Each individual of FEA use an Enlarged Genetic Code (EGC) which is evolved by the algorithm. The EGC includes the variables and other information about the process that can be useful in order to guide the optimization. A code with three sets of variables has been recently presented [68], where the structure of the population is $P(\vec{W}) = P(\vec{x}, \vec{y}, \vec{z})$, where P denotes the population of $\vec{W}$ individuals, $\vec{x}$ is the vector containing the variables, $\vec{y}$ the sampling methods and $\vec{z}$ some internal control parameters. The EGC is used by the *Decision Engine* through its rule based Control Mechanism (CM). The set of rules inside the CM allows managing the internal parameters and operators in order to prevent premature stagnation ensuring the efficiency of the whole process. FEA has been designed for optimization problems with real variables only and its possible impact on dependability problems were studied recently [68]. Several real coded dependability problems are suitable to use FEA to enhance the efficiency of the optimization process, among others Maintenance related Optimization problems [10,11,13,14,18,28,45,59] and Reliability related ones [26,55,57,69].

Under the Flexible Evolutionary framework no algorithm for integer or binary coded variables has been introduced as well as for multiple objectives optimization. In this work a new FEA for integer coded variables is introduced for the double loop multiple objectives algorithm described in Fig. 4. The algorithmic design of the developed FEA follows the scheme of the Fig. 8 being the main differences the Decision and Sampling Engines developed.

As Decision Engine and for the sake of simplicity a single Conservative/Explorative random choice has been used (Fig. 6) in this work. The Engine selects the type of decision to be used for each variable to be sampled. If the decision is Conservative, then the new variable value will be obtained using the same operator used to obtain the previous one. If the decision is Explorative a different operator will be used. The *Risk Front* is a value in the interval [0,1] which reflect the algorithm/user attitude for decision making along the optimization process. Higher values of *Risk Front* suggest more risky decisions while lower values suggest more conservative ones. Normal uses of *Risk Front* imply to adopt more Explorative decision when the optimization is starting and more Conservative decisions at the end, because if near optimum solutions have been found along the process probably the best operators have been identified too.

As Sampling Engine (Fig. 7), a set of operators designed for integer coded variables has been used. These operators are of the following types: Single Add/Deduct, Position Change and Control Variable. The first type adds or deducts integer values to the actual values of the variables, the second one changes the relative position of the actual variable value to a new one inside its range; the final one uses the value of a reference variable to change the actual variable value.

$$// \, Rnd(\,) = Uniform\ Random\ Number\ in\ [0,1]$$

$$// \, Risk\_Front = user's\ choice\ in\ [0,1]$$

$$Decision = Conservative$$

$$Attitude = Rnd(\,)$$

$$if\,(Attitude \geq Risk\_Front)\ Decision = Explorative$$

Fig. 6. Single Conservative/Explorative random choice Decision Engine for FEA

// Single Add / Deduct Operators

$$X_i = X_{i-1} \pm k \left[1 \le k \le \left(INT\ \frac{Range\_of\_X}{2} \right) \right]$$

// Position Change Operators

$$X_i = M\ ,\ M \in \left[X_{Min}, INT\frac{X_{Max} - X_{Min}}{2}, X_{Max} \right] \text{ // Singular Values Operator}$$

$$X_i = X_{Max} - \left(X_i - X_{Min} \right),\ X_i < \frac{X_{Max} - X_{Min}}{2} \text{ // Relative Position Change Operator}$$

$$X_i = X_{Min} + \left(X_{Max} - X_i \right),\ X_i > \frac{X_{Max} - X_{Min}}{2} \text{ // Relative Position Change Operator}$$

// Control Variable Operators

$$X_i = X_{i-1} + INT\left[Rnd[0,1]\left(X_j - X_i \right) \right]\ ,\ X_j > X_i\ ,\ X_j = Control\ Variable$$

$$X_i = X_{i-1} - INT\left[Rnd[0,1]\left(X_i - X_j \right) \right]\ ,\ X_j < X_i\ ,\ X_j = Control\ Variable$$

Fig. 7. A FEA Sampling Engine for integer coded variables

// **Algorithm parameters**
// *NG =Number of Generations*
// $Pop\_P\_size = Size\ of\ Population\ P$
// ***Other definitions***
// $NumVar = Number\ of\ Variables$
// $P_{NumVar}^{Pop\_P\_size} = Main\ Population$

```
Initialize (P_NumVar^Pop_P_size)_0
Evaluate & Sort {(P_NumVar^Pop_P_size)_0} //Eval. Obj. Func. & sort
While (i<=NG)  //Main loop
        Learning Engine
        // Use Evo. Alg. to generate "n_size" Children  R_NumVar^n_size
                Selection Engine
                One Point Crossover
                Sampling Engine & Decision Engine
        (P_NumVar^Pop_P_size+n_size)_i = (P_NumVar^Pop_P_size)_i ∪ (R_NumVar^n_size)_i
        Evaluate & Sort (P_NumVar^Pop_P_size+n_size)_i  //Eval. Obj. Func. & sort
        (P_NumVar^Pop_P_size)_i ← (P_NumVar^Pop_P_size+n_size)_i
        i++
End While
```

Fig. 8. FEA-based algorithm design for Integer Variables

5.3 Integrated Safety Systems Optimization

The Safety Systems mission is to operate when certain events or conditions occur, preventing hazardous situations or mitigating their consequences. Where possible, design of such systems must ensure the mission success even in the case of single failure of their components, or when external events can prevent entire subsystems from functioning. The designer(s) must optimize the overall system performance deciding the best compromise among: Component Selection, Component/Subsystems allocation and Time Intervals between Preventive Maintenance. In SSDO problems the system performance measure is normally the System Unavailability, being a second important variable the total System Cost. The whole optimization task is restricted, because practical considerations place limits on available physical and economic resources.

System components are chosen from possible alternatives, each of them with different nature, characteristics or manufacturers. Component/Subsystems allocation involves deciding the overall system arrangement as well as the levels of redundancy, diversity and physical separation. Redundancy duplicates certain system components; diversity is incorporated achieving certain function by means of two, or more, components/subsystems totally different. Physical separation among subsystems guarantees the mission success, even in the case of critical common causes of malfunctioning. Finally the modern safety system designs include the consideration of the time intervals between preventive maintenance as a very important variable to consider. On the one hand significant gains are to be made by considering maintenance frequency at the design stage [54], on the other hand good safety systems designs can become unsafe if the maintenance frequency is unrealistic (the maintenance team cannot keep such frequency).

In automated SSDO problems, a computer program will be in charge to obtain the optimum design considering all abovementioned alternatives and restrictions. There are two main components in such software: The system model and the optimization method.

Due to the inherent described characteristics of the optimum safety systems design problem, evolutionary algorithms are ideal for solving and obtaining optimum solution designs.

5.3.1 The System Model

The system model must reflect all possible design alternatives, normally using as starting point the knowledge of the design team and the physical restrictions on resources. In order to perform an automated computer-based design search, the typical approach consists in creating a system model and to define, inside this, a set of binary variables ("indicator variables") which permit to select (switch "on") or reject (switch "off") different configurations, components and available market models.

The Fault Tree Analysis (FTA) is the most widely accepted methodology to model safety systems for computer-based design optimization purposes. A Fault Tree is a Boolean logic diagram used to identify the causal relationships leading to a specific system failure mode [63]. The system failure mode to be considered is termed the "top event" and the fault tree is developed using logic gates below this event until component failure events, termed "basic events", are encountered. H.A. Watson of the Bell Telephone Labs developed the Fault Tree Analysis in 1961-62, and the first published paper was presented in 1965 [39], since these dates the use of the methodology has become very widespread. Important theoretical [60,63], methodological [44] and practical [44,53] advances have been introduced. For ISSDMO problems one important contribution was the possibility to consider design alternatives (redundancy and market models) into the fault tree [1], by means of associate the "house events" used in Fault Tree logic to "indicator variables" of the system design, whose values were managed by an evolutionary process. The Fig. 9B shows a fault tree section which permits to consider from one to three redundant components for a given system design.

In order to allow the possibility of adding and removing components to and from a given design along the computer-based integrated optimization, we introduce here a NAND logic gate (Fig. 9A) besides the traditional logic gates used to decide the redundancy level (Fig. 9B)[1]. The logic of the fault tree is arranged in a single manner using the logic gates showed in Fig. 9, permitting to create the whole system model. Such model will describe both the component decision (select/reject) and the redundancy level (if the component is selected) for all the design alternatives.

The logic complexity is simpler than the used in previous works [1,3,54,37,38] because the market available model selection for each component is performed by the evolutionary search, so the associated logic gates and house events no need to be included in the system fault tree, which result now of easy develop and construction. This is especially important when large/complex systems are under study were the size of the fault tree becomes a bottleneck of the process, opening new possibilities for both research & practitioners which use other systems models.

Among other system measures [53], the FTA permits to calculate the system unavailability for every configuration of the "indicator variables" (house events), which is of special interest in ISSDMO. Along the time many methods have been developed for quantitative evaluation of fault trees. During the 90's a very important contribution to this task was presented and developed, which use the FT conversion to Binary Decision Diagrams (BDD) [2,12,19], suitable for easy computation of the top event measure. Despite the widespread develop and use of the methodology, one significant inconvenient still remains unresolved: the basic event ordering necessary to obtain the simplest BDD suitable for easy computation (among the set of BDDs associated to one Fault Tree).

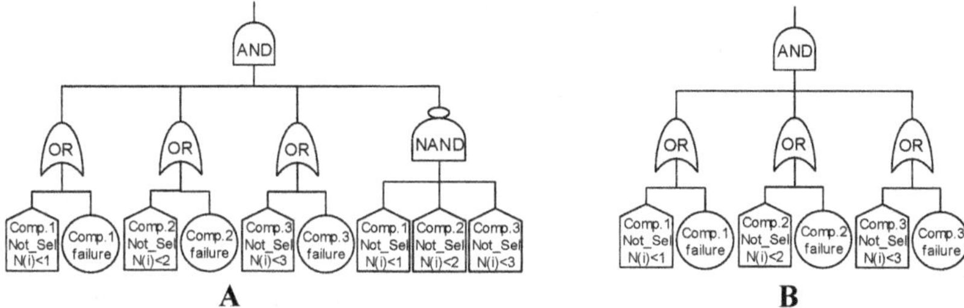

Fig. 9. Fault Tree for Remove/Redundancy-level (A) and Redundancy-level (B) of a system component

Many contributions have been presented in the field [2,8,9], but up to date only a few heuristics support the selection of the adequate basic event ordering. That is the reason why to evaluate fault trees, we use here an efficient direct algorithm based on a Monte Carlo Variance Reduction Technique called Restricted Sampling (RS) (paragraph 3.3), which provides accuracy estimates for the system unavailability as well as exact intervals and, in many cases, the exact solution. This method was introduced and tested a few years ago [30], having the advantage of working directly with the system fault tree, without any previous conversion to other structure (like Binary Decision Diagrams) nor simplification (like Minimal Cut Sets). The RS efficiency to evaluate large/complex fault trees was demonstrated too [30] using a wide test bed.

5.3.2 Components Unavailability and Cost Models

Since the FTA provides the tool to assess the unavailability of the whole system, it is only necessary to select a suitable model for calculating the unavailability of each single component or basic event of the FT. In that sense we adopt the model proposed by Martorell et al. [49] which is specially formulated for nuclear plants safety systems.

For this model, the unavailability of each component u(x) is expressed as the sum of four basic contributions:

$$u(x) = u_r(x) + u_t(x) + u_m(x) + u_c(x) \tag{1}$$

where each contribution is calculated according to the expressions listed in Table 1. The variables involved in this unavailability model are listed in Table 2. Likewise, the cost model is formulated as a sum of basic cost contributions for each basic event, thus the cost of the whole design is

$$\sum_{basic-events} c_t(x) + c_m(x) + c_c(x) + c_u(x) + c_r(x) \tag{2}$$

where each contribution is assessed according to the expressions in Table 4 for the following parameters showed in Table 3.

Table 1. Basic contributions for the component unavailability according to [49]

Basic Contribution	Source or Description	Functional expression
$u_r(x)$	Random failures	$u_r \approx \rho + \lambda T/2$
$u_t(x)$	Testing	$u_t = t/T$
$u_m(x)$	Preventive maintenance	$u_c = (\rho + \lambda T)\mu\left(1 - e^{-(D/\mu)}\right)/T$
$u_c(x)$	Corrective maintenance	$u_m = m/M$
λ	PAR and PAS models	$PAS = \lambda_0 + \alpha M \Psi^2(z)(2-\varepsilon)/2\varepsilon$
		$PAR = \lambda_0 + \alpha M \Psi^2(z)[1 + (1-\varepsilon)(L/M - 3)]/2$

Table 2. Basic contributions for the component unavailability according to [49]

T	Surveillance test interval (STI)
t	Mean time for testing
ρ	Cyclic or per-demand failure probability
λ	Stand-by failure rate considering either a PAR or PAS model [49]
α	Linear aging factor
$\psi(z)$	Effect of the working conditions. Is set to one under normal or nominal operational and environmental conditions
ε	Maintenance effectiveness $\in [0,1]$
m	Mean time for preventive maintenance
M	Period to perform time-directed preventive maintenance
L	Overhaul period to replace a component by a new one
μ	Mean time to repair when there are no time limitations for repairing
D	Allowed outage time

Table 3. Parameters considered in Table 4

c_{ht}	Hourly costs for conducting surveillance testing
c_{hm}	Hourly costs for conducting preventive maintenance
c_{hc}	Hourly costs for conducting corrective maintenance
c_{tu}	Total cost of loss of production due to the plant shutdown
c_{tr}	Total cost of replacing the component

Table 4. Basic contributions for the component cost according to [49]

Basic Contribution	Source or Description	Functional expression
$c_t(x)$	Testing	$c_t(x) = c_{ht} \cdot t / T$
$c_m(x)$	Preventive maintenance	$c_m(x) = c_{hm} \cdot m / M$
$c_c(x)$	Corrective maintenance	$c_c(x) = (\rho + \lambda T)\mu[1 - e^{-(D/\mu)}]c_{hc} / T$
$c_u(x)$	Plant shutdown	$c_u(x) = (\rho + \lambda T)\mu e^{-(D/\mu)} c_{tu} / T$
$c_r(x)$	Overhaul maintenance	$c_r(x) = c_{tr} / L$

5.3.3 Computing the System Model Using Restricted Sampling

The assessment of the Unavailability is made by a two stage method that works on the fault tree with the chosen design/redundancy alternatives. In general words, during the first stage the a deterministic procedure (the Weights Method, [36]) binds the real value into a proper interval; then a restricted sampling scheme makes use of the proper interval during the second stage to obtain accurate estimates.

The first stage works un this way: the so called Weights Method [30,35,36] is used to compute the upper and lower bounds Q_L, Q_U for the system unavailability Q_s. If these bounds are sufficiently close between them (e.g.: both values have equaled at least the first non-zero figure), the system unavailability is computed as $Q_s = (Q_L + Q_U)/2$ and the method is stopped. Otherwise Q_s is computed using the unbiased Monte Carlo estimator:

$$Q_S = \hat{Q}_R = [Q_U - Q_L]N^{-1} \sum_{j=1}^{k} \psi(m_j) + Q_L \quad \text{being: } 0 \le Q_L \le Q_s \le Q_U \le 1$$

where the suffix R means "Restricted" and m_j are fault tree basic event-samples restricted over the set of the fault tree cuts not used to obtain the bounds Q_L, Q_U, N is the total number of Monte Carlo samples used and $\sum_{j=1}^{k} \psi(m_j)$ is the sum of all fault tree outcomes (one per sample).

The efficiency of the first stage has been exhaustively tested [30] showing a great capacity finding exact solutions for many large/complex fault trees in short CPU times, so the second stage is used only when very large/complex fault trees are encountered by the ISSDMO procedure. The combined action of both stages gives the method a great adaptability to the fault tree to be evaluated, which is a key characteristic for the efficiency of the whole optimization process.

5.3.4 Coding Schemes

Once the fault tree with design alternatives has been constructed, the process to obtain an integrated safety system optimum design consists in finding the set of Indicator Variables and Test Intervals which defines the tree (system) with minimum Unavailability. The nature of the variables and the combinatorial character of the search, justify the application of heuristic methods based on evolutionary algorithms. However if the search process evolves pure binary (design & redundancy) and non-pure binary (Test Intervals) values, different coding alternatives can be considered:

- Binary coding of all variables (pure binary and real) and to employ a standard binary-based evolutionary algorithm.
- Binary coding for pure binary variables and real coding for real variables. To use an evolutionary search managing the operators and/or the algorithm to evolve properly.
- Integer coding of all variables utilizing a standard integer-based evolutionary algorithm.

For the first option is sufficient to have a standard binary-based evolutionary algorithm (e.g.: Binary Genetic Algorithm, BGA) and this is its main advantage, because both the convergence properties and the more efficient operators of BGA are well known. This option was explored in the past [54] but the authors reported premature convergence problems due to the different nature of the variables. The solution tested was to "delay" the convergence velocity by means of a penalty formula, but no convincing results were presented.

The second option requires using two optimizers: Binary and Real. The first devoted to evolve the designs and the second to find the optimal set of test intervals. The advantage of this option is that each group of variables is evolved by evolutionary processes developed specially for its nature. The way to link both optimizers is probably the main problem to solve for practical applications, but the double loop presented in this work seems to be a promising alternative to be tested.

The third option has the same algorithmic advantages than the first one, because only one type of optimizer performs the whole search, but the possibility of coding both binary and real variables as integers seems to suggest new advantages. On the one hand, it is easy and well known coding binary variables as integers. On the other hand, recent researches [49] demonstrate how practical considerations on available maintenance resources, permits to consider only integer values for test intervals.

No research results have been reported for the second and the third options at the moment. The first and the third options are explored and compared in this work.

In order to keep the evolutionary search as efficient as possible, the variable information coded in the chromosome should be structured in such a way that the information interchange between solutions using crossover or its modification using mutation does not produce an infeasible solution. The chromosome codification used here for ISSDMO allows the use of the conventional crossover and mutation operators. It considers one variable for each design component, which includes the number of redundancies of this component (number of parallel elements). Also it considers one variable for each component, which represents the test interval (TI) in hours. Finally, it includes one variable representing the model type for each possible parallel element of each component. In this last case, it is possible to encode redundant information when the number of redundancies of the component is less than the maximum number of redundancies. However using this encoding, has the advantages of generality and permits the use of standard mutation and crossover without producing infeasible solution candidates.

5.4. Application example: The Containment Spray System of a Nuclear Power Plant

5.4.1 System Description

The PWR Containment Spray System (CSS) is generally designed to provide particular and different functions inside the containment of a Pressured Water Reactor (PWR) of a Nuclear Power Plant (NPP), among others: To allow borated water flow to the spray rings, to remove heat (heat exchanger) following a loss of coolant accident (LOCA) or a steam line break, and Sodium Hydroxide chemical addition to improve the removal of iodine from the atmosphere. The particular design conditions on each PWR can avoid the necessity of some of the functions, so not all NPP designs include a heat exchanger or a sodium hydroxide addition sub-system into its CSS.

In this work we concentrate our attention in the borated water injection function because the associated sub-system design can be the most common to many different PWRs. The sub-system (Fig. 10) typically consists of two separate and complete trains, each with centrifugal pumps and valves to control the flow of borated water to the containment spray system from the Refueling Water Storage Tank (RWST). The main subsystem components are the Motor Operated Valves (MOV), Motor Driven Pumps (MDP), single Valves (V) and One Way Valves (OWV). The component failure modes used to model and evaluate the sub-system were the typical for PWRs [53].

Assuming the physical separation between both trains will be implemented during the building phase, the remaining reliability-based design concepts will be explored here: Redundancy (to include more than one component in parallel) and Diversity (to use different component models). Additionally the possibility of providing some connection between both trains in order to increase the reliability level will be explored too, that is the reason for the two connection paths (3 & 8) depicted on Fig. 10.

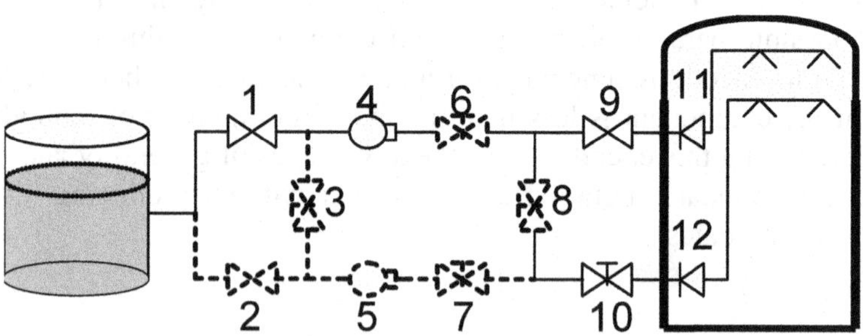

Fig. 10. CSS basic a priori design

In order to perform a computer-based design optimization, the following criteria has been adopted:

1. Only four components are considered: Valves (1,2,9), Motor Driven Pumps (4,5), Motor Operated Valves (3,6,7) and One Way Valves (11,12).

2. A maximum of three redundant components in parallel can be placed on each position.
3. Three different models of each component are available on the market
4. It will be possible to adopt different models for the redundancies of each component.
5. Only the following components are mandatory for each design: Valves (1, 9), Motor Operated Valves (10, 11, 12), Motor Driven Pump (4), pipe through MOV 8.
6. The pipe between the MDP 5 and the MOV 7 will be included only if the MDP 5 take part of the design.
7. The MOV 3 will be considered only if the MDP 5 and the MOV 7 are included in the design.

5.4.2 Models and Parameter Selected

The parameter values were adapted from [49] for simulating the situation when the designer has several alternatives in the market from where to choose the final model of the components. Table 5 lists the values for the parameters of the unavailability model whereas Table 6 presents the values for the parameters of the cost model.

Table 5. Parameters for the component unavailability model [49]

	Class of component											
Parameter	Pumps			Valves			MOV			OWV		
	M 1	M 2	M 3	M 1	M 2	M 3	M 1	M 2	M 3	M 1	M 2	M 3
$P(10^{-3})$	0.53	0.65	0.55	1.82	1.72	2.33	2.26	1.9	2.07	1.76	1.74	1.63
$\alpha(10^{-11}/h^2)$	2.20	2.436	2.1901	45.7	44.3	48.9	53.5	57.9	61.2	32.0	33.8	37.8
ε	0.5	0.5	0.5	0.5	0.5	0.5	0.5	0.5	0.5	0.5	0.5	0.5
$\Psi(z)$	1	1	1	1	1	1	1	1	1	1	1	1
$\mu(h)$	24	24	24	2.6	2.6	2.6	3.1	3.1	3.1	2.6	2.6	2.6
$\lambda_0(10^{-6}/h)$	3.89	3.89	3.89	5.83	5.83	5.83	5.90	5.90	5.90	5.79	5.79	5.79
λ	PAS	PAS	PAS	PAR	PAR	PAR	PAR	PAR	PAR	PAR	PAR	PAR
$t(h)$	4	4	4	0.75	0.75	0.75	0.75	0.75	0.75	0.75	0.75	0.75
$m(h)$	4	4	4	0.75	0.75	0.75	0.75	0.75	0.75	0.75	0.75	0.75
$M(h)$	4320	4320	4320	4320	4320	4320	4320	4320	4320	4320	4320	4320
$D(h)$	72	72	72	8	8	8	8	8	8	8	8	8
$L(h)$	43200	43200	43200	34560	34560	34560	34560	34560	34560	34560	34560	34560
$TI\ max(h)$	2184	2184	2184	2184	2184	2184	2184	2184	2184	2184	2184	2184

Table 6. Parameters for the component cost model [49]

Param.	Pumps			Valves			MOV			OWV		
	M 1	M 2	M 3	M 1	M 2	M 3	M 1	M 2	M 3	M 1	M 2	M 3
c_{ht}	20	20	20	20	20	20	25	25	25	20	20	20
c_{hc}	15	15	15	15	15	15	20	20	20	15	15	15
c_{hm}	15	15	15	15	15	15	20	20	20	15	15	15
c_{tr}	360	300	400	120	140	215	200	250	200	100	110	130
c_{tu}	1500	1500	1500	1500	1500	1500	1500	1500	1500	1500	1500	1500

Class of component

5.4.3 System Model

The Fault Tree (FT) has been constructed using as Top Event "Low Fluid Level on Spray Rings" and following the logic rules explained in section 3.1 in order to reflect the design alternatives and the redundancy levels. Figs. 11 to 13 contain the whole resulting FT. The house events in Figs. 11 and 12 are used to adapt the FT to each design explored by the evolutionary search, while the house events in Fig. 13 are used to adapt the FT to each redundancy level considered.

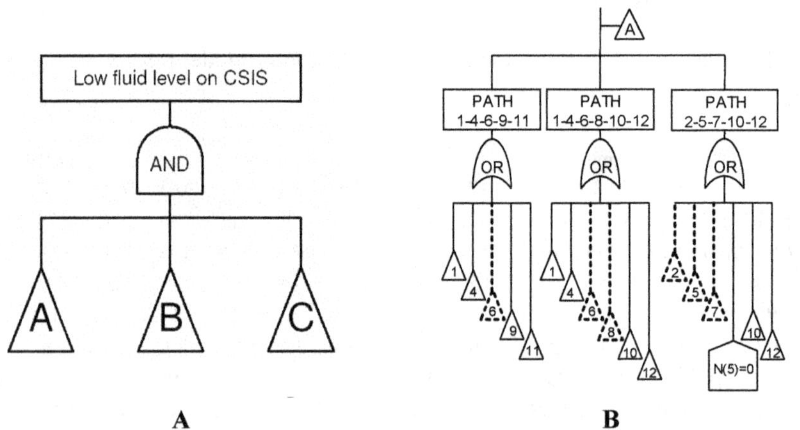

Fig. 11. Fault Tree Top Event (A) and First Sub Tree (B)

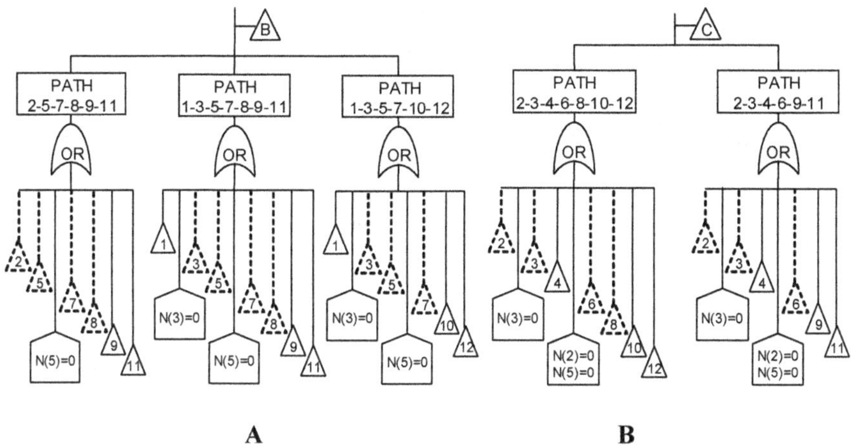

Fig. 12. Second (A) and third (B) Sub-Tree for the System Faul Tree

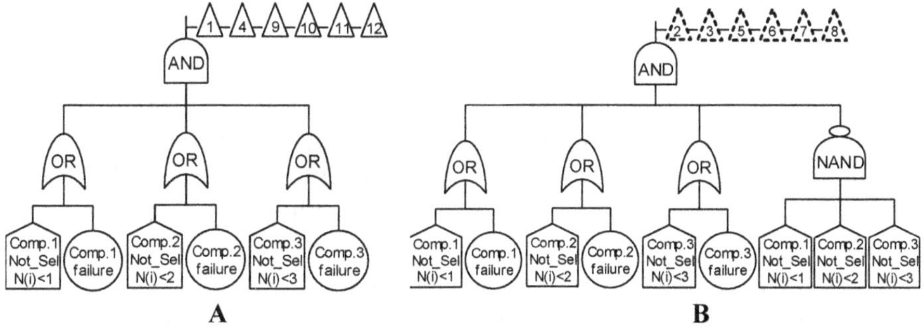

Fig. 13. Sub-Trees for obligatory (A) and non-obligatory (B) components

5.5 Results

The results of a set of ten runs with each multiple objective algorithm have been collected and compared. The general experimental settings for all the algorithms at each run were: 50 Individuals and 100 generations. In order to compare the single vs. the double loop algorithms all the runs were stopped when 5000 FT were evaluated. Previously, the optimal crossover and mutation parameters for the best performance of each algorithm were independently established. In the binary codification, four mutation rates were compared: 0.004, 0.008, 0.015 and 0.03. In the FE double loop approach, a variable strategy for the number of iterations of the inner loop has been considered, varying from lower to higher. The table 7 shows the final set of parameters for all the algorithms and runs.

Table 7. Parameter settings for all the algorithms

	Binary Algorithm	Integer Algorithm	Double Loop Algorithm
Number of Generations	100	100	100
Population Size	50	50	50
External File size	50	50	50
Crossover type	Uniform	One-point	One point
Mutation Type	Uniform	Triangular	Uniform
Crossover Probability	1.0	0.90	0.07
Mutation Probability	0.004	0.10	0.01
Number of Monte Carlo Samples	100	100	100
Internal loop: Number of Generations			3
Internal loop: Population Size			8
Internal loop: Pareto File size			4
Internal loop: Crossover type			One point
Internal loop: Mutation Type			Uniform
Internal loop: Crossover Probability			0.07
Internal loop: Mutation Probability			0.416
Restricted Sampling: Max cut order	6	6	6
Restricted Sampling: Number of samples	100	100	100

The main goal has been to assess the capacity of each algorithm for contributing with solutions to the Pareto Front. Fig. 14 depicts the best front obtained and the contribution of each algorithm. As can be seen, the binary approach contributes with 79 solutions, the Integer contributes with 44 solutions and the double loop contributes with 24 solutions. So, all the approaches shows capacity to found non-dominated solutions, but with different search characteristics. These characteristics can be seen with more detail in the Figs. 15 to 17. On the other hand, the Fig.14 shows the solution of each Pareto Front "corner" labeled with numbers from one to ten. These solutions are depicted in Figs. 18 to 22 showing the different layouts of the achieved designs.

Notice that both the unavailability and the cost axis are represented in logarithmic scale. It gives the opportunity to appreciate better the two mechanisms that constitute the heuristic search: on one hand we have the introduction of minor changes that conserve the basic layout by modifying only some models and surveillance intervals. These slight variations produce unavailability values of the same magnitude which are characterized by the horizontal swathes of the front. On the other hand we have the introduction of major modifications that reduce drastically the unavailability by placing key components in the layout, generating the vertical strips.

The above comment can be understood better by giving a glance to the labeled solutions from Fig. 18 and keeping in mind that a single pump has a higher impact in the system unavailability that a valve for our particular design. Thus, designs with the same number of pumps like solutions 5 and

6 (Figs. 22, 23), as well as 7 and 8 (Figs. 24, 25) have the same order of magnitude, whereas the displacements from 6 to 7 and 8 to 9 that entail reductions in one order of magnitude are characterized by the introduction of one pump into the layout.

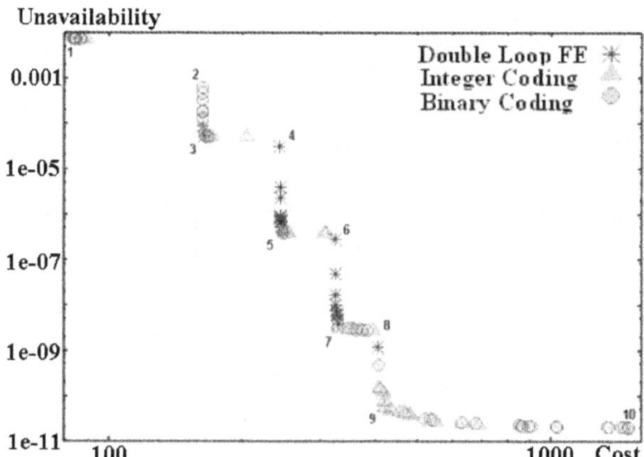

Fig. 14. Best Non-Dominated Front found considering all the codifications, including the labeling of ten characteristics designs

The labeled designs (Figs. 18 to 27) show a smooth transition in the design layout that is mainly governed by the number of parallel branches and the total number of pumps. Therefore, considering the designs from lower cost and higher unavailability to higher cost and lower unavailability, if we expose consecutively the labeled designs from 1 to 10, then the parallel number of pumps that are included in each of the two possible branches are the following (total number of pumps: number of first branch pump + number of second branch pump): (1:1+0), (2:2+0), (2:1+1), (3:2+1), (3:3+0), (3:2+1), (4:3+1), (4:3+1), (5:2+3), (6:3+3).

The binary algorithm is capable to obtain the most extreme solutions of the best non-dominated front found. Even more, they are not isolated points, but a set of solutions in both the minimum unavailability and minimum cost corresponding to those extreme portions of the front. Also the binary algorithm has located 79 non-dominated designs over a total of 167 proposed solutions. However, in some intermediate parts of the front, the prevalence of the other codifications is evident, lacking of binary coded solutions.

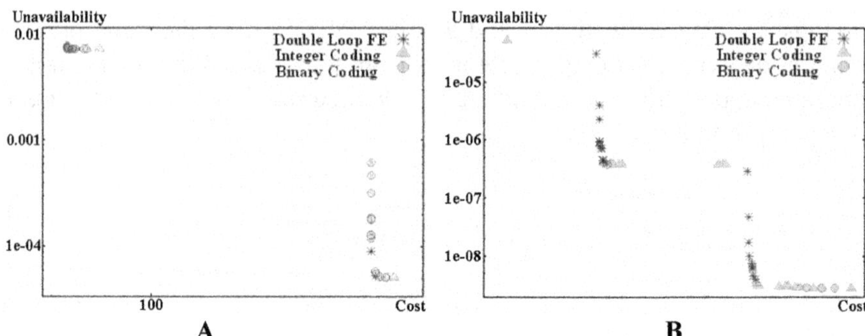

Fig. 15. Top (A) and medium (B) Part of Best Non-Dominated Front found considering all the codifications

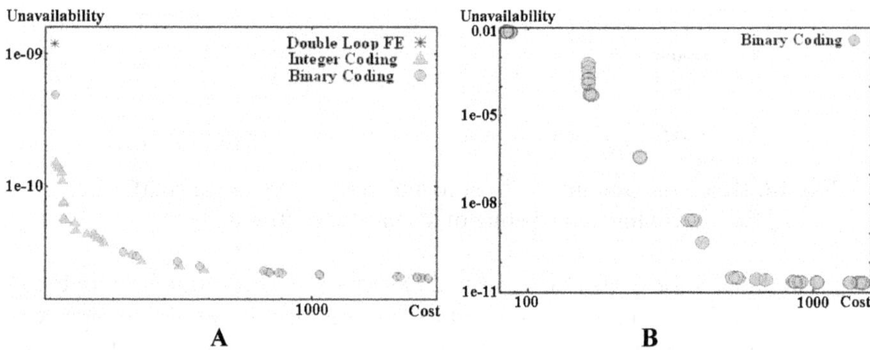

Fig. 16. Lower section of best Non-Dominated front found considering all the codifications (A) and Binary coded Single Loop Non-Dominated Solutions belonging to the best Front found (B)

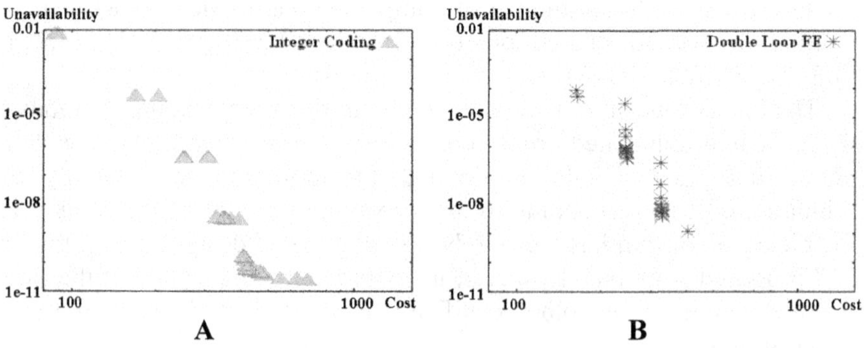

Fig. 17. Integer coded Single Loop (A) and Integer coded Double Loop Flexible Evolution (B) Non-Dominated Solutions belonging to the best Front found

The integer codification is the natural option for representing combinatorial problems like the one studied here. Since one hour is a good step for discretizing time and the remaining variables are integer in nature, the integer representation looks like a good option “a priori”. Nevertheless, the recombination operators and the parameters setting must be selected carefully in order to assure diversity during the search. Figs. 14, 15 and 16 show that the NSGA-II Single Loop with integer representation found a large number of non-dominated solutions in the area of best trade-off, that is the zone with high cost and low unavailability (labeled solution n° 9. This leads to consider that this implementation scores over the others in terms of DM’s desires; however a wider coverage of the whole non-dominated front is also important, and in this point the integer representation showed difficulties to succeed.

The NSGA-II Double Loop Flexible Evolutionary Algorithm with integer representation found non-dominated solutions in a very difficult area of the Pareto front. The main characteristic of this area is the existence of many solutions, allocated in two vertical strips, with significant differences in Unavailability (one or two orders of magnitude) but very similar in Costs. This result suggests the Double Loop take advantage of the loops structure to find deeper solutions in this difficult area. However the whole front coverage was not of the same quality, so results seem to indicate that the approaches are complementary but, due the duplicate number of parameters with regard to single loop algorithms, a more exhaustive work on parameter settings is needed.

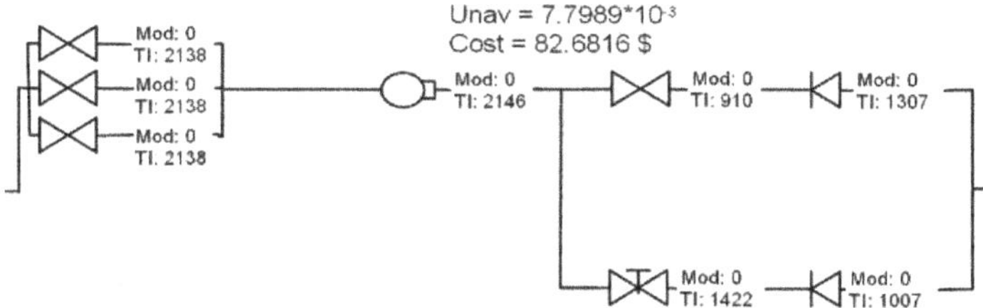

Fig. 18. Detailed Solutions Label 1

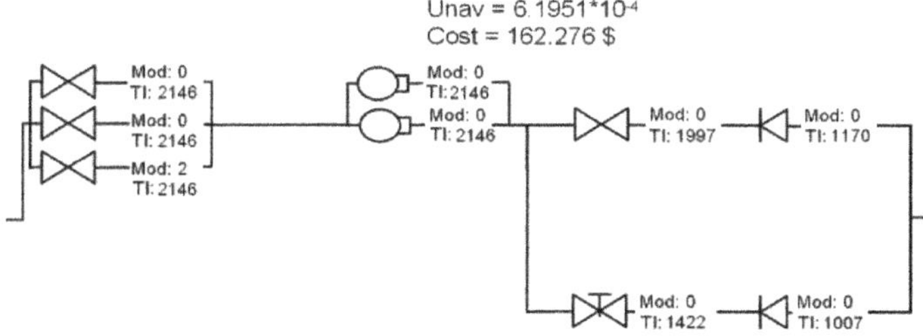

Fig. 19. Detailed Solutions Label 2

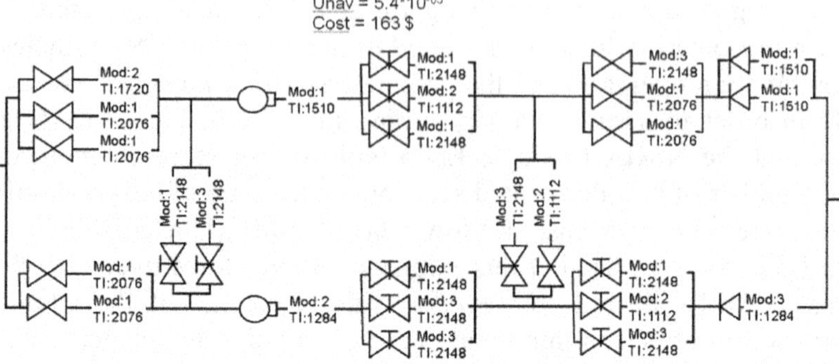

Fig. 20. Detailed Solution Label 3

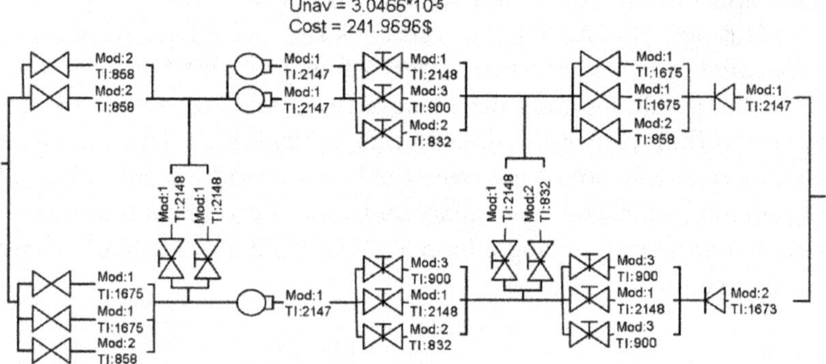

Fig. 21. Detailed Solution Label 4

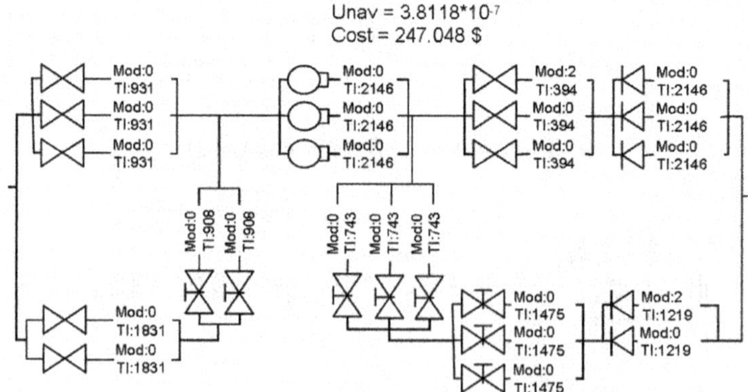

Fig. 22. Detailed Solution Label 5

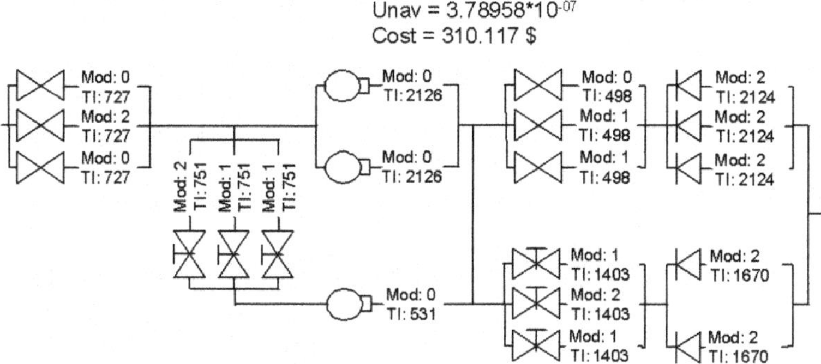

Fig. 23. Detailed Solution Label 6

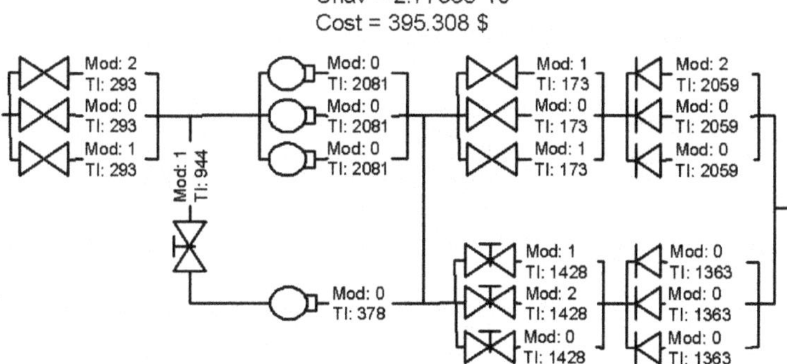

Fig. 24. Detailed Solution Label 7

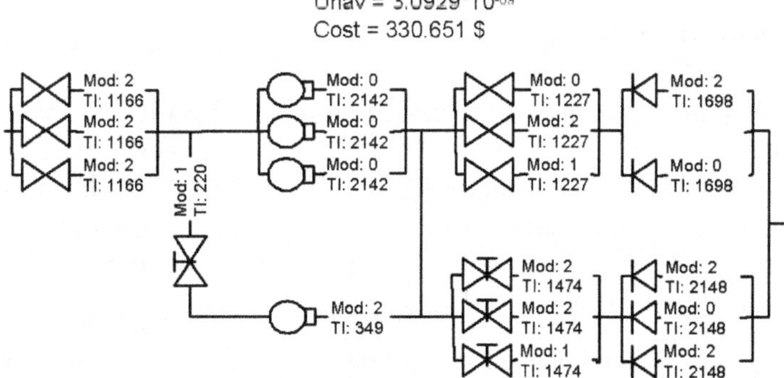

Fig. 25. Detailed Solution Label 8

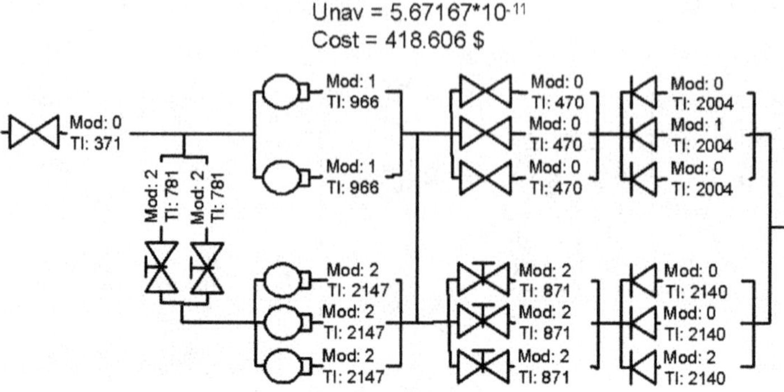

Fig. 26. Detailed Solution Label 9

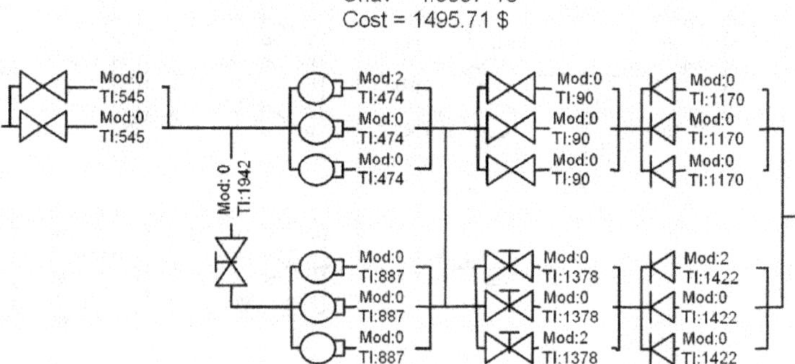

Fig. 27. Detailed Solution Label 10

5.6 Conclusions

In this chapter new methodologies for Integrated Safety Systems Design and Maintenance Optimization (ISSDMO) have been introduced and tested. The development and/or adaptation process of each one has been performed taken into account, as main objective, their use into a computer-based ISSDMO problem. Considering the previous methodologies, the main characteristics introduced are:

- The use of Fault Trees with less number of logic gates and events. This is due to the fact that available component models are not included inside the fault tree (only design changes and redundancy levels). Moreover the enhanced tree logic allows adding/deducting components to the system automatically during the evolutionary search. The combined effect of both characteristics convey in building a smaller tree that fully represents the whole domain of layout designs.

- The use of a powerful and self-adaptable two-stage direct method for fault tree quantitative evaluation. The first stage finds proper bounds for the system unavailability using a fast deterministic process. The high efficiency of this stage computes exact solutions (the bounds reach the same value) for many large/complex fault trees. Only if the first stage cannot approach the bounds with the required precision, the second stage performs a Restricted Monte Carlo search using the proper bounds into an unbiased estimator, which produces accurate estimates of the system unavailability. The combined effect of both methods permits to adapt the computing effort to the size/complexity of the fault tree under evaluation "on the fly".
- The joint use of the two abovementioned methods constitutes a powerful tool for computer-based ISSDMO, avoiding the necessity of Binary Decision Diagrams or Minimal Cut Sets.
- The selection of appropriate component models among those available in the market has been transferred to the Evolutionary Optimizer, designing an internal process to assign their characteristics to the related fault tree basic events.
- Single and Double Loop Evolutionary Multiobjective algorithms have been developed. The coding alternatives for computer-based ISSDMO as well as the desirable evolutionary search characteristics were considered.

From an evolutionary point of view, a methodology for the simultaneous minimization of cost and unavailability for ISSDMO is exposed by using evolutionary multiple objective algorithms, taking into account the maintenance strategy of each design component (Test Intervals, TI). The best non-dominated solutions obtained represent a complete set of optimum candidate designs where not only the TI, and number of parallel components are defined, but also the existence or inexistence of some components. So, each solution constitutes an optimum different design as can be seen in Figs. 18 to 27 for the test case used in the chapter. In this case, solutions vary from a single branch design (one pump) with some parallel valves (Fig. 18), corresponding to the minimum cost, to an interconnected double branch design (six pumps in Fig. 27) corresponding to the minimum unavailability.

The NSGA-II has been compared with binary and integer encodings versus a flexible evolution approach FEA-NSGAII with double loop strategy, showing the capability of each algorithm to succeed in the resolution of the optimum design problem. Given the complexity of this kind of task in real life problem scale, we have tested a well known multiple-objective evolutionary algorithm with two different encodings and we have introduced a new heuristic algorithm with a double-loop structure in order to investigate a diversity of optimization alternatives and get a feedback of the performance of those algorithms over real design problems. The ability of Double Loop Evolutionary Multiple objective Algorithms to find Non Dominated solutions not found by Single Loop ones has been demonstrated in practice. The multiple objective algorithmic features of such methods have been introduced and detailed too.

In general, the three heuristics show the ability of solving the problem, but none was able to overcome the others; moreover the results seem to indicate that the approaches are complementary, which encourages us to test new ways to profit the good features of each one.

References

1. Andrews JD (1994) Optimal safety system design using fault tree analysis. In Proc. Instn. Mech. Engrs. Vol. 208, pp. 123-131.
2. Andrews JD, Bartlett LM (1998) Efficient Basic Event Orderings for Binary Decision Diagrams, IEEE Proceedings Annual Reliability and Maintainability Symposium, pp. 61-68.
3. Andrews JD, Pattison RL (1997) Optimal Safety-system performance. In Proceedings of the 1997 Reliability and Maintainability Symposium, Philadelphia, Pennsylvania, January, pp. 76-83.
4. Andrews JD, Pattison RL (1999) Genetic Algorithms in optimal safety system design, Proceedings Instn. Mech. Engrs., vol. 213, Part E, pp. 187-197.
5. Back T (1996) Evolutionary Algorithms in Theory and Practice. Oxford university Press, New York.
6. Back T, Naujoks B (1998) Innovative methodologies in evolution strategies. INGENET Project Report D 2.2, Center for Applied Systems Analysis, Informatik Centrum Dortmund, June.
7. Barros A, Berenguer C, Grall A (2003) Optimization of Replacement Times Using Imperfect Monitoring Information. IEEE Trans on Reliab, Vol. 52-4.
8. Bartlett LM, Andrews JD (2001) An ordering heuristic to develop the binary decision diagram based on structural importance. Reliab Engng & Syst Safety, Vol. 72, issue 1, 2001, pp. 31-38.
9. Bouissou M (1996) An ordering Heuristic for building Binary Decisión Diagrams from Fault Trees. Proceedings RAMS 96, Las Vegas, January.
10. Brezavscek A, Hudoklin A (2003) Joint Optimization of Block-Replacement and Periodic-Review Spare-Provisioning Policy. IEEE Trans on Reliab, Vol. 52, No. 1.

11. Bris R, Chatelet E, Yalaoui F (2003). New method to minimize the preventive maintenance cost of series-parallel systems. Reliab Engng and Syst Safety 82:247-255.
12. Bryant RE (1992) Symbolic boolean manipulation with ordered binary-decision diagrams, *ACM Computing Surveys* 24(3):293-318
13. Busacca PG, Marseguerra M, Zio E (2001) Multiobjective optimization by genetic algorithms: application to safety systems, Reliab Engng & Syst Safety, Vol. 72, issue 1, pp. 59-74
14. Chen D, Trivedi KS (2005) Optimization for condition-base maintenance with semi-Markov decision process. Reliab Engng & Syst Safety Vol. 90:25-29.
15. Chen H, Zhou J (1991) Reliability Optimization in Generalized Stochastic-Flow Networks. IEEE Trans on Reliab, 40-1.
16. Coello Coello CA, Van Veldhuizen D, Lamont G (2002) Evolutionary Algorithms for solving multi-objective problems. Kluwer Academic Publishers - GENA Series
17. Coit DW, Jin T, Wattanapongsakorn N (2004) System Optimization With Component Reliability Estimation Uncertainty: A Multi-Criteria Approach. IEEE Trans on Reliab, Vol. 53-3, September.
18. Coolen FPA, Dekker R (1995) Analysis of a 2-phase Model for Optimization of Condition-Monitoring Intervals. IEEE Trans on Reliab, 44-3.
19. Coudert O, Madre JC (1993) Fault Tree analysis: 10^{20} Prime Implicants and beyond, *IEEE Proceedings Annual Reliability and Maintainability Symposium*, pp. 240-245.
20. Cui L, Kuo W, Loh HT and Xie M (2004) Optimal Allocation of Minimal & Perfect Repairs Under Resource Constraints. IEEE Trans on Reliab, Vol. 53-2, June.
21. Deb K, Pratap A, Agrawal S, Meyarivan T (2002) A fast and elitist multiobjective genetic algorithm: NSGA-II", IEEE Trans on Evol Comput 6 (2), pp. 182-197.
22. Dhingra AK (1992) Optimal Apportionment of Reliability & Redundancy in Series Systems Under multiple Objectives. IEEE Trans on Reliab, Vol. 41, no. 4, December.
23. Dugan JB, Bavuso SJ, Boyd MA (1992). Dynamic Fault-Tree models for fault-tolerant computer systems, IEEE Trans on Reliab 41(3):363-377.
24. Dugan JB, Lyu MR (1994) Reliability and Sensitivity analysis of hardware and software fault tolerant systems. In Proceedings of the Fourth IFIP Working Conference on Dependable Computing for Critical Applications, January.
25. Dugan JB, Venkataraman B, Gulati R (1997). A software package for the analysis of dynamics fault tree models, IEEE Proceedings Annual Reliability and Maintainability Symposium, pp. 64-70.
26. Elegbede AOC, Chu C, Adjallah KH, Yalaoui F (2003) Reliability Allocation Through Cost Minimization. IEEE Trans on Reliab, Vol. 52-4.
27. Fiori de Castro H, Cavalca KL (2005) Maintenance resources optimization applied to a manufacturing system. Reliab Engng & Syst Safety. In press.
28. Fry AJ (2003) Integrity-Based Self-Validation Test Scheduling". IEEE Trans on Reliab, Vol. 52-2.

29. Galván BJ (2000) Evolución Flexible. Informe CEA-04-001. Centro de investigación de Aplicaciones Numéricas en Ingeniería. CEANI. Universidad de Las Palmas de Gran Canaria. Islas Canarias, España.
30. Galván BJ (1999) Contributions to Fault Tree Quantitative Evaluation., *PhD Thesis*, Physics Dep., Las Palmas de Gran Canaria University (Canary Islands-Spain)(In Spanish).
31. Gen M, Cheng R (1997) Genetic Algorithms & Engineering Design. Wiley Interscience, John Wiley & Sons, USA.
32. Gersht A, Kheradpir S, Shulman A (1996). "Dynamic Bandwidth-Allocation and Path-Restoration in SONET Self-Healing Networks". IEEE Trans on Reliab, 45-2.
33. Goel HD, Grievink J, Herder PM, Weijnen MPC (2002) Integrating reliability optimization into chemical process synthesis. Reliab Engng & Syst Safety, 78:247-258.
34. Goldberg DE (1989) Genetic Algorithms in Search, Optimization, and Machine Learning. Reading, MA: Addison-Wesley.
35. González L, García D, Galván BJ (1995) Sobre el análisis computacional de funciones Booleanas estocásticas de muchas variables. EACA95-Actas del primer encuentro de Álgebra computacional y Aplicaciones - Santander (Spain), Sept, pp. 45-55.
36. González L, García D, Galván BJ (2004) An Intrinsic Order Criterion to Evaluate Large, Complex Fault Trees, IEEE Trans on Reliab, Vol 53 (3) September, pp. 297-305
37. Greiner D, Galván B, Winter G (2003) Safety Systems Optimum Design using Multicriteria Evolutionary Algorithms. In: Evolutionary Multicriterion Optimization, Lecture Notes in Computer Science 2632 pp. 722-736. Springer Verlag.
38. Greiner D, Winter G, Galván B (2002) Multiobjective Optimization in Safety Systems: A Comparative between NSGA-II and SPEA2. In: Proceedings of the IV Congress on Reliability: Dependability. Spain.
39. Haasl DF (1965) Advanced concepts on Fault Tree Analysis. The Boeing company System Safety Symposium, USA.
40. Holland JH (1975) Adaptation in natural and artificial systems. Ann Arbor: University of Michigan Press.
41. Kuo W, Prasad VR (2000) An Annotated of System-Reliability Optimization. IEEE Trans on Reliab, Vol. 49, No. 2, pp. 176-187.
42. Lapa CMF, Pereira CMNA, Paes de Barros M (2005) A model for preventive maintenance planning by genetic algorithms based in cost and reliability. Reliab Engng & Syst Safety. In press.
43. Lapa CMF, Pereira CMNA, Frutuoso e Melo PF (2003) Surveillance test policy optimization through genetic algorithms using non-periodic intervention frequencies and considering seasonal constraints". Reliab Engng & Syst Safety 81:103-109.
44. Lee WS, Grosh DL, Tillman FA, Lie CH (1985) Fault Tree Analysis, Methods and Applications – A review. IEEE Trans on Reliabi, R-34(3), pp. 194-203.

45. Lhorente B, Lugtigheid D, Knights PF, Santana A (2004) A model for optimal armature maintenance in electric haul truck wheel motors: a case study. Reliab Engng & Syst Safety 84:209-218.
46. Liang YC, Smith AE (2004) An Ant Colony Optimization Algorithm for the redundancy Allocation Problem (RAP) IEEE Trans on Reliab, Vol. 53-3 September.
47. Martorell S, Carlos S, Villanueva JF, Sánchez AI, Galván B, Salazar D, Čepin M (In Press) Special Issue: "Use of Multiple Objective Evolutionary Algorithms in Optimizing Surveillance Requirements". Reliab Engng & Syst Safety.
48. Martorell S, Muñoz A, Serradell V (1996) Age-dependent models for evaluating risks and costs of surveillance and maintenance of components. IEEE Trans on Reliab, Vol. 45, No. 3, pp. 433-442.
49. Martorell S, Sánchez A, Carlos S, Serradell V (2002) Comparing effectiveness and efficiency in technical specifications and maintenance optimization. Reliab Engng Syst Safety 77:281-289.
50. Martorell S, Sánchez A, Carlos S, Serradell V (2000) Constrained optimization of test intervals using steady-state genetic algorithm. Reliab Engng & Syst Safety, Vol. 67, 2000, pp. 215-232.
51. Myerson RB (1999) "Nash Equilibrium and the History of Economic theory". Journal of Economic Literature, 36:1067-1082.
52. Nourelfath M, Dutuit Y (2004) A combined approach to solve the redundancy optimization problem for multi-state systems under repair policies. Reliab Engng & Syst Safety 86:205-213.
53. NUREG (1975), Reactor Safety Study-An Assessment of Accident Risks in U.S.Commercial Nuclear Power Plants, WASH-1400, U.S. Nuclear Regulatory Commission NUREG-75/014, Washington, D.C.
54. Pattisson RL, Andrews JD (1999) Genetic algorithms in optimal safety system design. In Proc. Instn. Mech. Engrs. Vol. 213, Part E, pp. 187-197.
55. Prasad VR, Kuo W (2000) Reliability Optimization of Coherent Systems. IEEE Trans on Reliab, 49-3.
56. Quagliarella D, Vicini A (1999) A genetic algorithm with adaptable parameters. IEEE Systems, Man and Cibernetics, pp. III-598-III-602.
57. Ravi V, Murty BSN, Reddy PJ (1997) Nonequilibrium Simulated Annealing-Algorithm Applied to reliability Optimization of Complex systems. IEEE Trans on Reliab, Vol. 46-2.
58. Rubinstein RY, Levitin G, Lisnianski A, Ben-Haim H (1997) Redundancy Optimization of static Series-Parallel Reliability Models Under Uncertainty. IEEE Trans on Reliab, Vol. 46, No. 4, December.
59. Samrout M, Yalaoui F, Chatelet E, Chebbo N (2005) New methods to minimize the preventive maintenance cost of series-parallel systems using ant colony optimization. Reliab Engng & Syst Safety 89:346-354.
60. Schneeweiss WG (1989) Boolean Functions with Engineering Applications and Computer Programs. Springer-Verlag.

61. Sefrioui M, Periaux J, "Nash Genetic Algorithms: examples and applications", Proceedings of the 2000 Congress on Evolutionary Computation, CEC-2000, Vol. 1, pp. 509-516.
62. Soni S, Narasimhan S, LeBlanc LJ (2004) Telecommunication Access Network Design UIT Reliability Constraints. IEEE Trans on Reliab, Vol. 53-4.
63. Vesely WE, Goldberg FF, Roberts NH, Haals DF (1981) Fault Tree Handbook. Systems and Reliability research office of Nuclear Regulatory Research, U.S. Nuclear Regulatory Commission, Washington D.C., January.
64. Winter G, Gonzalez B, Galvan B, Benitez E, "Numerical Simulation of Transonic Flows by a Double Loop Flexible Evolution", Parallel Computacional Fluid Dynamics 2005, Washington, U.S.A.
65. Wattanapongsakorn N, Levitan S (2004) Reliability Optimization Models for Embedded Systems With Multiple Applications. IEEE Trans on Reliab, Vol. 53-3, September.
66. Winter G, Galván B, Alonso S, González B, Jiménez JI, Greiner D (2005) A Flexible Evolutionary Agent: cooperation and Competition among real-coded evolutionary operators, Soft Computing Journal, Vol. 9 No. 4:299-323
67. Winter G, Galvan B, Alonso S and González B (2002) Evolving From Genetic Algorithms to Flexible Evolution Agents. In the *Late-Breaking Papers Book* of the Genetic and Evolutionary Computation Conference, Erik Cantú-Paz (Lawrence Livermore National Laboratory) Editor, New York. GECCO 2002, pp. 466-473.
68. Winter G, Galván B, Alonso S, Mendez M (2005) New Trends In Evolutionary Optimization And Its Impact On Dependability Problems. Evolutionary and Deterministic Methods for Design, Optimization and Control with Applications to Industrial and Societal Problems, EUROGEN 2005, R. Schilling, W.Haase, J. Periaux, H. Baier, G. Bugeda (Eds).
69. Yalaoui A, Chu C, Chatelet E, (2005) Reliability allocation problem in a series-parallel system. Reliab Engng & Syst Safety 90:55-61.
70. Zitzler E, Laumanns M, Thiele L "SPEA2: Improving the Strength Pareto Evolutionary Algorithm for Multiobjective Optimization", Evolutionary Methods for Design, Optimization and Control with Applications to Industrial Problems, (EUROGEN 2001), September 2001, Athens, Greece.
71. Ziztler E (1999) Evolutionary algorithms for multiobjective optimization: methods and applications. Swiss Federal Institute of Technology (ETH) Zurich, TIK-Schriftenreihe Nr. 30, Diss ETH No. 13398. Germany: Shaker Verlag, ISBN 3-8265-6831-1, December.

Optimal Redundancy Allocation of Multi-State Systems with Genetic Algorithms

Zhigang Tian, Ming J Zuo

Department of Mechanical Engineering, University of Alberta, Canada

Hong-Zhong Huang

School of Mechanical and Electronic Engineering, University of Electronic Science and Technology of China, P.R. China

6.1 Introduction

6.1.1 Optimal Redundancy Allocation of Multi-state Systems

Traditional reliability theory assumes that a system and its components may take only two possible states, working or failed. However, many practical systems can actually perform their intended functions at more than two different levels, from perfectly working to completely failed. These kinds of systems are known as multi-state systems (Kuo and Zuo, 2003). A multi-state system (MSS) reliability model provides more flexibility for modeling of system conditions than a binary reliability model. A typical example of a multi-state component is the generator in a power station (Lisnianski and Levitin, 2003). Such a generator can work at full capacity, which is its nominal capacity, when there are no failures at all. Certain types of failures can cause the generator to be completely failed, while other failures will lead to the generator working at some reduced capacities. Another typical example of multi-state systems is wireless communication systems. Suppose that a wireless communication system consists of several transmission stations arranged in a sequence. Each station can work at different levels of capacity. A station is in state 0 if it can not transmit any signals, it is in state 1 if it can transmit signals to the next station only, and it is in state 2 if it can transmit signals to the next two stations.

Z. Tian et al.: *Optimal Redundancy Allocation of Multi-State Systems with Genetic Algorithms*, Computational Intelligence in Reliability Engineering (SCI) **39**, 191–214 (2007)
www.springerlink.com

Thus, such a wireless system is actually a multi-state system with components that can be in three possible states.

For the purpose of maximizing system performance and cost effectiveness, we need to consider the issue of optimal design of multi-state systems. Some complex engineering systems, e.g. a space shuttle, consist of hundreds of thousands of components. These components functioning together form a system. The reliable performance of the system depends on the reliable performance of its components and the system configuration. Basically, there are three ways to enhance reliability of multi-state systems (Lisnianski and Levitin, 2003): (1) to provide redundancy, such as adding redundant components in parallel and using redundancy in the form of *k*-out-*n* systems; (2) to adjust the system configuration while keeping the constituent components the same, such as optimal arrangement of the existing components; (3) to enhance the reliability or performance of the components, either by selecting components with better performance, or by building components with high performance in the design stage which will be built into the manufacturing process. Reliability has become a more and more important index for complex engineering systems so as to ensure these systems to perform their targeted functions reliably. It is almost impossible to enhance the reliability of such a complex system only based on experience. Sophisticated and systematic approaches for component reliability measuring, system reliability analysis and optimization are necessary. In reliability optimization of multi-state systems, we aim at maximizing system reliability while satisfying physical and economical constraints, or minimizing the cost in development, manufacturing and operation while satisfying other constraints.

El-Neweihi et al (1988) first dealt with the problem of optimal allocation of multi-state components. Meng (1996) analyzed the optimal allocation of components in multi-state coherent systems, and gave the principle for interchanging components. Chen et al (1999) did the selective maintenance optimization for multi-state systems, and used the shortest path method to solve the optimization model. Zuo et al (2000) investigated the replacement-repair policy for multi-state deteriorating products under warranty. They examined the optimal value of the deterioration degree of the item and the length of the residual warranty period, so as to minimize the manufacturer's expected warranty servicing cost per item sold. A simple heuristic algorithm was used to implement the optimization. A heuristic method was also used to deal with the design optimization of multi-state series-parallel systems (Zuo et al. 1999). Gurler and Kaya (2002) proposed a maintenance policy for a system with multi-state components, using an approximation of the average cost function to reduce the problem complexity and using a numerical method to solve the formulated optimization problem. Hsieh and Chiu (2002) investigated the optimal maintenance policy in a multi-state deteriorating standby system by determining the optimal number of standby components required in the system and the optimal state in which the replacement of deteriorating components shall be made. A heuristic algorithm was developed to solve the problem. Ramirez-Marquez and Coit (2004) proposed a heuristic approach for solving the redundancy allocation problem where the system utility was evaluated using the universal generating function approach (Levitin et al. 1998). Generally speaking, the optimization techniques used above are basically heuristic approaches. As is known, the key characteristic of a heuristic approach is that its specific procedure is different with respect to different problems. Thus, it is not a universal approach, and sometimes approximation has to be made to get the final solution.

6.1.2 Optimal Redundancy Allocation of Multi-state Systems with Genetic Algorithms

Advancement in the optimization area provides us new tools for optimal design of multi-state systems. Genetic Algorithm (GA) is a very powerful optimization approach with two key advantages (Tian and Zuo, 2005, Davis, 1991). The first advantage is that it has flexibility in modeling the problem. GA has no strict mathematical requirements, such as derivative requirement, on the objective functions and constraints. The only requirement is that the objective functions and constraints can be evaluated in some way. GA is also suitable for dealing with those problems including discrete design variables. The second advantage of GA is its global optimization ability. GA has been recognized as one of the most effective approaches in searching for the global optimal solution. Of course GA does not guarantee to find the optimal optimum. But compared with other optimization techniques, GA has more chance to find solutions that are closer to global optimum. There are important discrete decision variables, such as redundancies in redundancy allocation problems, in the optimization of multi-state systems. Thus, GA is very suitable to deal with the optimal design issue of multi-state systems due to its flexibilities in building the optimization model. GA is also a universal tool that can deal with any type of optimization problems regarding multi-state systems.

Levitin and Lisnianski have produced a series of reports on using Universal Generating Function (UGF) approach and GA for the optimization of multi-state systems (Lisnianski and Levitin, 2003). They pointed out that UGF has two advantages in evaluating MSS reliability: (1) it is fast enough to be used in these problems due to its effectiveness in dealing with high-dimension combinatorial problems; (2) it allows us to calculate the entire MSS performance distribution based on the performance distributions of its components. Levitin et al (1998) systematically presented the UGF approach for redundancy optimization for series-parallel multi-state systems, and described in details how GA was used as an optimization tool. Later, Levitin and his colleagues applied the UGF and GA approach to the optimization of multi-state systems with other types of structures, such as structure optimization of a power system with bridge topology (Levitin and Lisnianski, 1998a). They dealt with system optimization considering maintenance, such as joint redundancy and maintenance optimization (Levitin and Lisnianski, 1998b), imperfect preventive maintenance (Levitin and Lisnianski, 2000a), and optimal replacement scheduling (Levitin and Lisnianski, 2000b). They also investigated such optimization problems by taking into account more practical constraints, such as time redundancy (Lisnianski et al. 2000), and fixed resource requirements and unreliable sources (Levitin, 2001b). They also studied the dynamic optimization of multi-state systems such as optimal multistage modernization (Levitin and Lisnianski, 1999) and system expansion scheduling (Levitin, 2000).

Tian and Zuo (2005a) studied the joint reliability-redundancy allocation for multi-state series-parallel systems, where both redundancy and component reliability are treated as decision variables for system utility enhancement. GA was used to solve the formulated optimization problem including both discrete and continuous design variables. In practical situations involving reliability optimization, there may exist mutually conflicting goals such as maximizing system utility and minimizing system cost and weight (Kuo et al. 2001). Tian and Zuo (2005b) proposed to use the physical programming approach (Messac, 1996) and GA to deal with redundancy allocation of multi-state systems while considering multiple conflicting objectives.

Besides GA, other computational intelligence techniques have been investigated for the optimal design of multi-state systems. Nourelfath and Nahas (2004a) applied the ant colony method, another effective optimization method of computational intelligence, to the redundancy allocation of multi-state systems. They also developed a heuristic approach combining the universal generating function approach and the stochastic Petri nets to solve the redundancy optimization problem for multi-state systems under certain repair policies (Nourelfath and Nahas, 2004b), and mentioned that either GA or the ant algorithm can be used to solve the integrated optimization problems.

A multi-state consecutively-connected system is an important special type of multi-state system. Most reported studies on this system structure focus on its reliability or performance evaluation (Hwang and Yao, 1989, Zuo and Liang, 1994, Kuo and Zuo, 2003). Malinowski and Preuss (1996) studied how to increase system reliability through optimal ordering of components using a heuristic method. Levitin (2002a, 2003b) investigated the optimal allocation of components in a linear multi-state sliding window system, which was generalized from consecutive *k*-out-of-*r*-from-*n*:F system, using the UGF and GA approaches. He also used such approaches for the optimal allocation of multistate components in a linear consecutively-connected system (Levitin, 2003a). A multi-state network is another important special type of multi-state system with practical applications. Levitin (2002b, 2002c) has developed approaches based on UGF and GA for the optimal design of multi-state transmission networks.

6.1.3 Content of this Chapter

In this chapter, we will limit our discussions to redundancy allocation of multi-state systems. In Section 2, we will present several multi-state system structures involved in redundancy allocations, and how to evaluate their system utilities when implementing the optimization. The presented multi-state system structures include the multi-state series-parallel system by Barlow and Wu (1978), multi-state load sharing series-parallel system by Levitin et al (1998), and the multi-state *k*-out-of-*n* systems (Huang et al. 2000). In Section 3, we will describe the optimization models for redundancy allocation of multi-state systems, both the single-objective optimization model and the optimization model considering multiple objectives. In Section 4, how to implement GA in optimization of multi-state series-parallel systems is presented in details, including encoding of design variables and how to implement GA operators. In Section 5, we will use a specific example to illustrate how GA is applied to the redundancy allocation of multi-state systems. Finally, conclusions and future research are presented in Section 6.

6.2 System Utility Evaluation

Notation

$\boldsymbol{x}$:	A vector representing the states of all components in the multi-state system
$\phi(\boldsymbol{x})$:	State of the system. $\phi(\boldsymbol{x}) = 0, 1, ..., M$
U:	Expected system utility

u_s: The utility when the system is in state s

In optimal design of binary-state systems, system reliability is a very important performance measure. It should be dealt with as an optimization objective or a constraint so that we can meet the requirements on system reliability. In multi-state systems, the concept equivalent to the concept "reliability" in binary-state systems is "utility". Utility is the most widely used performance measure for multi-state systems (Aven, 1993). A multi-state system can be in different discrete states, in contrast to that a binary system can only be in two possible states. When a multi-state system is in a certain state s, it can produce a certain amount of benefit. Such benefit is called the system utility when it is in state s, and is denoted by u_s. A multi-state system can be in different states with different probabilities. Thus, the measure we are interested in is the expected system utility. In optimization of multi-state systems, we usually use the term "system utility" to indicate "expected system utility".

System utility can be calculated as follows (Aven, 1993)

$$U = \sum_{s=0}^{M} u_s \cdot \Pr(\phi(x) = s) \tag{1}$$

where U is the expected system utility, and u_s is the utility when the system is in state s. In practical applications, u_s is usually known. Thus, the key issue in multi-state system utility evaluation is to evaluate the probabilities of the system in different possible states.

In this section, we will show how to calculate the state distributions of several typical multi-state system structures, including the multi-state series-parallel system by Barlow and Wu (1978), multi-state load sharing series-parallel system by Levitin et al (1998), and the multi-state *k*-out-of-*n* systems. These system structures are the most commonly used ones in the study of redundancy allocation of multi-state systems.

6.2.1 Multi-state Series-parallel Systems by Barlow and Wu (1978)

Notation

N Number of subsystems (stages)

S_i Subsystem (stage) i, $1 \le i \le N$

n_i Redundancy in S_i

p_{ij} Probability of an i.i.d. component of S_i in state j

The multi-state series-parallel system defined by Barlow and Wu (1978) has been widely studied. Its structure is shown in Fig. 1. A multi-state series-parallel system consists of N subsystems, S_1 to S_N, connected in series. Each subsystem, say S_i, has n_i identical components connected in parallel. The following assumptions are used: (1) The components in a subsystem are identical and independently distributed (i.i.d.). (2) The components and the system may be in M+1 possible states, namely, 0, 1,..., M. (3) The multi-state series-parallel systems under consideration are coherent systems.

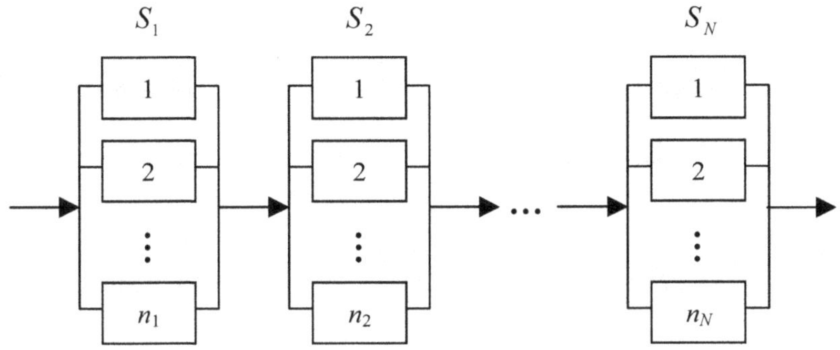

Fig. 1. Structure of a multi-state series-parallel system

According to the multi-state system definition of Barlow and Wu (Barlow and Wu, 1978), the state of a parallel system is defined to be the state of the best component in the system, and the state of a series system is defined to be the state of the worst component in the system. Hence, the system tem state of the series-parallel system shown in Fig. 1 is

$$\phi(\boldsymbol{x}) = \min_{1\le i\le N} \max_{1\le j\le n_i} x_{ij} \,. \tag{2}$$

Consider an example of multi-state system definition of Barlow and Wu, as shown in Fig. 2. There are two subsystems, with two components in each subsystem. The numbers in the box indicate the states of the corresponding components. Thus, according to Barlow and Wu's definition, the state of subsystem S_1 is 2, the state of subsystem S_2 is 4, and the state of this system is 2.

The probability that the system is in state s or above ($s = 0, 1, ..., M$) is

$$\Pr\left(\phi(x) \geq s\right) = \prod_{i=1}^{N} \left[1 - \left(1 - \sum_{k=s}^{M} p_{ik} \right)^{n_i} \right]. \tag{3}$$

Using this formula, we can get probabilities of the system in different states, and thus get the expected system utility.

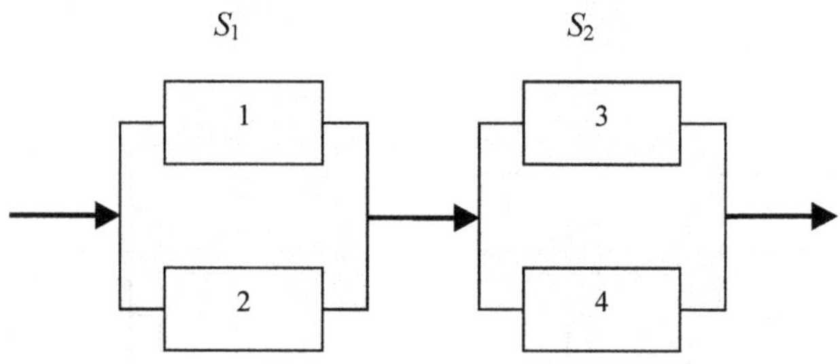

Fig. 2. An example of multi-state series-parallel system

6.2.2 The Multi-state Load Sharing Series-parallel Systems by Levitin et al (1998)

The difference between this system structure and the one discussed in Section 2.1 is that the state of a parallel system in this structure is the sum of the states of its components. The state of a series system is still the state of the worst component. Thus the state of the multi-state load sharing series-parallel system defined by Levitin et al (1998) is:

$$\phi(x) = \min_{1 \leq i \leq N} \left\{ \sum_{j=1}^{n_i} x_{ij} \right\}. \tag{4}$$

If we use the example in Fig. 2 to illustrate this type of systems, the state of subsystem S_1 is 3, the state of subsystem S_2 is 7, and thus the state of the system is 3.

A practical example of a multi-state load sharing series-parallel system is a power station coal transportation system (Lisnianski and Levitin, 2003). Such a system consists of five subsystems connected in series: primary feeder, primary conveyor set, stacker-reclaimer, secondary feeder, and secondary conveyor set. The capacity of a subsystem, say the primary conveyor set, is equal to the sum of the capacities of the conveyers in it which are connected in parallel. The capacity of the whole system is determined by the bottle neck, which is the subsystem with the lowest capacity.

A general method for the state distribution evaluation of multi-state load sharing series-parallel systems is the UGF approach. The UGF approach is referred to as a generalized sequence generating approach from the mathematical perspective. The approach is based on the definition of a UGF of discrete random variables and composition operators over UGFs (Levitin et al. 1998). Some additional notation is listed as follows.

Notation

$U(z)$: UGF representation.

p_k: Probability that a component is in state k when they are iid.

G_k: Output performance of a component (or system) in state k.

M_i: The number of possible states of component i minus 1.

Ω: The UGF operator.

The following assumptions are used: (1) The components in a subsystem are identical and independently distributed. (2) The components and the system may have discrete state set. (3) The multi-state load sharing series parallel systems under consideration are coherent systems.

The UGF of component i in MSS is defined as

$$U(z)=\sum_{k=0}^{M_i} p_k z^{G_k}, \tag{5}$$

Suppose that we have two components i and j with the following UGF functions respectively:

$$U_i(z)=\sum_{k=0}^{M_i} p_k z^{G_k},\ U_j(z)=\sum_{k=0}^{M_j} p_k z^{G_k}, \tag{6}$$

where M_i+1 and M_j+1 denote the number of possible states for component i and j respectively. If these two components are connected in parallel and are sharing load to form a system, the UGF of the system will be

$$U_S(z)=\Omega[U_i(z),U_j(z)]=\sum_{k=0}^{M_i}\sum_{l=0}^{M_j} p_k\cdot p_l\cdot z^{G_k+G_l}. \tag{7}$$

The common terms of the resulting polynomial should be combined if there are any. If the two components are connected in series to form a system, the UGF of the system will be

$$U_S(z) = \Omega\left[U_i(z), U_j(z)\right] = \sum_{k=0}^{M_i}\sum_{l=0}^{M_j} p_k \cdot p_l \cdot z^{\min\{G_k+G_l\}}. \tag{8}$$

Actually, a subsystem can be treated as a component in the UGF approach. From the UGF of the system, we can know the probability of the system in different possible states, i.e., the system state distribution. When there are more than two components connected in parallel sharing load and more subsystems connected in series, the same approach is applicable. Refer to Levitin et al (1998) for details.

We need to note that the UGF approach is a general approach for state distribution evaluation of multi-state systems. It is applicable not only to multi-state load sharing series-parallel systems, but also to other structures such as bridge systems, consecutively connected systems, etc.

6.2.3 Multi-state *k*-out-of-*n* Systems

The *k*-out-of-*n* redundancy is a very important and widely used form of redundancy. In binary reliability theory, both series systems and parallel systems are special cases of the *k*-out-of-*n* system structure. The issues of *k*-out-of-*n* systems in the multi-state context are much more complicated. The structure of a multi-state *k*-out-of-*n* system is shown in Fig. 3.

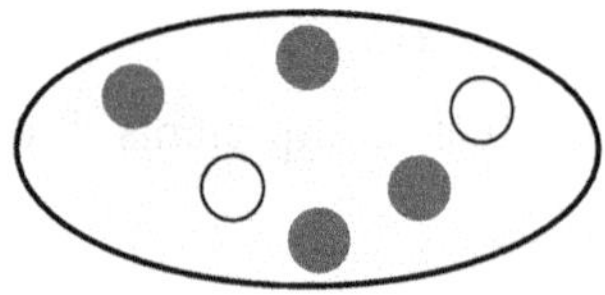

Fig. 3. A multi-state *k*-out-of-*n* system

Assume that both the components and the system may be in M+1 possible states, namely, 0, 1, 2, …, M. A system is called a constant multi-state *k*-out-of-*n*:G if the system is in state j or above when at least k components are in state j or above. Constant multi-state *k*-out-of-*n*:G systems can be evaluated efficiently using recursive algorithms (Huang et al. 2000). If the k values are allowed to be different for different states, we have the generalized multi-state *k*-out-of-*n* systems. Efficient algorithms are available for the evaluation of such generalized systems as well (Tian et al. 2005b).

The redundancy optimization issue in such systems involves the determination of the optimal number of components in the system. Several such multi-state k-out-of-n systems can also be connected in series to make a mixed multi-state system. Then the redundancy optimization issue in such a mixed system is the determination of the optimal number of components for each subsystem.

6.2.4 General Comments

System utility evaluation is an integral part of optimal redundancy allocation. The system utility will be evaluated repetitively during the optimization process. In each iteration of GA, the utility evaluation is required for each individual solution of the solution population. Thus, a suitable system utility evaluation approach has to be efficient. Specifically, the state distribution evaluation method for the considered multi-state system should be efficient. Currently, intensive researches are ongoing to find more efficient methods for the performance evaluation of multi-state load-sharing systems and multi-state k-out-of-n systems. Progresses in these areas will be very helpful to the efficient application of GA to multi-state system optimizations.

6.3 Optimization Models for Multi-state Systems

In a multi-state system, if we add more redundant components, the system utility will be improved. However, system cost and system weight will increase as well. Therefore, there is the issue of optimal redundancy allocation. In this section, we will discuss the general optimization framework of redundancy allocation of multi-state systems. There are many design objectives when designing a multi-state system, such as the system utility, cost, weight, etc. If we pick one of these objectives as the objective function while using others as constraints, we will have the single-objective optimization model. If we want to consider all of these objectives simultaneously, we will be dealing with a multi-objective optimization model. The elements of an optimization model, i.e. the objective functions, the constraints and the design variables are discussed in this section. Physical programming (Messac, 1996) is presented as an effective multi-objective optimization approach in this section as well.

When presenting the optimization models, the multi-state systems being considered are multi-state series-parallel systems, multi-state load sharing series-parallel systems, and the mixed multi-state *k*-out-of-*n* systems where several multi-state *k*-out-of-*n* systems are connected in series.

6.3.1 Single-objective Optimization Models

Notation

N : Number of subsystems in the considered multi-state system.

$\boldsymbol{h}$: Component versions of the subsystems. $\boldsymbol{h}=\left(h_1,h_2,\ldots h_N\right)$.

$\boldsymbol{n}$: Number of components in the subsystems. $\boldsymbol{n}=\left(n_1,n_2,\ldots n_N\right)$.

H_i : Number of available versions of components for subsystem S_i

C_S : System cost.

W : System weight.

U_0 : Minimum system utility required

C_0 : Maximum system cost allowed

W_0 : Maximum system weight allowed

The important elements in an optimization model are the objectives, the constraints and the design variables. As mentioned in Section 2, the concept "utility" in multi-state systems is like the concept "reliability" in binary-state systems. In this section, we should note that system utility U, is an important performance measure in the optimization of multi-state systems. Other system performance measures include system cost, system weight, system volume, etc. When designing a multi-state system, the system utility may be maximized, system cost may be minimized, system weight and volume may be minimized, and so on.

The design variables are the parameters that we can manipulate so as to influence the performance measures of the considered system. In redundancy allocation problems, the most important type of design variables are the redundancies. That is, the number of components n_i for each subsystem i. Increasing the number of identical components for subsystem S_i will increase the utility of the subsystem, but will also increase the cost and weight simultaneously. Redundancies for all subsystems make up of a vector $\boldsymbol{n}=\left(n_1,n_2,\ldots n_N\right)$. Another type of design variables is the versions of components we choose for each subsystem (Levitin et al, 1998). Suppose that for each subsystem, there are different versions of components with different state distributions, different costs, different weights, etc. Choosing the optimal component version for each subsystem is also an optimization issue. We use h_i to represent the component version for subsystem i, and use $\boldsymbol{h}$ to represent the vector $\left(h_1, h_2, \ldots h_N\right)$.

As mentioned, there are different design objectives such as system utility, system cost, system weight, etc. A regular and widely used way to deal with these different and usually conflicting design objectives is using one design objective as the objective function while using others as constraints. If system cost is used as the objective function, the formulated single-objective optimization model is as follows:

$$\begin{aligned} & \min C_S \\ & \text{s.t.} \\ & \quad U \geq U_0 \\ & \quad W \leq W_0 \\ & \quad 0 < h_i \leq H_i, \; i = 1,2,\ldots,N \\ & \quad 0 < n_i, \qquad i = 1,2,\ldots,N \\ & \quad n_i, h_i \text{ are integers} \end{aligned} \tag{9}$$

If system utility is used as the objective function, the formulated single-objective optimization model is as follows:

$$\begin{aligned} & \max U \\ & \text{s.t.} \\ & \quad C_S \leq C_0 \\ & \quad W \leq W_0 \\ & \quad 0 < h_i \leq H_i, \; i = 1,2,\ldots,N \\ & \quad 0 < n_i, \qquad i = 1,2,\ldots,N \\ & \quad n_i, h_i \text{ are integers} \end{aligned} \tag{10}$$

6.3.2 The Multi-objective Optimization Model

The multi-objective optimization model seeks to maximize system utility and minimize system cost and system weight simultaneously. Physical programming (Messac, 1996) may be used to formulate the multi-objective optimization model. In addition to the notation defined in previous sections, we provide the following additional notion:

Notation

$\overline{g}_i$ Class function of objective i

g_{ij} Boundary value of preference ranges for objective i. j=1, 2, …,5

$\overline{g}_U$ Class function of system utility

$\overline{g}_C$ Class function of system cost

$\overline{g}_W$ Class function of system weight

Physical Programming is a multi-objective optimization tool that explicitly incorporates the Decision Maker's (DM's) preferences on each design goal into the optimization process (Messac, 1996). It eliminates the typical iterative process involving the adjustment of the physically meaningless weights, which is required by virtually all other multi-objective optimization methods, and thus substantially reduces the computational intensities. The DM's preferences are specified individually on each goal through physically meaningful values, which makes the physical programming method easy to use and advantageous in dealing with a large number of objectives.

Physical programming captures the designer's preferences using class functions. Class functions are functions of the corresponding objective values, and they may be classified into four classes: smaller is better (i.e. minimization), larger is better (i.e. maximization), value is better (i.e. close to a value), and range is better (as long as in a range). There are two so-called class functions, one soft and one hard, with respect to each class. Soft class functions will become constituent parts of a single objective function, the so-called aggregate objective function, which is to be minimized in the physical programming optimization model (Messac, 1996). There are four types of soft class functions, Class-1S to Class-4S, with respect to the four classes of class functions. Class-1S is a monotonously increasing function, and it is used to represent the objectives to be minimized. Class-2S is a monotonously decreasing function, and it is used to represent the objectives to be maximized. In the reliability-redundancy allocation problem, the utility objective has a Class-2S class function, and the cost objective has a Class-1S class function.

What a designer needs to do is just to specify ranges of different degrees of desirability (highly desirable, desirable, tolerable, undesirable, highly undesirable, and unacceptable) for the class function of each objective. For example, the DM can specify ranges of degrees of desirability for the cost of a system, which is to be minimized, as follows: (1) the cost is considered to be unacceptable if it is over \$200, (2) the cost is considered to be highly undesirable if it is between \$150 and \$200, (3) the cost is considered to be undesirable if it is between \$100 and \$150, (4) the cost is considered to be tolerable if it is between \$80 and \$100, (5) the cost is considered to be desirable if it is between \$50 and \$80, and (6) the cost is considered to be highly desirable if it is below \$50. Such ranges are specified by the DM based on experience and design purpose.

Here we assume the DM knows the rough ranges of variation of each objective a priori, which is usually true in practical engineering applications. If for an objective, the DM does not have prior knowledge or experiences, he/she can try several groups of design variable inputs to see the objective values, so as to have some ideas of the range of this objective. Another approach is first use a much wider variation range for the objective and run the optimization model, after which the DM will have a better idea of the objective and be able to define a more precise range of variation of the objective.

The physical programming approach solves a multi-objective optimization problem by transforming it into a single-objective optimization problem. The class functions of design objectives are combined into the aggregate objective function. The multi-objective optimization model of the multi-state system may be formulated as

$$
\begin{aligned}
&\min_{\boldsymbol{n},\boldsymbol{h}} g(\boldsymbol{n},\boldsymbol{h}) \\
&\quad = \log_{10}\left\{\frac{1}{3}[\bar{g}_U(U(\boldsymbol{n},\boldsymbol{h}))+\bar{g}_C(C(\boldsymbol{n},\boldsymbol{h}))+\bar{g}_W(W(\boldsymbol{n},\boldsymbol{h}))]\right\} \\
&\text{s.t.} \\
&\quad U(\boldsymbol{n},\boldsymbol{h}) \geq U_0 \\
&\quad C(\boldsymbol{n},\boldsymbol{h}) \leq C_0 \\
&\quad W(\boldsymbol{n},\boldsymbol{h}) \leq W_0 \\
&\quad 0 < h_i \leq H_i,\ i=1,2,\ldots,N \\
&\quad 0 < n_i, \qquad i=1,2,\ldots,N \\
&\quad n_i, h_i \text{ are integers}
\end{aligned}
\tag{11}
$$

where $g(\boldsymbol{n},\boldsymbol{h})$ is the aggregate objective function, $\bar{g}_U$, $\bar{g}_C$ and $\bar{g}_W$ are the class functions of system utility, system cost and system weight, respectively, H_i is the number of types available for component C_i, and U_0, C_0 and W_0 are the constraint values, which are equal to the boundaries of the unacceptable ranges of the corresponding design metrics.

6.4 Implementation of Genetic Algorithms

6.4.1 General Framework of GA

GA is a very powerful optimization approach with two key advantages. (1) Flexibility in modeling the problem. GA has no strict mathematical requirements, such as derivative requirement, on the objective functions or the constraints. The only requirement is that the objective functions and constraints can be evaluated in some way. GA is also suitable for dealing with those problems including discrete design variables. (2) Global optimization performance. GA has been recognized as one of the most effective approaches for searching for the global optimal solution.

In GA, an individual solution is called chromosome, which include all the design variables. First the design variables are encoded to form a chromosome. Encoding methods including binary encoding, decimal encoding and real encoding. For example, if binary encoding is used, a design variable is represented by several adjacent binary digits in a chromosome. In each iteration of GA, all the chromosomes form a population. Each chromosome is evaluated by calculating its corresponding objectives values to get a single fitness value. The current population is updated to generate the new population for next iteration using GA operators including selection, crossover, mutation and duplication.

The procedure of GA is as follows.

(1) Initialization. Set the size of population, i.e. how many chromosomes there are in the population. Set the length of the chromosome and how each design variable is represented in a chromosome. Set $k = 0$, and generate the initial population $P(0)$.

(2) Evaluation. Calculate the fitness value of each chromosome, based on its corresponding objective function values, of the current population $P(k)$. Save the chromosome $B(k)$ with the best fitness value.

(3) Selection. Select chromosomes from the current population based on their fitness values to form a new population $P(k+1)$.

(4) Crossover. One point crossover is used. In $P(k+1)$, choose several pairs of chromosomes. For each pair of chosen chromosomes, generate a random position in a chromosome, and exchange the parts on the left (or right) side of the position of the chromosomes to form a pair of new chromosome. For example, suppose we have a pair of decimal encoded chromosomes: $\boldsymbol{x}_1 = (1, 3, 5, 7)$ and $\boldsymbol{x}_2 = (2, 4, 6, 8)$. The crossover operation

point is randomly chosen as the point after the second digit. Thus the resulting new chromosomes are $\boldsymbol{x}_1' = (2, 4, 5, 7)$ and $\boldsymbol{x}_2' = (1, 3, 6, 8)$.

(5) Mutation. Implement even mutation on chromosomes in $P(k+1)$. That is, for any randomly selected digit in the population, use a randomly generated possible value to replace the old value.

(6) Duplication. Use $B(k)$ to replace the first chromosome in $P(k+1)$.

(7) If the maximal iteration is reached, terminate the procedure and output the result. Otherwise, set $k = k+1$, and go to step (2).

6.4.2 Encoding and Decoding

Encoding deals with how to represent the design variables in a chromosome. In a redundancy allocation problem, the numbers of components $\boldsymbol{n} = (n_1, n_2, \ldots n_N)$ and the component versions $\boldsymbol{h} = (h_1, h_2, \ldots h_N)$ are the commonly used design variables. The decimal encoding method is usually used. A simple example will be used to illustrate the decimal encoding and the corresponding decoding process.

Suppose that the considered system consists of two subsystems, S_1 and S_2. Each subsystem will not have more than 100 components. There are 10 versions of components available for subsystem S_1, and 4 versions of components available for subsystem S_2. Thus, the chromosome representing the 4 design variables, n_1, n_2, h_1 and h_2, will have a length of 6. Ch is used to represent the encoded chromosome string. The first two elements represent variable n_1, the third and fourth elements represent variable n_2, the fifth element represents variable h_1 and the sixth element represents variable h_2. In the decoding process, we will have:

$$\begin{aligned} n_1 &= 10 \cdot ch_1 + ch_2 + 1 \\ n_2 &= 10 \cdot ch_3 + ch_4 + 1 \\ h_1 &= ch_5 + 1 \\ h_2 &= Floor\left(\frac{4 \cdot ch_6}{10}\right) + 1 \end{aligned} \tag{12}$$

where function $Floor(x)$ rounds x down to the nearest integer.

6.4.3 Fitness Function Value

First the objective function and all the constraints are integrated into an aggregate function denoted by f, which needs to be minimized. Suppose $f^{\max}$ and $f^{\min}$ are the maximum and minimum values of the population in the current iteration. The fitness value for a candidate solution with aggregate objective function value f is

$$F = \frac{f^{\max} - f + 0.3\cdot\left(f^{\max} - f^{\min}\right)}{(1+0.3)\cdot\left(f^{\max} - f^{\min}\right)} \tag{13}$$

where the number 0.3 is the so-called "selection pressure". The selection pressure is used to make sure that the solution with the worst fitness value still has a chance to be selected to form the new population. It is usually selected from the range 0.1~0.5, and the exact value of selection pressure will not impact the performance of GA too much.

6.4.4 Fitness Function Value

GA operators include selection operator, crossover operator, and mutation operator. In each type of these operators, there are several options we can choose. In redundancy allocation problems of multi-state systems, we can simply choose the simple roulette-wheel selection scheme, one-point cross operator, and even mutation operator. For more details, see Davis (1991).

6.5 An Application Example

In this section, we will investigate an example of redundancy allocation of a multi-state series-parallel system as defined by Barlow and Wu (1978) while considering multiple objectives. This example is taken from Tian et al (2005a), and it will illustrate how GA is applied to the redundancy allocation of multi-state systems.

We consider a multi-state series-parallel system with four subsystems. The state space of the component and the system is $\{0, 1, 2, 3\}$. Suppose we have different versions of components for each subsystem, with their characteristics listed in Table 1. For a component with version h_i in subsystem S_i, the table gives component cost $c_i(h_i)$, component weight $w_i(h_i)$, and the state distribution for each component, p_{i0} to p_{i3}. The system utility u_s with respect to the corresponding system state s is given in Table 2.

We aim at maximizing the expected system utility while minimizing system cost and system weight simultaneously. The expected system utility can be calculated using equation (1) to (3). For given h_1, h_2,..., h_N and n_1, n_2,..., n_N values, the total cost of the system is expressed as (Dhingra, 1992).

$$C = \sum_{i=1}^{N} c_i(h_i)\left[n_i + \exp(n_i/4)\right]. \tag{14}$$

Table 1. Characteristics of available components

S_i	h_i	p_{i0}	p_{i1}	p_{i2}	p_{i3}	$c_i(h_i)$	$w_i(h_i)$
1	1	0.100	0.450	0.250	0.200	0.545	7
	2	0.060	0.450	0.200	0.290	0.826	3
	3	0.045	0.400	0.150	0.405	0.975	10
	4	0.038	0.350	0.160	0.452	1.080	8
2	1	0.050	0.450	0.300	0.200	0.550	12
	2	0.040	0.400	0.300	0.260	0.630	5
	3	0.040	0.300	0.320	0.340	0.740	8
	4	0.030	0.250	0.250	0.470	0.900	10
	5	0.025	0.215	0.180	0.580	1.150	12
3	1	0.145	0.625	0.130	0.100	0.250	10
	2	0.110	0.400	0.250	0.240	0.380	12
	3	0.065	0.450	0.230	0.255	0.494	13
	4	0.080	0.300	0.300	0.320	0.625	15
	5	0.050	0.250	0.250	0.450	0.790	13
	6	0.038	0.235	0.240	0.487	0.875	17
4	1	0.115	0.535	0.200	0.150	0.545	10
	2	0.074	0.550	0.186	0.190	0.620	15
	3	0.045	0.440	0.215	0.300	0.780	12
	4	0.035	0.330	0.250	0.385	1.120	14

Table 2. System utility with respect to each system state

s	0	1	2	3
u_s	0.0	0.5	0.8	1.0

The system weight is (Dhingra, 1992)

$$W = \sum_{i=1}^{N} w_i(h_i) n_i \exp(n_i/4). \tag{15}$$

Now we will build the physical programming based optimization model, as shown in Equation (11). There are two groups of design variables: the component version variables (h_1 to h_4) and the numbers of components (n_1 to n_4). There are three objectives: the utility objective has Class-2S class function (larger is better), while the cost and weight objectives have Class-1S class functions (smaller is better). Assume the variations of the objectives are known, and we are clear about our requirements on different objectives. We set the class function settings for the three objectives as shown in Table 3. The constraint values U_0, C_0 and W_0 are equal to the boundaries of the unacceptable ranges, g_{i5}, of the corresponding objectives.

Table 3. Physical programming class functions setting

	g_{i1}	g_{i2}	g_{i3}	g_{i4}	g_{i5}
Utility	0.99	0.98	0.95	0.92	0.9
Cost	15	20	25	30	45
Weight	400	500	600	800	1000

GA is used to solve the formulated physical programming based optimization model. The GA parameter settings we used in this example are as follows. The decimal encoding is used. The population size is chosen to be 100. The chromosome length is 12: the first 4 digits are used to represent the component version variables h_1 to h_4, and the rest 8 digits are used to represent the numbers of components n_1 to n_4, each with 2 digits. We use the roulette-wheel selection scheme, one-point cross operator with cross rate of 0.25, and even mutation operator with a mutation rate of 0.1.

Solving the physical programming optimization model, we get the optimization results shown in Table 4. The optimal solution is $\boldsymbol{h}$ = (4, 5, 5, 4) and $\boldsymbol{n}$ = (4, 3, 4, 5). Using this optimal structure, the state distribution of the resulting multi-state system is:

$$\Pr(\phi(\boldsymbol{x}) = 0) = 2.40 \times 10^{-5};\ \Pr(\phi(\boldsymbol{x}) = 1) = 0.05;$$
$$\Pr(\phi(\boldsymbol{x}) = 2) = 0.25;\ \Pr(\phi(\boldsymbol{x}) = 3) = 0.70$$

Thus, based on Table 2, the resulting optimal system utility is

$$U = \sum_{s=0}^{3} u_s \cdot \Pr(\phi(x) = s) = 0.9245$$

The optimization results in Table 4 represent the best tradeoff among system utility, cost and weight, and they are the best results we can get under the preference setting of the designer.

Table 4. Optimization results using different methods

	U	C	W	h	n
Optimal value	0.9245	27.9569	548.8717	(4, 5, 5, 4)	(4, 3, 4, 5)

6.6 Concluding Remarks

Today's engineering systems are becoming more and more complex. In addition, to make the optimization results more useful, we need to include in the optimization model more practical factors that may influence the system design, manufacturing and operation. Reliability or system utility is just one performance measure that may be used when designing a system. For example, when designing a laptop, we need to consider its cost, reliability, size, computing performance, etc. In the design process of such a product, we need to integrate tools from different disciplines. Such a design is also known as Multi-disciplinary Design Optimization (Sobieski, 1990). Based on its flexibilities in building optimization models, GA is suitable for integrating analysis and evaluation tools from different disciplines into a universal optimization framework. The only requirement on a GA optimization model is that the system performance can be evaluated in a certain way and produce a numerical index. Therefore, GA is a desirable tool when including more practical factors into the design process and dealing with more complex systems.

The only disadvantage of GA is that it needs intensive objective function evaluations, and thus it is sort of slow compared to conventional optimization methods. However, with easier access to high performance computing equipment, this disadvantage will become trivial. Furthermore, in case of problems we can not use conventional optimization methods to solve due to mathematical difficulties, GA would be the only choice. However, GA is usually not suitable to deal with real time optimization problems. In addition, GA has the best global optimization property though not guaranteeing global optimality.

The following issues regarding the optimal design of multi-state systems using GA need to be further investigated:

(1) Consider more practical systems that can be modeled as multi-state systems, and use GA for the optimization.

(2) The evaluation of performance distribution of multi-state systems is the bottleneck when implementing the optimization using GA. Thus, more efficient performance evaluation algorithms for multi-state systems are desirable.

(3) Some special types of multi-state systems, such as consecutively-connected systems and k-out-of-n systems, have many optimal design issues. These issues are worth further investigation as well.

References

T. Aven, (1993) On performance-measures for multistate monotone systems. Reliability Engineering & System Safety, Vol. 41, No. 3, pp. 259-266.

C. Chen, M. Meng and M. Zuo, (1999) Selective Maintenance Optimization for Multi-state Systems. Electrical and Computer Engineering, IEEE Canadian Conference, Vol. 3, pp. 1471-1476.

R.E. Barlow and A.S. Wu, (1978) Coherent systems with multi-state components. Mathematics of Operations Research, Vol. 3, No. 4, pp. 275-81.

L. Davis, (1991) Handbook of genetic algorithms. New York: Van Nostrand Reinhold.

E. El-Neweihi, F. Proschan and J. Sethuraman, (1988) Optimal allocation of multistate components. In: Handbook of Statistics, Vol.7: Quality Control and Reliability. Edited by P.R.Krishnaiah, C.R.Rao. North-Holland, Amsterdam, pp. 427-432.

U. Gurler and A. Kaya, (2002) A maintenance policy for a system with multi-state components: an approximate solution, Reliability Engineering & System Safety, Vol. 76, pp. 117-127.

D. Hsieh and K. Chiu, (2002) Optimal maintenance policy in a multistate deteriorating standby system, European Journal of Operational Research, Vol. 141, pp. 689-698.

J. Huang, M.J. Zuo and Y.H. Wu, (2000) Generalized multi-state k-out-of-n:G systems. IEEE Transactions on Reliability, Vol. 49, pp. 105-111.

J. Huang, M.J. Zuo and Z. Fang, (2003) Multi-state consecutive-k-out-of-n systems. IIE Transactions, Vol. 35, pp. 527-534.

F.K. Hwang and Y.C. Yao, (1989) Multistate consecutively-connected systems. IEEE Transactions on Reliability, Vol. 38, October No. 4, pp. 472-474.

W. Kuo and M.J. Zuo. (2003) Optimal reliability modeling: Principles and Applications. Hoboken: John Wiley & Sons, Inc.

G. Levitin, (2000) Multistate series-parallel system expansion scheduling subject to availability constraints, IEEE Transactions on Reliability, Vol. 49, pp. 71-79.

G. Levitin, (2001a) Optimal allocation of multi-state retransmitters in acyclic transmission network, Reliability Engineering & System Safety, Vol. 75, pp. 73-82.

G. Levitin, (2001b) Redundancy optimization for multi-state system with fixed resource-requirements and unreliable sources, IEEE Transactions on Reliability, Vol. 50, pp. 52-59.

G. Levitin, (2002a) Optimal allocation of elements in linear multi-state sliding window system, Reliability Engineering & System Safety, Vol. 76, pp. 247-255.

G. Levitin, (2002b) Optimal reliability enhancement for multi-state transmission networks with fixed transmission time, Reliability Engineering & System Safety, Vol. 76, pp. 289-301.

G. Levitin, (2002c) Maximizing survivability of acyclic transmission networks with multi-state retransmitters and vulnerable nodes, Reliability Engineering & System Safety, Vol. 77, pp. 189-199.

G. Levitin, (2003a) Optimal allocation of multi-state elements in a linear consecutively-connected system, IEEE Trans. Reliability, Vol. 52, pp. 192-199.

G. Levitin, (2003b) Linear multi-state sliding window systems, IEEE Trans. Reliability, Vol. 52, pp. 263-269.

G. Levitin and A. Lisnianski, (1998a) Structure optimization of power system with bridge topology, Electric Power Systems Research, Vol. 45, pp. 201-208.

G. Levitin and A. Lisnianski, (1998b) Joint redundancy and maintenance optimization for multistate series-parallel systems, Reliability Engineering & System Safety, Vol. 64, No. 1, pp. 33-42.

G. Levitin and A. Lisnianski, (1999) Optimal multistage modernization of power system subject to reliability and capacity requirements, Electric Power Systems Research, Vol. 50, pp. 183-190.

G. Levitin and A. Lisnianski, (2000a) Optimization of imperfect preventive maintenance for multi-state systems, Reliability Engineering & System Safety, Vol. 67, pp. 193-203.

G. Levitin and A. Lisnianski, (2000b) Optimal replacement scheduling in multi-state series-parallel systems (short communication), Quality and Reliability Engineering International, Vol. 16, pp. 157-162.

G. Levitin, A. Lisnianski, H. Ben Haim and D. Elmakis, (1998) Redundancy optimization for series-parallel multistate systems, IEEE Transactions on Reliability, Vol. 47, No. 2, pp. 165-172.

A. Lisnianski and G. Levitin, (2003) Multi-state System Reliability: Assessment, Optimization and Applications. World Scientific, Singapore.

A. Lisnianski, G. Levitin and H. Ben Haim, (2000) Structure optimization of multi-state system with time redundancy, Reliability Engineering & System Safety, Vol. 67, pp. 103-112.

J. Malinowski and W. Preuss, (1996) Reliability increase of consecutive-k-out-of-n:F and related systems through components' rearrangement, Microelectronics and Reliability, Vol. 36, pp. 1417-1423.

F. Meng, (1996) More on optimal allocation of components in coherent systems. Journal of Applied Probability , Vol. 33, pp. 548-556.

A. Messac, (1996) Physical Programming: Effective Optimization for Computational Design. AIAA Journal, Vol. 34, No. 1, pp. 149-158.

M. Nourelfath and N. Nahas, (2004a) Ant Colony Optimization to Redundancy Allocation for Multi-state Systems, Fourth International Conference on Mathematical Methods in Reliability Methodology and Practice, June 21-25, Santa Fe, New Mexico.

M. Nourelfath and Y. Dutuit, (2004b) A combined approach to solve the redundancy optimization problem for multi-state systems under repair policies, Reliability Engineering & System Safety, Vol. 86, No. 3, pp. 205-213.

J.E. Ramirez-Marquez and D.W. Coit, (2004) A heuristic for solving the redundancy allocation problem for multi-state series-parallel systems, Reliability Engineering & System Safety, Vol. 83, pp. 341-349.

J. Sobieski, (1990) Multidisciplinary design optimization. Aerospace America, Vol. 28, No. 12, pp. 65-65.

Z. Tian, M.J. Zuo and H. Huang, (2005a) Reliability-redundancy allocation for multi-state series-parallel systems. In Advances in Safety and Reliability, Kołowrocki (ed.) Taylor & Francis Group, London. pp. 1925-1930.

Z. Tian, M.J. Zuo and RCM Yam. (2005b) Performance evaluation of generalized multi-state k-out-of-n systems. Proceedings of 2005 Industrial Engineering Research Conference, Atlanta, Georgia, USA.

Z. Tian and M.J. Zuo, (2005) Redundancy allocation for multi-state systems using physical programming and genetic algorithms. Reliability Engineering and System Safety, Special issue on GA in Reliability. Accepted.

M.J. Zuo, L. Choy and R. Yam, (1999) A Model for Optimal design of Multi-state parallel-series systems, Electrical and Computer Engineering, IEEE Canadian Conference, Vol. 3, pp. 1770-1773.

M.J. Zuo and M. Liang, (1994) Reliability of multistate consecutively-connected systems, Reliability Engineering & System Safety, Vol. 44, pp. 173-176.

M.J. Zuo, B. Liu and D. Murthy, (2000) Replacement-repair policy for multi-state deteriorating products under warranty, European Journal of Operational Research, 123, pp. 519-530.

Intelligent Interactive Multiobjective Optimization of System Reliability

Hong-Zhong Huang

School of Mechanical and Electronic Engineering, University of Electronic Science and Technology of China, P.R. China

Zhigang Tian, Ming J Zuo

Department of Mechanical Engineering, University of Alberta, Canada

7.1 Introduction

In most practical situations involving reliability optimization, there are several mutually conflicting goals such as maximizing system reliability and minimizing cost, weight and volume. Sakawa [1] considered a multiobjective formulation to maximize reliability and minimize cost for reliability allocation by using the surrogate worth trade-off method. Inagaki *et al.* [2] solved another problem to maximize reliability and minimize cost and weight by using an interactive optimization method. The multiobjective reliability apportionment problem for a two component series system has been analyzed by Park [3] using fuzzy logic theory. Dhingra [4] and Rao and Dhingra [5] researched the reliability and redundancy apportionment problem for a four-stage and a five-stage overspeed protection system, using crisp and fuzzy multiobjective optimization approaches respetively. Ravi *et al.* [6] modeled the problem of optimizing the reliability of complex systems as a fuzzy multiobjective optimization problem and studied it.

It is very difficult for designers to specify accurately their preference on the goals a priori in multiobjective reliability optimization problems. The most effective methods have been interactive procedures [7], which typically include alternately solution generation and solution evaluation phases. There are three key issues in the interactive multiobjective optimization

H.-Z. Huang et al.: *Intelligent Interactive Multiobjective Optimization of System Reliability*, Computational Intelligence in Reliability Engineering (SCI) **39**, 215–236 (2007)
www.springerlink.com

method [8]: (1) how to elicit preference information from the designer over a set of candidate solutions, (2) how to represent the designer's preference structure in a systematic manner, (3) how to use the designer's preference structure to guide the search for improved solutions.

Current interactive multiobjective optimization methods mainly include STEM, the Geoffrion-Dyer-Feinberg procedure, the Visual Interactive Approach, the Tchebycheff method, the Zionts-Wallenius method, the Reference Point Method [7] and the Interactive FFANN Procedure [8, 9]. Most of these methods do not make full use of the designer's preference information on the generated solutions, and therefore can't build the model of the designer's preference structure systematically. In Sun *et al.* [8], an Artificial Neural Network (ANN) model of the designer's preference structure is built with the objective function value vector as input and the corresponding preference value as desired output. An optimization problem is solved with the ANN model representing the objective to search for improved solutions. Nevertheless, the improved solutions found in this way are not guaranteed to be Pareto solutions, and nor is the final solution. In Sun *et al.* [9], the ANN model of the designer's preference structure is built in the same way. It is just used to evaluate the Pareto solutions generated with the Augment weighted Tchebycheff programs (AWTPs) in order to pick half of them with the highest preference values and present them to the designer for evaluation. The ANN model can reduce the designer's burden of evaluating generated solutions, but it can not help to search for improved solutions, which is vital in interactive multiobjective optimization procedures.

This chapter reports a new effective multiobjective optimization method, Intelligent Interactive Multiobjective Optimization Method (IIMOM), which is characterized by the way the designer's preference structure model is built and used in guiding the search for improved solutions [16]. In IIMOM, the general concept of the model parameter vector, which refers to the parameter vector determined by the designer in the multiobjective optimization model (such as the weight vector in the weighted-sum method), is proposed. From a practical point of view, the designer's preference structure model is built using Artificial Neural Networks (ANN) with the model parameter vector as input and the preference information articulated by designers over representative samples from the Pareto frontier as desired output. Then with the ANN model of the designer's preference structure as the objective, an optimization problem is solved to search for improved solutions. Two key advantages of IIMOM are: (1) the ANN model of the designer's preference structure can guide the designer to explore the interesting part of the Pareto frontier efficiently and accurately; (2) the improved solutions generated at each iteration are Pareto solutions, which is in stark contrast to the method presented in Sun *et al.* [8].

IIMOM is applied to the reliability optimization problem of a multistage mixed system. Five different value functions are used to simulate the designer in the solution evaluation process. The results illustrate that IIMOM is effective in capturing different kinds of preference structures of the designer, and it is an effective tool for the designer to find the most satisfying solution.

7.2 Multiobjective Optimization Problem

7.2.1 Problem Formulation

A general multiobjective optimization problem is to find the design variables that optimize m different objectives over the feasible design space. A mathematical formulation of the multiobjective optimization problem is

$$\begin{aligned} &\text{minimize} \quad \boldsymbol{f}(\boldsymbol{x}) = \{f_1(\boldsymbol{x}), f_2(\boldsymbol{x}), \ldots, f_m(\boldsymbol{x})\} \\ &\text{Subject to} \quad \boldsymbol{x} \in \mathrm{X} \end{aligned} \tag{1}$$

where $\boldsymbol{x}$ is an n-dimensional vector of design variables, X is the feasible design space, $f_i(x)$ is the objective function of the ith design objective and $\boldsymbol{f}(\boldsymbol{x})$ is the design objective vector.

7.2.2 Pareto Solution

A design variable vector $\boldsymbol{x}^{\mathrm{P}}$ is said to be a Pareto solution if there exists no feasible design variable vector $\boldsymbol{x}$ that would decrease some objective functions without causing a simultaneous increase in at least one objective function. Mathematically, a solution $\boldsymbol{x}^{\mathrm{P}}$ is said to be a Pareto solution if for any $\boldsymbol{x} \in \mathrm{X}$ satisfying $\boldsymbol{f}_j(\boldsymbol{x}) < \boldsymbol{f}_j(\boldsymbol{x}^{\mathrm{P}})$, $\boldsymbol{f}_k(\boldsymbol{x}) > \boldsymbol{f}_k(\boldsymbol{x}^{\mathrm{P}})$ for at least one other objective $k \neq j$.

The set of all Pareto solutions of a multiobjective optimization problem is known as the Pareto frontier (or Pareto set), which is denoted by N. It is evident that the final solution of a multiobjective optimization problem should be selected from the Pareto frontier.

7.2.3 Weighted-sum Method

The Weighted-sum Method is one of the most widely used solution methods for multiobjective optimization problems. It converts a multiobjective optimization problem into a single-objective optimization problem using a weighted sum of all the objective functions as the single objective. The mathematical model of the Weighted-sum Method takes the form

$$\begin{aligned} &\text{minimize} \quad f = \sum_{i=1}^{m} w_i f_i(\boldsymbol{x}) \\ &\text{Subject to} \quad \boldsymbol{x} \in \mathrm{X} \end{aligned} \tag{2}$$

where w_i is the weight of objective i, and

$$\sum_{i=1}^{m} w_i = 1, \; w_i \geq 0, \; i = 1,2,...,m$$

7.2.4. Augment Weighted Tchebycheff Programs (AWTP)

AWTPs is another widely used solution method for multiobjective optimization problems. The mathematical model of AWTPs takes the form

$$\begin{aligned} \min \quad & \alpha + \rho \sum_{i=1}^{m} (1 - z_i) \\ \text{s.t.} \quad & \alpha \geq \lambda_i (1 - z_i), \; \forall i \\ & z_i = \frac{f_i(x) - f_i^{\text{nadir}}}{f_i^{\text{ideal}} - f_i^{\text{nadir}}}, \; \forall i \\ & x \in \mathrm{X} \end{aligned} \tag{3}$$

where ρ is a small positive scalar; f^{ideal} is the utopian point, that is, f_i^{ideal} is the optimization result with the *i*th design objective as the only objective function and $x \in \mathrm{X}$ as constraints; λ_i is the weight of design objective i, and satisfies $\sum_{i=1}^{m} \lambda_i = 1$, and $\lambda_i \geq 0$, $i = 1,2,...,m$; f_i^{nadir} is the worst value of the *i*th objective function (the worst value is the maximum value since we desire to minimize this objective function) among all the points in the Pareto frontier. For multiobjective linear programming problems, f_i^{nadir} can be evaluated with the method provided by Korhonen *et al.* [10]. For nonlinear programming problems, f_i^{nadir} can be estimated by the optimization result with the minus of the *i*th design objective as the only objective function and $x \in X$ as constraints, or it can be estimated by simply being assigned a value based on experience.

Some interactive multiobjective optimization methods, such as the Tchebycheff method, WIERZ and SATIS, are based on AWTPs [7]. By changing the weight vector λ in AWTPs, each point of N can be reached [10]. Therefore, AWTPs is effective in constructing interactive multiobjective optimization methods.

7.3 Designer's Preference Structure Model

7.3.1 Model Parameter Vector

The general concept of the model parameter vector is proposed in this section. The model parameter vector refers to the parameter vector determined by the designer in the multiobjective optimization model, such as the weight vector $[w_1, w_2,..., w_m]$ in the weighted-sum method and the weight vector $[\lambda_1, \lambda_2,..., \lambda_m]$ in AWTPs.

In a specific multiobjective optimization problem, the design variables, objective functions and constraints have already been determined through analysis and modeling. What the designer can manipulate is only the model parameter vector. That is, in a specific multiobjective optimization problem, once the model parameter vector is determined, the final solution of the problem will also be determined.

7.3.2 General Multiobjective Optimization Procedure

The general multiobjective optimization procedure is shown in Fig. 1. First, the problem is analyzed and the design variables, objective functions and the constraints are determined. Then the model parameter vector is set by the designer. Optimization is conducted and the solution is obtained. If the designer is satisfied with the obtained solution, the optimization procedure is terminated. Otherwise, the designer will modify the model parameter vector and conduct the next iteration of optimization.

There is one problem in the general multiobjective optimization procedure. The designer can control the model parameter vector, but he can not control the generated objective function vector of the optimization solution with respect to the model parameter vector. The objective function vector of the optimization solution determines the designer's preference on the solution. Therefore, the general multiobjective optimization procedure typically requires many iterations on the choice of the model parameter vector and often provides no clear guidance as to how to converge to the right model parameter vector. For instance, consider the case where the weighted-sum method is used in the multiobjective optimization procedure. If the designer is not satisfied with the obtained solution and desires to improve objective i, the weight w_i should be increased relatively. However, how much w_i should be relatively increased is unknown. The convergence of the optimization procedure could not be assured. Therefore, an interactive multiobjective optimization method is needed which is able to intelligently guide the designer to explore his interesting part of the Pareto frontier efficiently and accurately and to converge to the satisfying solution finally.

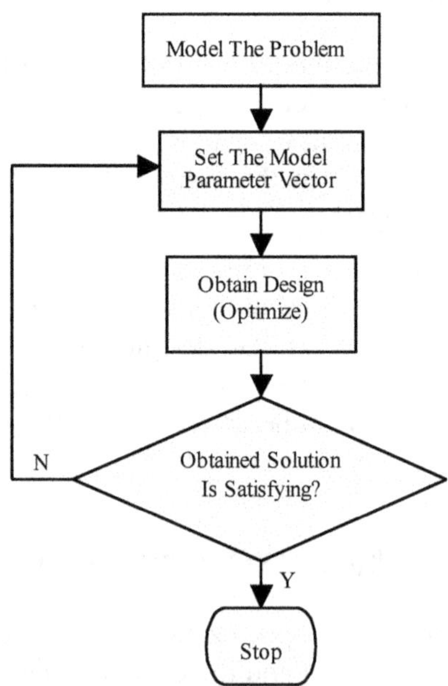

Fig. 1. General multiobjective optimization procedure

Another important point implied in Fig. 1 is the vital role the model parameter vector plays in the general multiobjective optimization procedure. The model parameter vector is the only item the designer can manipulate. The designer modifies the model parameter vector to express his preference on the generated solution and to generate new solutions.

7.3.3 Designer's Preference Structure Model

The designer's preference structure represents the designer's preferences on the design objectives and their trade-off. After a solution is generated, the designer could express his preference information (e.g. assigned preference value) on the solution based on its objective functions' values. The process of how the preference information is elicited is depicted in Fig. 2.

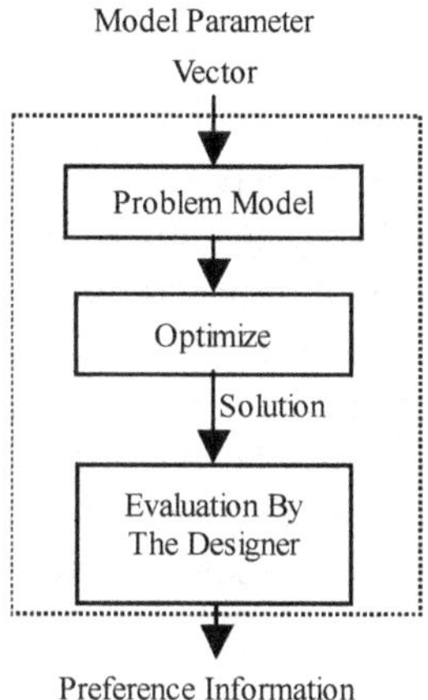

Fig. 2. The preference information elicitation process

A direct approach to model designer's preference structure is to build a model mapping the objective function value vector to the preference information, that is, to simulate the function block of "Evaluation By The Designer" in Fig. 2. ANNs have been used to build the designer's preference structure model in this way by Sun *et al.* [8, 9], Stam *et al.* [11] and Lu *et al.* [12]. The method is easy to understand, but the designer's preference structure model built in this way is difficult to use to guide the designer effectively in the following optimization iterations, because the input of the model (the generated solution) can not be controlled directly. For example, in the method proposed by Sun *et al.* [8], an optimization problem is solved with the ANN model of the designer's preference structure model as the objective to search for improved solution. Nevertheless, the improved solutions found in this way are not guaranteed to be Pareto solutions, and nor is the final solution.

For the preference information elicitation process in Fig. 2, it can be seen that the model parameter vector is the input and the preference information is the output. And the designer can modify the model parameter vector in order to achieve the solution that best satisfies his preference.

Therefore, from a practical point of view, it is proposed in this chapter that the designer's preference structure model should be built using an ANN that maps the model parameter vector to the preference information, that is, simulates the function blocks in the dotted frame in Fig. 2.

ANN has proved its ability to represent complex nonlinear mapping using a set of available data, so it is used to model the designer's preference structure, that is, to represent the mapping from the model parameter vector to the preference information. The ANN model of the designer's preference structure is trained with the Pareto solutions generated during the interactive optimization procedure and their corresponding preference information as the training set.

The model parameter vector is the vital factor in interactive optimization and the vital factor to build the designer's preference structure model, but it has always been overlooked in previous research. The ANN model of the designer's preference structure in this chapter is easy to use because the input of the model (the model parameter vector) can be directly controlled. And there is no evidence that the ANN model built in this way is more complex than the ANN model mapping the objective function value vector to the preference information, because both of them are using a set of generated data to approximate nonlinear relationships using ANNs. The ANN model of the designer's preference structure plays a vital role in the IIMOM procedure presented in the following section.

7.3.4 Preference Information Elicitation

As the output of the ANN model of the designer's preference structure, the preference information could be elicited in two ways [8]. The designer determines a Pareto solution's preference information by either directly assigning a preference "value" or making pairwise comparisons among the generated Pareto solutions. The elicited preference information could be represented by a numerical value, so it is called preference value.

7.4 IIMOM Procedure

IIMOM developed in this chapter is based on AWTPs formulated in Eq. (3). The weight vector $[\lambda_1, \lambda_2, ..., \lambda_m]$ is the model parameter vector in AWTPs. The IIMOM procedure is shown in Fig. 3 and is specified step-by-step below, followed by comments about its different steps.

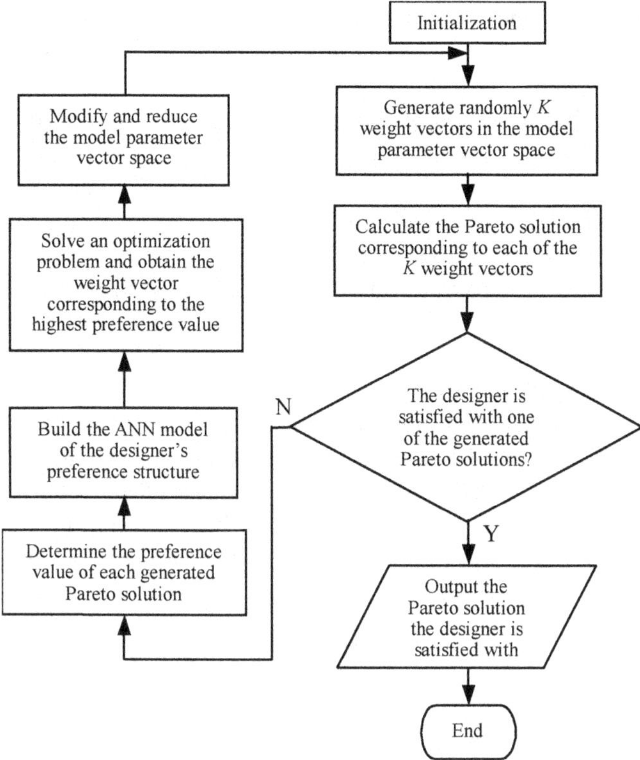

Fig. 3. The flow chart of IIMOM

Step 1: $l_i^{(h)}$ and $u_i^{(h)}$ denote respectively the lower and upper boundaries of weight λ_i at iteration h. $\left[l_i^{(h)}, u_i^{(h)}\right] \subseteq [0,1], \forall i, \forall h$. Let $\left[l_i^{(1)}, u_i^{(1)}\right] = [0,1], \forall i$, and a more specific $\left[l_i^{(1)}, u_i^{(1)}\right] \subset [0,1], \forall i$ will be helpful for efficient convergence of the IIMOM procedure. $\Lambda^{(h)}$ denotes the weight vector space at iteration h

$$\Lambda^{(h)} = \left\{ \lambda \mid \sum_{i=1}^{m} \lambda_i = 1, \lambda_i \in \left[l_i^{(h)}, u_i^{(h)}\right], \forall i \right\}$$

Specify the weight vector space reduction factor r, the number of Pareto solutions K to be evaluated at each iteration, and the structure of the ANN model of the designer's preference structure. Calculate f^{ideal}, and determine f_i^{nadir}.

Step 2: In the weight vector space, K weight vectors are randomly generated. If $h > 1$, the first weight vector is replaced by the best weight vector $\lambda^{(\text{Opt}, h-1)}$ obtained at the previous iteration. The best weight vector $\lambda^{(\text{Opt}, h-1)}$ is obtained by solving an optimization problem with the ANN model of the designer's preference structure as the objective. It refers to the weight vector with respect to the highest preference value in the ANN model at iteration h.

Step 3: Solve one AWTPs problem for each of the K weight vectors to obtain K Pareto solutions.

Step 4: If the designer is satisfied with one of the Pareto solutions, that Pareto solution is output and the interactive optimization procedure is terminated. Otherwise, go to step 5.

Step 5: The Pareto solutions obtained at the current iteration are presented to the designer, and their preference values are evaluated.

Step 6: With the Pareto solutions obtained in the latest several iterations as the training set, the weight vector as input, and the corresponding preference value as desired output, a feed-forward neural network is trained to obtain the ANN model of the designer's preference structure.

Step 7: With the ANN model of the designer's preference structure as the objective function, the optimization problem shown in Eq. (4) is solved to obtain the best weight vector $\lambda^{(\mathrm{Opt},h)}$ with respect to the highest preference value.

$$\begin{array}{ll} \max & \mathrm{ANN}(\lambda) \\ \text{s.t.} & \lambda \in \Lambda^{(h)} \end{array} \tag{4}$$

where $\mathrm{ANN}(\lambda)$ represents the preference value calculated using the ANN model of the designer's preference structure when the input weight vector is λ.

Step 8 : Let $P^{(\mathrm{Best},h)}$ denote the Pareto solution with respect to the highest preference value among all the Pareto solutions that have been generated up to the current iteration, and $P^{(\mathrm{Opt},h)}$ denote the Pareto solution obtained in Step 7. $\lambda^{(\mathrm{Best},h)}$ and $\lambda^{(\mathrm{Opt},h)}$ are the weight vectors with respect to $P^{(\mathrm{Best},h)}$ and $P^{(\mathrm{Opt},h)}$ respectively. Let

$$\lambda^{(\mathrm{Center},h)} = \frac{\lambda^{(\mathrm{Best},h)} + \lambda^{(\mathrm{Opt},h)}}{2} \tag{5}$$

Modify $l_i^{(h+1)}$ and $u_i^{(h+1)}$ for each design objective i to determine the new weight vector space $\Lambda^{(h+1)}$:

If $\left|\lambda_i^{(\mathrm{Opt},h)} - \lambda_i^{(\mathrm{Best},h)}\right| > r^h$

$$\left[l_i^{(h+1)}, u_i^{(h+1)}\right] = \left[\min\left(\lambda_i^{(\mathrm{Opt},h)}, \lambda_i^{(\mathrm{Best},h)}\right), \max\left(\lambda_i^{(\mathrm{Opt},h)}, \lambda_i^{(\mathrm{Best},h)}\right)\right] \tag{6}$$

$$\text{if } \left|\lambda_i^{(\text{Opt},h)} - \lambda_i^{(\text{Best},h)}\right| \le r^h \text{ and } \left(\lambda_i^{(\text{Center},h)} - \frac{r^h}{2}\right) \le 0$$

$$\left[l_i^{(h+1)}, u_i^{(h+1)}\right] = \left[0, \lambda_i^{(\text{Center},h)} + \frac{r^h}{2}\right]$$

$$\text{if } \left|\lambda_i^{(\text{Opt},h)} - \lambda_i^{(\text{Best},h)}\right| \le r^h \text{ and } \left(\lambda_i^{(\text{Center},h)} + \frac{r^h}{2}\right) \ge 1$$

$$\left[l_i^{(h+1)}, u_i^{(h+1)}\right] = \left[\lambda_i^{(\text{Center},h)} - \frac{r^h}{2}, 1\right]$$

otherwise

$$\left[l_i^{(h+1)}, u_i^{(h+1)}\right] = \left[\lambda_i^{(\text{Center},h)} - \frac{r^h}{2}, \lambda_i^{(\text{Center},h)} + \frac{r^h}{2}\right]$$

Then go to step 2 and conduct the next iteration of IIMOM.

IIMOM has two key advantages: (1) the ANN model of the designer's preference structure can guide the designer to explore the part of the Pareto frontier of interest to him efficiently and accurately. IIMOM has outstanding convergence performance. During the IIMOM procedure, the ANN model of the designer's preference structure will become more and more accurate around the part of the Pareto frontier of interest to the designer; (2) the improved solutions generated at each iteration are Pareto solutions, which is in stark contrast to the method presented by Sun *et al.* [8].

7.5 Application of IIMOM to Reliability Optimization Problem

In this section, IIMOM is applied to the reliability optimization problem of a multi-stage mixed system. Five different value functions are used to simulate the designer in the solution evaluation process in order to illustrate the effectiveness of IIMOM in capturing different kinds of preference structures of the designer and finding the most satisfying solution.

7.5.1 Problem Definition

The multiobjective reliability optimization problem is taken from Sakawa [13] and Ravi *et al.* [6] and the problem is modeled as a fuzzy multiobjective optimization problem. A multistage mixed system is considered, where the problem is to allocate the optimal reliabilities $r_i, i = 1, 2, 3, 4$ of four components whose redundancies are specified. The multiobjective optimization model of the problem takes the following form [6].

$$\text{Maximize } R_S, \text{ Minimize } C_S, \text{ Minimize } W_S$$
$$\text{Subject to} \tag{7}$$
$$V_S = \sum_{j=1}^{4} V_j n_j \le 65\,;\; P_S \le 12000$$

where R_S, C_S, W_S, V_S are the reliability, cost, weight and volume of the system.

$$P_S = W_S \cdot V_S\,;$$
$$R_S = \prod_{j=1}^{4}\left[1-\left(1-r_j\right)^{n_j}\right];\; C_S = \sum_{j=1}^{4} C_j n_j;\; W_S = \sum_{j=1}^{4} W_j n_j\,; \tag{8}$$

$$C_j = \alpha_j^c\left[\log_{10}\left(\frac{\beta_j^c}{1-r_j}\right)\right]^{\gamma_j^c};\; W_j = \alpha_j^w\left[\log_{10}\left(\frac{\beta_j^w}{1-r_j}\right)\right]^{\gamma_j^w};$$
$$V_j = \alpha_j^v\left[\log_{10}\left(\frac{\beta_j^v}{1-r_j}\right)\right]^{\gamma_j^v} \tag{9}$$

$$\alpha_j^c = 8.0,\; \alpha_j^w = 6.0,\; \alpha_j^v = 2.0\,;$$
$$\gamma_j^c = 2.0,\; \gamma_j^w = 0.5,\; \gamma_j^v = 0.5\,;$$
$$\beta_1^c = 2.0,\; \beta_2^c = 10.0,\; \beta_3^c = 3.0,\; \beta_4^c = 18.0\,; \tag{10}$$
$$\beta_1^w = 3.0,\; \beta_2^w = 2.0,\; \beta_3^w = 10.0,\; \beta_4^w = 8.0\,;$$
$$\beta_1^v = 2.0,\; \beta_2^v = 2.0,\; \beta_3^v = 6.0,\; \beta_4^v = 8.0\,;$$
$$n_1 = 7,\; n_2 = 8,\; n_3 = 7,\; n_4 = 8\,.$$

where all the values use the corresponding SI units.

IIMOM is applied to the formulated multiobjective reliability optimization problem. Let $\lambda_i^{(1)} \in [0,1], i = 1, 2$, $\lambda_3^{(h)} = 1 - \lambda_1^{(h)} - \lambda_2^{(h)}$. The ideal and nadir values of the three objectives are determined as follows:

$$R_S^{\text{ideal}} = 1,\; R_S^{\text{nadir}} = 0.9$$
$$C_S^{\text{ideal}} = 0,\; C_S^{\text{nadir}} = 550 \tag{11}$$

$$W_S^{\text{ideal}} = 0\,,\ W_S^{\text{nadir}} = 350$$

In the framework of AWTPs formulated in Eq. (3), we have

$$z_1 = \frac{R_S - R_S^{\text{nadir}}}{R_S^{\text{ideal}} - R_S^{\text{nadir}}}\,;\ z_2 = \frac{C_S - C_S^{\text{nadir}}}{C_S^{\text{ideal}} - C_S^{\text{nadir}}}\,;\ z_3 = \frac{W_S - W_S^{\text{nadir}}}{W_S^{\text{ideal}} - W_S^{\text{nadir}}} \tag{12}$$

7.5.2 The Mapping from Weight Vector to Preference Value

Through numerical experiments in this section, we try to make sure that a specific weight vector will result in a corresponding specific preference value, that is, the preference value is a function of the weight vector.

The value function is used to simulate the designer in the solution evaluation process. Assume that the value function takes the form

$$V = w_1 z_1 + w_2 z_2 + w_3 z_3 \tag{13}$$

where w_1, w_2 and w_3 are equal to 0.5, 0.3 and 0.2 respectively.

Let $\lambda = [0.40, 0.25, 0.35]$. Genetic algorithm is used to solve the AWTPs model formulated in Eq. (3) five times, and the obtained Pareto solutions are evaluated with the value function. The results are shown in Table 1.

Table 1. Results with respect to a specific weight vector

	Design Objectives			Preference value
	R_S	C_S	W_S	
1	0.9573	295.36	170.98	2.5274
2	0.9573	295.80	170.95	2.5273
3	0.9573	294.75	170.94	2.5276
4	0.9573	296.12	170.96	2.5273
5	0.9573	296.25	170.97	2.5272

It can be concluded from Table 1 that a specific weight vector will result in corresponding specific objective function values and a corresponding preference value. There are still small variations in the obtained preference values because genetic algorithm may not obtain accurately the same optimal solution in a limited number of generations. The ANN model of the designer's preference structure is built with the weight vector as input and the preference value as desired output. The variations are too small to impact the function of the ANN model to intelligently guide the multiobjective optimization procedure.

7.5.3 Results and Discussions

Value functions are specified in order to simulate the designer in the generated solution evaluation process during the IIMOM procedure. The following five value functions [9] are used in order to illustrate the effectiveness of IIMOM to capture different kinds of preference structures. $w = [0.5, 0.3, 0.2]$ and $KV = 2$ for all the value functions.

(1) Linear value function:

$$V = \sum_{i=1}^{3} w_i z_i \tag{14}$$

(2) Quadratic value function:

$$V = KV - \sqrt{\sum_{i=1}^{3} \left[w_i (1 - z_i)^2 \right]} \tag{15}$$

(3) L_4-metric value function:

$$V = KV - \left[\sum_{i=1}^{3} \left[w_i (1 - z_i)^4 \right] \right]^{\frac{1}{4}} \tag{16}$$

(4) Tchebycheff metric value function:

$$V = KV - \max_{1 \le i \le 3} \{ w_i (1 - z_i) \} \tag{17}$$

(5) Combined value function:

$$V = KV - \frac{1}{2} \left\{ \sqrt{\sum_{i=1}^{3} \left[w_i (1 - z_i)^2 \right]} + \max_{1 \le i \le 3} [w_i (1 - z_i)] \right\} \tag{18}$$

The combined value function is obtained by combining the quadratic value function and the Tchebycheff metric value function.

Genetic algorithm is used to solve the AWTPs problems, shown in Eq. (3), in Step 3 of the IIMOM procedure. Compared with standard nonlinear programming techniques, genetic algorithm is computationally more expensive, but it has much better capability in finding the global optimum, while standard nonlinear programming techniques are easily trapped in local optima. The preference values based on the optimization results obtained by solving these AWTPs problems will be used to train the ANN to represent the designer's preferences. In this problem, the population size is chosen to be 100. The decimal encoding is used and the chromosome length is set to be 20, that is, each of the four design variables is represented by a five-digit segment of the chromosome. We use the roulette-wheel selection scheme, one-point crossover operator with a crossover rate of 0.25, and uniform mutation operator with a mutation rate of 0.1.

The model parameter vector space reduction factor r is 0.7. Ten Pareto solutions are evaluated at each iteration, in order to make the trained ANN model accurate enough while not requiring too many Pareto solution generating procedures. The Pareto solutions generated in the latest five iterations are used to train the ANN model of the designers' preference structure, so that the data used to train the ANN model will focus gradually on the region that the designers are interested in, and make the ANN model more accurate in this region. Except for the first four iterations, fifty training pairs in total are used to train the ANN at each iteration. The numbers "ten" and "five" are chosen based on computational experience, and there are no definite criteria on how to choose these numbers.

In this problem, the ANN used for building the model of the designers' preference structure is a three layered feedforward neural network, with two neurons in the input layer, one neuron in the output layer, and mostly three neurons in the hidden layer. The reason for selecting three hidden neurons is that an ANN with three hidden neurons is believed to be a parsimonious model which can model the nonlinear relationship without overfitting the data when there are two input neurons, one output neuron and, in most cases, fifty training pairs in the training set [14]. Because there are ten training pairs in the first iteration and twenty training pairs in the second iteration, we use two hidden neurons in these two iterations so that there will not be too many free parameters in the ANN model.

The case of the linear value function is considered first. IIMOM is run for ten iterations. The weight vector space, average preference value of the generated Pareto solutions at the current iteration, the weight vectors and preference values of $P^{(\mathrm{Best},h)}$ and $P^{(\mathrm{Opt},h)}$ are shown in Table 2.

In Table 2, $\left[l_1^{(h)},u_1^{(h)}\right]$ and $\left[l_2^{(h)},u_2^{(h)}\right]$ are the range of the weights λ_1 and λ_2 at iteration h, $\lambda_3=1-\lambda_1-\lambda_2$. P^{Best} is the best Pareto solution generated up to the current iteration, P^{opt} is the optimization result obtained by solving the optimization problem with the ANN model of the designer's preference structure as the objective function at the current iteration.

The ANN models of the designer's preference structure at iterations 2, 5, 8 and 10 are depicted in Fig. 4. During the process of IIMOM, the model parameter vector space is reduced, focusing step by step on the part of the Pareto frontier the designer is interested in. The new model parameter vector space $\Lambda^{(h+1)}$ is determined by $P^{(\mathrm{Best},h)}$ and $P^{(\mathrm{Opt},h)}$.

Table 2. The IIMOM process in the case of linear value function

h	Weight vector space		Average preference value	$P^{(\mathrm{Best},h)}$		$P^{(\mathrm{Opt},h)}$	
	$\left[l_1^{(h)}, u_1^{(h)}\right]$	$\left[l_2^{(h)}, u_2^{(h)}\right]$		Weight vector $[\lambda_1, \lambda_2]$	Preference value	Weight vector $[\lambda_1, \lambda_2]$	Preference value
1	[0, 1]	[0, 1]	2.3123	[0.7388, 0.1461]	2.6579	[0, 0]	2.3187
2	[0.0000, 0.7694]	[0, 0.4731]	2.4070	[0.7388, 0.1461]	2.6579	[0.7694, 0.1459]	2.6596
3	[0.4341, 1.0000]	[0, 0.4660]	2.5796	[0.7694, 0.1459]	2.6596	[0.7659, 0.2341]	2.6367
4	[0.5116, 1.0000]	[0, 0.4460]	2.6314	[0.8246, 0.0840]	2.6727	[0.8261, 0.1738]	2.6552
5	[0.6206, 1.0000]	[0, 0.3337]	2.6360	[0.8246, 0.0840]	2.6727	[0.8883, 0.0694]	2.6750
6	[0.6926, 1.0000]	[0, 0.2405]	2.6582	[0.8883, 0.0694]	2.6750	[0.8703, 0.0587]	2.6751
7	[0.7482, 1.0000]	[0, 0.1951]	2.6647	[0.8800, 0.0576]	2.6755	[0.8790, 0.0402]	2.6745
8	[0.7747, 0.9844]	[0, 0.1537]	2.6701	[0.8800, 0.0576]	2.6755	[0.8947, 0.0499]	2.6753
9	[0.8035, 0.9712]	[0, 0.1376]	2.6713	[0.8800, 0.0576]	2.6755	[0.8885, 0.0555]	2.6754
10	[0.8171, 0.9514]	[0, 0.1236]	2.6723	[0.8800, 0.0576]	2.6755	[0.8942, 0.0433]	2.6745

As can be seen from Table 2, the average preference value increases in general during the process of IIMOM, which means that the model parameter vector space is converging to the model parameter vector that best satisfies the designer's preference. The model parameter vector space is reduced in most iterations, though there might be an iteration in which the model parameter vector space isn't reduced. The model parameter vector space is sure to converge at the end. IIMOM captures the hidden tendency of the designer's preference structure through the discrete generated Pareto solutions, and uses all the information the generated Pareto solutions can provide. $P^{(\mathrm{Opt},h)}$ is obtained by solving an optimization problem with the ANN model of the designer's preference structure as the objective. Although the preference value of $P^{(\mathrm{Opt},h)}$ may not be superior to that of $P^{(\mathrm{Best},h)}$, $P^{(\mathrm{Opt},h)}$ does lead the designer to the Pareto frontier part he is interested in. On the other hand, the IAWTPs method presented by Sun *et al.* (2000) only uses the best generated Pareto solution to adjust the weights, and it overlooks some important information provided by other generated Pareto solutions.

The linear value function is used to simulate the designer in evaluating the generated Pareto solutions in IIMOM in this case. If the linear value function formulated in Eq. (14) is used as the single objective for the reliability optimization problem, the obtained solution must be the Pareto solution that best satisfies the designer's preference. For the purpose of comparison, the Pareto solution obtained in this way, termed the comparing result, is compared with the result obtained using IIMOM in Table 3.

It can be seen that the objectives values and the preference value of the result obtained using IIMOM are very close to those of the comparing result. These numerical results illustrate that IIMOM is effective in capturing the designer's preference structure and obtaining the most satisfying Pareto solution if the designer's preference structure can be approximated as a linear value function.

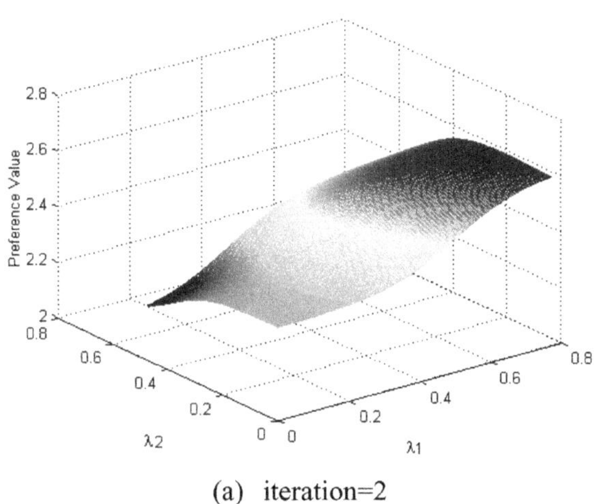

(a) iteration=2

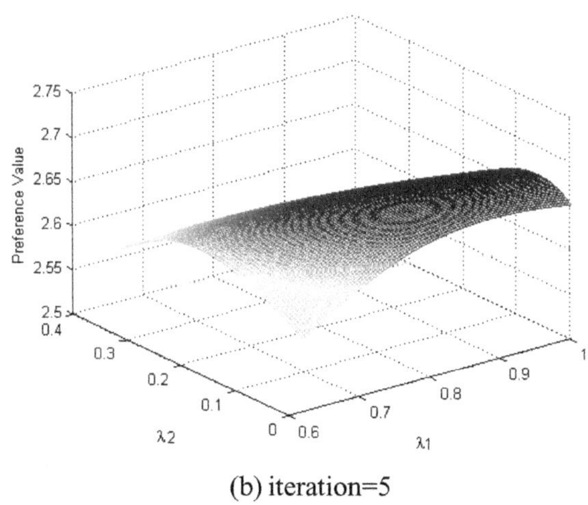

(b) iteration=5

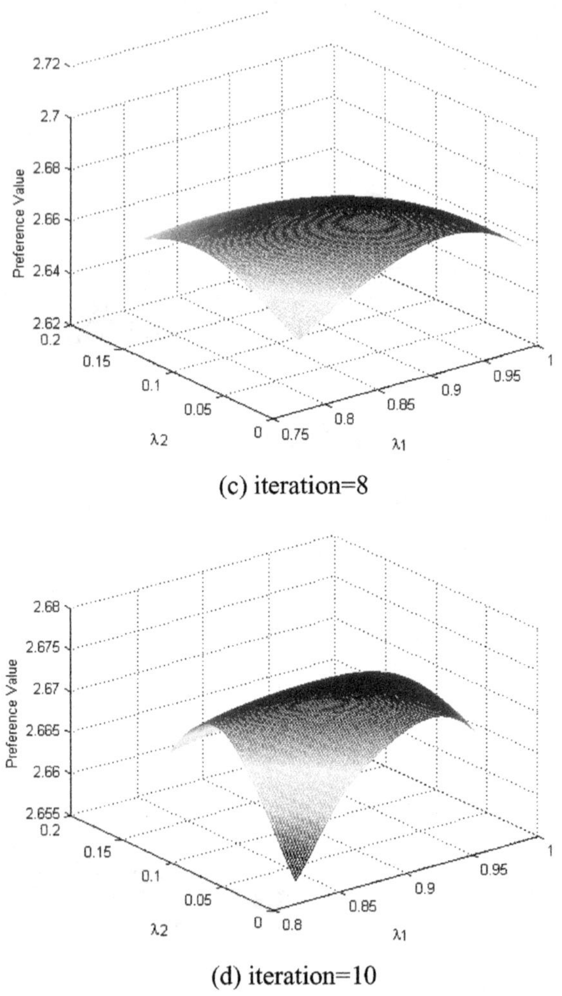

(c) iteration=8

(d) iteration=10

Fig. 4. The ANN model of the designer's preference structure in the case of the linear value function

The IIMOM program we used is based on the MATLAB platform. Here the IIMOM procedure which runs for ten iterations takes about ten minutes, that is, each iteration of IIMOM takes about one minute. It takes about five seconds to solve a single AWTPs problem. And it takes about four seconds to train the ANN model, and five seconds to solve the optimization problem in Eq. (4), at each iteration. Generally, for this problem, the IIMOM procedure can be completed in acceptable time.

In the cases of the quadratic value function, L_4-metric value function, Tchebycheff metric value function, and combined value function, the optimization results are shown in Table 4-Table 7. These cases illustrate the same trends as found in the case of the linear value function. The results obtained using IIMOM are very close to those of the comparing results obtained by solving the reliability optimization problems with the corresponding value functions as the single objectives.

Table 3. The results in the case of the linear value function

	IIMOM	Comparing result
r_1	0.6394	0.6442
r_2	0.5680	0.5680
r_3	0.6400	0.6375
r_4	0.5551	0.5612
r_s	0.9957	0.9959
C_S	362.9294	364.3245
W_S	183.6814	183.8811
Preference Value	2.6755	2.6756

Table 4. The results in the case of the quadratic value function

	IIMOM	Comparing result
r_1	0.5726	0.5736
r_2	0.4892	0.4900
r_3	0.5682	0.5650
r_4	0.4813	0.4834
r_s	0.9848	0.9849
C_S	323.4813	323.7218
W_S	176.9803	177.0066
Preference Value	1.5920	1.5920

Table 5. The results in the case of the *L*4-metric value function

	IIMOM	Comparing result
r_1	0.5403	0.5350
r_2	0.4503	0.4484
r_3	0.5219	0.5250
r_4	0.4365	0.4398
r_s	0.9717	0.9719
C_S	304.8938	305.0382
W_S	173.6443	173.5886
Preference Value	1.5430	1.5430

Table 6. The results in the case of the Tchebycheff metric value function

	IIMOM	Comparing result
r_1	0.5246	0.5214
r_2	0.4355	0.4402
r_3	0.5090	0.5118
r_4	0.4342	0.4298
r_s	0.9672	0.9672
C_S	300.6453	300.5795
W_S	172.6808	172.6974
Preference Value	1.8360	1.8360

Table 7. The results in the case of the combined value function

	IIMOM	Comparing result
r_1	0.5615	0.5620
r_2	0.4787	0.4760
r_3	0.5470	0.5511
r_4	0.4698	0.4670
r_s	0.9814	0.9812
C_S	317.3904	316.9782
W_S	175.8632	175.8256
Preference Value	1.7087	1.7087

The results illustrate that IIMOM is effective in capturing different kinds of preference structures of the designer, including linear, quadratic, L_4-metric, Tchebycheff metric, and the combined modes, and it can obtain the Pareto solution that best satisfies the designer's preference finally.

7.5.4 Discussion on the Performances of IIMOM

(1) Effectiveness in the multiobjective optimization process [15]. IIMOM is very effective in capturing different kinds of preference structures of the designer, and it can obtain the Pareto solution that best satisfies the designer's preference finally.

(2) Ease in actual use. What the designer needs to do in IIMOM is to evaluate the generated Pareto solutions. Therefore, the designer's cognitive burden is not too heavy, and it is not too complex to use IIMOM in actual problems. How to evaluate the generated solutions and determine their preference values is a key problem.

(3) Convergence. The ANN model of the designer's preference structure guides the designer to explore the part of the Pareto frontier of interest.

The numerical experiments indicate that the IIMOM procedure has good convergence performance.

(4) Change of the designer's preference structure. The designer's knowledge of the handled problem increases during the optimization process. After examining some generated solutions, the designer's preference structure may change gradually. IIMOM uses Pareto solutions generated in the latest several iterations to train the ANN model of the designer's preference structure. The training strategy could resolve the problem of the change of the designer's preference structure.

7.6 Conclusions

This chapter reports a new effective multiobjective optimization method, IIMOM, and applies it to the reliability optimization problem of a multistage mixed system. In IIMOM, the general concept of the model parameter vector is proposed. From a practical point of view, the designer's preference structure model is built using an ANN with the model parameter vector as input and the preference information articulated by designers over representative samples from the Pareto set as desired output. Then with the ANN model of the designer's preference structure as the objective, an optimization problem is solved to search for improved solutions.

Two key advantages of IIMOM are: (1) the ANN model of the designer's preference structure can guide the designer to explore the part of the Pareto frontier of interest efficiently and accurately; (2) the improved solutions generated at each iteration are Pareto solutions.

In the reliability optimization problem, five value functions are used to simulate the designer in the solution evaluation process of IIMOM. The results illustrate that IIMOM is very effective in capturing different kinds of preference structures of the designer, and it can obtain the Pareto solution that best satisfies the designer's preference finally.

Acknowledgement

This research was partially supported by the National Natural Science Foundation of China under the contract number 50175010, the Excellent Young Teachers Program of the Ministry of Education of China under the contract number 1766, the National Excellent Doctoral Dissertation Special Foundation of China under the contract number 200232, and the Natural Sciences and Engineering Research Council of Canada.

References

[1] Sakawa, M. (1978) Multiobjective optimization by the surrogate worth trade-off method. IEEE Transactions on Reliability, 27, 311-314.

[2] Inagaki, T., Inoue, K. and Akashi, H. (1978) Interactive optimization of system reliability under multiple objectives. IEEE Transactions on Reliability, 27, 264-267.

[3] Park, K.S. (1987) Fuzzy apportionment of system reliability. IEEE Transactions on Reliability, 36, 129-132.

[4] Dhingra, A.K. (1992) Optimal apportionment of reliability and redundancy in series systems under multiple objectives. IEEE Transactions on Reliability, 41, 576-582.

[5] Rao, S.S. and Dhingra, A.K. (1992) Reliability and redundancy apportionment using crisp and fuzzy multiobjective optimization approaches. Reliability Engineering and System Safety, 37, 253-261.

[6] Ravi, V., Reddy, P.J. and Zimmermann, H-J. (2000) Fuzzy global optimization of complex system reliability. IEEE Transactions on Fuzzy System, 8, 241-248.

[7] Gardiner, L.R. and Steuer, R.E. (1994) Unified interactive multiple objective programming. European Journal of Operational Research, 74, 391-406.

[8] Sun, M.H., Stam, A. and Steuer, R.E. (1996) Solving multiple objective programming problems using feed-forward artificial neural networks: the interactive FFANN procedure. Management Science, 42, 835-849.

[9] Sun, M.H., Stam, A. and Steuer, R.E. (2000) Interactive multiple objective programming using Tchebycheff programs and artificial neural networks. Computers & Operations Research, 27, 601-620.

[10] Korhonen, P., Salo, S. and Steuer, R.E. (1997) Heuristic for estimating nadir criterion values in multiple objective linear programming. Operations Research, 45, 751-757.

[11] Stam, A., Sun, M.H. and Haines, M. (1996) Artificial neural network representations for hierarchical preference structures. Computers & Operations Research, 23, 1191-1201.

[12] Lu, Q.Z., Sheng, Z.H. and Xu, N.R. (1995) A multiobjective decision-making algorithm based on BP-Hopfield neural networks. Journal of Decision Making and Decision Support Systems, 5, 93-103.

[13] Sakawa, M. (1982) Interactive multiobjective optimization by the sequential proxy optimization technique. IEEE Transactions on Reliability, 31, 461-464.

[14] Rojas, R. (1996) Neural networks: a system introduction. Berlin: Springer.

[15] Shin, W.S. and Ravindran, A. (1991) Interactive multiple objective optimization: survey 1-continuous case. Computers & Operations Research, 18, 97-114.

[16] Huang, H.Z., Tian, Z. and Zuo, M.J. (2005) Intelligent interactive multiobjective optimization method and its application to reliability optimization. IIE Transactions, 37(11), 983-993.

Reliability Assessment of Composite Power Systems Using Genetic Algorithms

Nader Samaan

EnerNex Corp., Knoxville, TN, USA

Chanan Singh

Department of Electrical and Computer Engineering,
Texas A&M University, College Station, TX, USA

8.1 Introduction

Reliability is a measure of the ability of a system to perform its designated functions under the conditions within which it was designed to operate. Given this concept, power system reliability is a measure of the ability of a power system to deliver electricity to all points of utilization at acceptable standards and in the amount desired.

Power systems reliability assessment, both deterministic and probabilistic, is divided into two basic aspects: system adequacy and system security [1]. System adequacy examines the availability of sufficient facilities within the system to satisfy the customer load demand without violating system operational constraints. These include the facilities necessary to generate sufficient energy and the associated transmission and distribution facilities required to transport the energy to consumer load points. Adequacy is therefore associated with static conditions which do not include system disturbances. System security is the ability of the system to respond to sudden shocks or disturbances arising within the system such as the loss of major generation and/or transmission facilities and short circuit faults. Under such condition, security studies show system's ability to survive without cascading failures or loss of stability.

N. Saman and C. Singh: *Reliability Assessment of Composite Power System Using Genetic Algorithms*, Computational Intelligence in Reliability Engineering (SCI) **39**, 237–286 (2007)
www.springerlink.com

Power system reliability evaluation is important for studying the current system to identify weak points in the system, determining what enforcement is needed to meet future demand and planning for new reliable power system, i.e., network expansion. Reliability studies are vital to avoid economic and social losses resulting from power outages.

Adequacy analysis of power systems essentially consists of identification and evaluation of failure states, states in which the power system cannot satisfy customer demand and load-shedding action is needed to maintain the system integrity. Since the number of possible states can run into millions, straightforward enumeration and evaluation is not feasible even for moderate sized networks. Monte Carlo simulation is currently the most common method used in sampling states, yet in its basic form it suffers from three major drawbacks. The first one is the excessive simulation time. The second one is the lack of information about outage scenarios that can happen and the contribution of different system components to these outages. The third one is the difficulty to sample failure states when system reliability is very high which the case is in most practical systems.

Adequacy assessment methods in power systems are mainly applied to three different hierarchical levels [1]. At Hierarchical level I (HLI), the total system generation is examined to determine its adequacy to meet the total system load requirements. This is usually termed "generating capacity reliability evaluation". The transmission system and its ability to transfer the generated energy to the consumer load points is taken for granted in HLI. The only concern is estimating the necessary generation capacity to satisfy the demand and to have sufficient capacity to perform corrective and preventive maintenance on the generating facilities.

In HLII studies, the adequacy analysis is usually termed composite system or bulk transmission system evaluation. HLII studies can be used to assess the adequacy of an existing or proposed system including the impact of various reinforcement alternatives at both the generation and transmission levels. In HLII, two sets of indices can be evaluated; the first set includes individual load point indices and the second set includes overall system indices. These indices are complementary, not alternatives. The system indices give an assessment of the overall adequacy and the load-point indices indicate the reliability of individual buses and provide input values to the next hierarchical level.

The HLIII studies include all the three functional zones of the power system, starting at generation points and terminating at the individual consumer load points. To decrease complexity of these studies, the distribution functional zone is usually analyzed as a separate entity using the HLII load-point indices as the input. The objective of the HLIII study is to obtain suitable adequacy indices at the actual consumer load points.

Power system reliability has been an active research area for more than three decades. A recent comprehensive list of publications can be seen from bibliographies on power system reliability evaluation [2], [3]. A survey of the state-of-art models and analysis methods used in power system reliability assessment is given in [4].

Genetic algorithms (GAs) have shown a rapid growth of applications in power systems. An area which has not yet been investigated for their application is power system reliability. Application of GAs to power systems is found in areas such as economic dispatch, power system planning, reactive power allocation and the load flow problem. In all these applications, GA is used primarily as an optimization tool. The authors of this chapter have successfully used GA for the evaluation of generation system reliability [5].

8.2 Reliability Evaluation of Composite Generation-Transmission Systems

Adequacy assessment of composite generation-transmission systems is a more complex task. It is divided into two main parts, state sampling and state evaluation. Each sampled state consists of the states of generating units and transmission lines, some of them are in the up state and others are in the down state. The purpose of state evaluation is to judge if the sampled state represents a failure or success state. After state sampling stops, data from evaluated states is used to calculate adequacy indices of the composite power system. A wide range of techniques has been proposed for composite system reliability evaluation. These techniques can be generally categorized as either analytical or simulation.

Analytical techniques represent the system by analytical models and evaluate the indices from these models using mathematical solutions. The most widely used analytical method is the contingency enumeration approach [6].

Monte Carlo simulation methods estimate the indices by simulating the actual process and random behavior of the system. Monte Carlo simulation methods are divided into random sampling methods and sequential methods [7]. In both techniques, Monte Carlo simulation is used for state sampling. Two types of flow calculation methods are used for state evaluation in composite system reliability studies. One of these is the linearized model, also called DC load flow. The other is the full load flow model or the optimal load flow. Examples of linearized flow equations to calculate the amount of load curtailment if needed can be found in [8], [9]. Monte Carlo techniques can take a considerable computation time for convergence. Convergence can be accelerated by using techniques such as variance reduction to reduce the number of the analyzed system states [10] or using intelligent system methods like self organized maps [11].

There are two main types of composite system adequacy indices. The first set of indices are called annualized adequacy indices in which the system maximum load only is considered, i.e., load value at each load bus is fixed at its maximum yearly value. The second set of indices are called annual adequacy indices in which the yearly chronological load curve at each bus is considered. Each set of indices has its own importance. Annualized indices are used to compare the reliability of two different systems while annual indices are used for detecting weak load points and as a planning criterion.

Both random sampling and sequential Monte Carlo simulation can be used for the assessment of composite system annual adequacy indices. Chronological load is aggregated into a certain number of steps or represented by a certain number of clusters when using Monte Carlo random sampling technique. On the other hand sequential Monte Carlo simulation is able to represent different chronological load curves of load buses on hourly basis, and hence it is the most suitable method for the assessment of annual adequacy indices. However, this technique requires more extensive computational effort than the sampling method.

This chapter presents an innovative state sampling method based on GA for state sampling of composite generation-transmission power systems. GA is used as an intelligent tool to truncate the huge state space by tracing failure states, i.e., states which result in load curtailment. States with failure probability higher than a threshold minimum value will be scanned and saved in a state array. Binary encoded GA is used as a state sampling tool for the composite power system network states. The key to the success of the proposed method is the appropriate choice of a GA fitness function, a scaling method for fitness function and GA operators. Each scanned state will be evaluated through a linearized optimization load flow model to determine if a load curtailment is necessary.

The developed approach has been extended to evaluate adequacy indices of composite power systems while considering chronological load at buses. Hourly load is represented by cluster load vectors using the k-means clustering technique. Two different approaches have been developed which are GA parallel sampling and GA sampling for maximum cluster load vector with series state revaluation.

The developed GA based method is also used for the assessment of annual frequency and duration indices of composite system. The conditional probability based method is used to calculate the contribution of sampled failure states to system failure frequency using different component transition rates. The developed GA approach has been generalized to recognize multi-state components such as generation units with derated states. It also considers common mode failure for transmission lines. The proposed method is superior to the conventional Monte Carlo method in its ability for intelligent search through its fitness function. In addition, it reports the most common failure scenarios and severity of different scanned states.

8.3 Genetic Algorithms Approach for the Assessment of Composite Systems Annualized Indices

A genetic algorithm is a simulation of evolution where the rule of survival of the fittest is applied to a population of individuals. In the basic genetic algorithm [13]-[15] an initial population is randomly created from a certain number of individuals called chromosomes. All of the individuals are evaluated using a certain fitness function. A new population is selected from the old population based on the fitness of the individuals. Some genetic operators, e.g., mutation and crossover are applied to members of the population to create new individuals. Newly selected and created individuals are again evaluated to produce a new generation and so on until the termination criterion has been satisfied.

The proposed method can be divided into two main parts. First GA searches intelligently for failure states through its fitness function using the linear programming module to determine if a load curtailment is needed for each sampled state. Sampled state data are then saved in a state array. After the search process stops, the second step begins by using all of the saved states data to calculate the annualized indices for the whole system and at each load bus. Each power generation unit and transmission line is assumed to have two states, up and down. The probability of any generation unit to be down is equal to its forced outage rate "*FOR*". The failure probability of any transmission line "i" is "PT_i," which is calculated from its failure rate "λ_i" and repair rate "μ_i" as follows:

$$PT_i = \frac{\lambda_i}{\lambda_i + \mu_i} \tag{1}$$

The total number of states "N_{states}" for all possible combinations of generating units and transmission lines installed is:

$$N_{states} = 2^{ng + nt} \qquad (2)$$

where "*ng*" is the total number of generation units and "*nt*" is the total number of transmission lines in the system. GA is used to search for failure states and save such states in the state array. This is achieved by making each chromosome represent a system state. Each chromosome consists of binary number genes. Each gene represents a system component. The first "*ng*" genes in the chromosome represent generation units while the remaining "*nt*" genes represent transmission lines. If any gene takes a zero value this means that the component it represents is in the down state and if it takes a one value that means its component is in the up state. To illustrate the chromosome construction, consider the small RBTS test system [16] shown in Fig. 1. It consists of 2 generator (PV) buses, 4 load (PQ) buses, 9 transmission lines and 11 generating units. Consider the state that all system components are up, the chromosome representing this state is shown in Fig. 2.

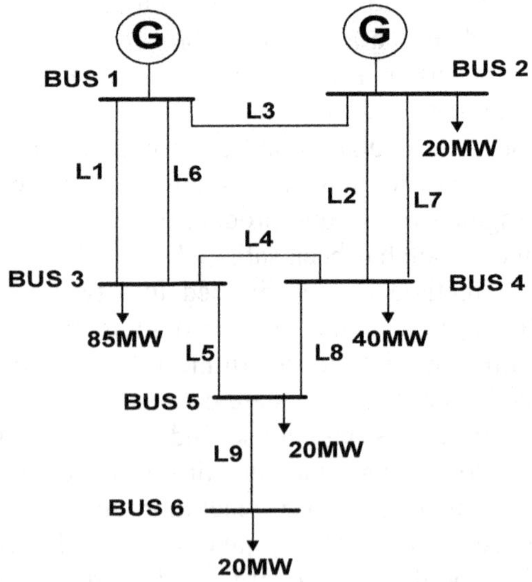

Fig. 1. Single line diagram of the RBTS test system

Each chromosome is evaluated through an evaluation function. The suitable choice for the evaluation function can add the required intelligence to GA state sampling. Many evaluation functions can be used. The simplest one returns zero, if it is a success state and the state probability if it is a failure state. The evaluation function then calls a linear programming optimization load flow model that returns the amount of load curtailment to satisfy power system constraints. If there is no load curtailment, the chromosome represents a success state; otherwise it represents a failure state. The fitness value for each chromosome will be the resultant value after linearized scaling of the evaluation function value. Scaling of the evaluation function enhances the performance of GA since it results in more diversity in the chromosomes of the new generations. After calculating the fitness value of all chromosomes in the current population, GA operators are applied to evolve a new generation. These operators are selection schema, cross over and mutation. There are many types of such operators and the ones used are explained later.

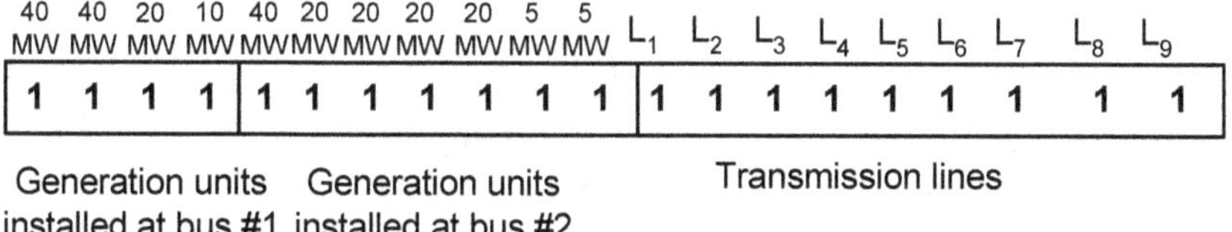

Fig. 2. Chromosome representation for composite system

For each chromosome produced with a state probability higher than a threshold value, the binary number it represents will be converted to its equivalent decimal number. For larger systems, it may be necessary to use more than one decimal number to represent a chromosome or binary representations may be used. A search for this number in the state array is performed and if such a number is found it means this state has been previously sampled and is not added again. There is also no need to call the linear programming module for this state as the load curtailment value for this state has been calculated and saved previously in state array. If the decimal number representing a state is not found in the state array, the linear programming module is then called to determine the load curtailment amount for the whole system and for each load bus, if necessary. All calculated data are saved in the state array. New generations are produced until reaching a stopping criterion. The main role of GA is to truncate state space searching for states that contribute most to system failure. The next phase is to calculate the full set of annualized adequacy indices for the whole system and for each load bus. This is achieved via the use of data stored in the state array.

8.3.1 Construction of System State Array

GA searches for failure states and saves sampled states with all their related data in the state array. This process can be summarized in the following steps:

1. Each chromosome represents a system state. The first "*ng*" binary genes represent generation units in the system. The last "*nt*" binary genes represent transmission lines.
2. Initial population is generated randomly. For each bit in the chromosome, a random binary number (0 or 1) is chosen, i.e., "*ng*+*nt*" random binary numbers for each chromosome. This process is repeated for all population chromosomes.
3. The state probability "SP_j" for each chromosome "*j*" is calculated.

$$SP_j = \prod_{i=1}^{ng} G_i \cdot \prod_{i=1}^{nt} T_i \tag{3}$$

   where $G_i = 1\text{-}FOR_i$ if its gene = 1 (up state) or $G_i = FOR_i$ if its gene = 0 (down state), and $T_i = 1\text{-}PT_i$ if its gene = 1 or $T_i = PT_i$ if its gene = 0.
4. A threshold probability value is set depending on the required accuracy. If the state probability calculated in step 3 is less than the threshold value this state is ignored and linear programming module is not called.
5. If the state probability is higher than the threshold value the binary number representing this state is converted into the equivalent decimal number. A search is carried out in the state array to find if this decimal number has been saved previously. If the equivalent decimal number is found, this means that this state has been scanned and evaluated previously. Hence, its evaluation function value is retrieved and the algorithm proceeds to step 9 otherwise, it goes to next step.
6. The linear programming optimization module for calculating load curtailment is called to evaluate the new state. The amount of load curtailment, if necessary to satisfy system constraints, for the whole system and for each load bus is obtained and saved in the state array. The state equivalent decimal number is also saved in the state array to prevent any state from being added to the state array more than once.
7. State contribution to system failure frequency is calculated using the conditional probability approach [17], [18] and the resultant value is also saved in the state array.

$$FS_j = SP_j \sum_{i=1}^{ng+nt} [(1-b_i)\mu_i - b_i\lambda_i] \tag{4}$$

where FS_j is state "*j*" contribution to system failure frequency, and b_i is the binary value of gene number "*i*" representing a generator unit or transmission line.

8. Expected Power not supplied "*EPNS*" for the new state is calculated and the result is saved in the state array.

$$EPNS_j = LC_j \cdot SP_j \tag{5}$$

where LC_j is the amount of load curtailment for the whole system calculated in step 6.

9. The chromosome is evaluated. Many evaluation functions can be used. Two of them are explained here. The first considers the state failure probability.

$$eval_j = \begin{cases} SP_j & \text{if new or old chromosome j represents a failure state.} \\ SP_j.\alpha & \text{if new or old chromosome j represents a success state.} \\ SP_j & \text{if chromosome probabilit y is less than the fixed threshold} \\ \text{value.} & \end{cases} \tag{6}$$

where "new chromosome" means it has not been previously saved in the state array, "old chromosome" means it has been found in the state array and α is a very small number, e.g., 10^{-30} to decrease the probability of success states to appear in next generations.

The second evaluation function considers the severity of the failure state which is represented by *EPNS*.

$$eval_j = \begin{cases} EPNS_j + \beta & \text{if new or old chromosome j represents a failure state} \\ \beta & \text{for all other chromosomes} \end{cases} \tag{7}$$

where β is a very small number, e.g., 10^{-20} to prevent obtaining a zero value for the evaluation function.

The first evaluation function guides GA to search for states with higher failure probabilities. The second evaluation function guides GA to search for more severe states that have high value of failure probability multiplied by the associated load curtailment.

10. The fitness of any chromosome "*j*" is calculated by linearly scaling its evaluation function value.

$$fitness_j = A \cdot eval_j + C \tag{8}$$

where A and C are fixed constant numbers. Scaling has the advantage of maintaining a reasonable difference between fitness values of different chromosomes. It also enhances the effectiveness of the search by preventing an earlier super-chromosome from dominating other chromosomes which decreases the probability of obtaining new more powerful chromosomes [15].

11. Repeat previous steps to calculate fitness value for all chromosomes in current population.
12. Apply GA operators to evolve a new population. These operators are selection, crossover and mutation. The suitable choice for the appropriate types of operators enhances the search performance of GA.
13. The evolution process is continued from one generation to another until a prespecified stopping criterion is reached.
14. Data saved in the state array are then used to calculate the full set of adequacy indices for the whole system and at each load bus.

Some of the previous steps are explained in more detail in the next sections.

8.3.2 Evolution of a New Generation

In the evolution of a new population from the old one in the simple GA, old population passes through three operations.

The first one is the selection from parents. There are many types of selection operators like roulette wheel selection, ranked selection and tournament selection. The three types have been tested and tournament selection is recommended as it improves the search process more than the other types. Tournament selection can be explained briefly as follows [15]: A set of chromosomes is randomly chosen. The chromosome that has the best fitness value, the highest in the proposed algorithm, is chosen for reproduction. Binary tournament is used in which the chosen set consists of two chromosomes. The probability of choosing any chromosome in the selected set is proportional to its fitness value relative to the whole population fitness value. Consider population size of GA is equal to *pop_size* chromosomes. Binary tournament selection is repeated *pop_size* times, i.e., until obtaining a new population.

The second step is to apply the crossover operator on the selected chromosomes. Single point cross over is used with cross over probability of P_c. For each pair of chromosomes in the new population a random number r from [0,1] is generated. If $r < P_c$ given chromosome pair is selected for crossover. At the end j pairs of chromosomes are eligible to apply crossover to them. Assume the pair X,Y is subjected to crossover. Generate a random number "*pos*" in the range $[1,ng+nt\text{-}1]$, the new two chromosomes genes are:

$$x_i` = x_i \text{ if } i < pos \text{ and } y_i \text{ otherwise (for } i=1 \text{ to } ng+nt)$$
$$y_i` = y_i \text{ if } i < pos \text{ and } x_i \text{ otherwise (for } i=1 \text{ to } ng+nt)$$

The third step is to apply the mutation operator. Uniform mutation with probability of P_m is used. For each gene in each chromosome in the newly created population after applying the previous two operators, generate a random number r from [0,1]. If $r < P_m$ convert that gene from one to zero or zero to one. Now a new population has been generated and the process is repeated until a stopping criterion is reached.

Now a new population is generated, and the process is repeated until a stopping criterion is reached. The main idea of the proposed method is that at each GA generation, more states are scanned, especially those with higher failure probabilities, i.e., have higher fitness values. Each of them is saved in the state array. If dealing with an ordinary optimization problem, the purpose is to obtain the maximum value of the fitness function and the decoded decimal value for its chromosome. But here, GA is used to scan or, in other words, to sample system states which have higher fitness values. Illustration of the GA search process is shown in Fig. 3.

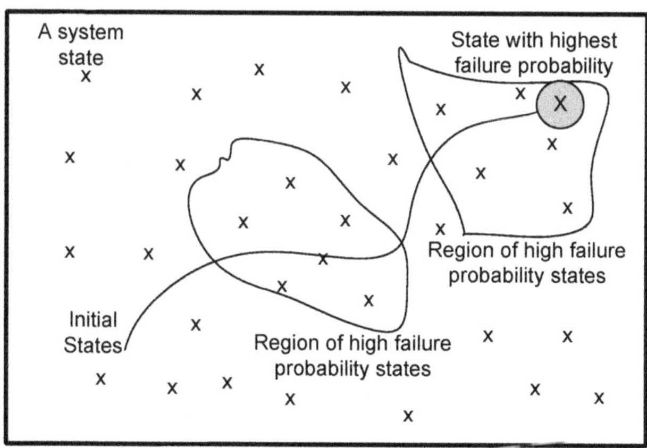

Fig. 3. The GA search in the state space

8.3.3 Stopping Criterion

Any of the following three criteria can be used to stop the algorithm:

1. The first stopping criterion is to stop the algorithm after reaching a certain number of generations. If a small number of generations has been used this will lead to inaccurate results as not enough states would have been sampled.

2. The second one is to stop when the number of new states that has been added to state array is less than a specified value within certain number of GA generations.
3. The third stopping criterion is updating the value of system Loss of Load Probability "*LOLP*" for each new failure state added to the state array. The algorithm will stop when the change of *LOLP* is below a specified value within certain number of GA generations.

A flowchart for GA sampling procedures is shown in Fig. 4.

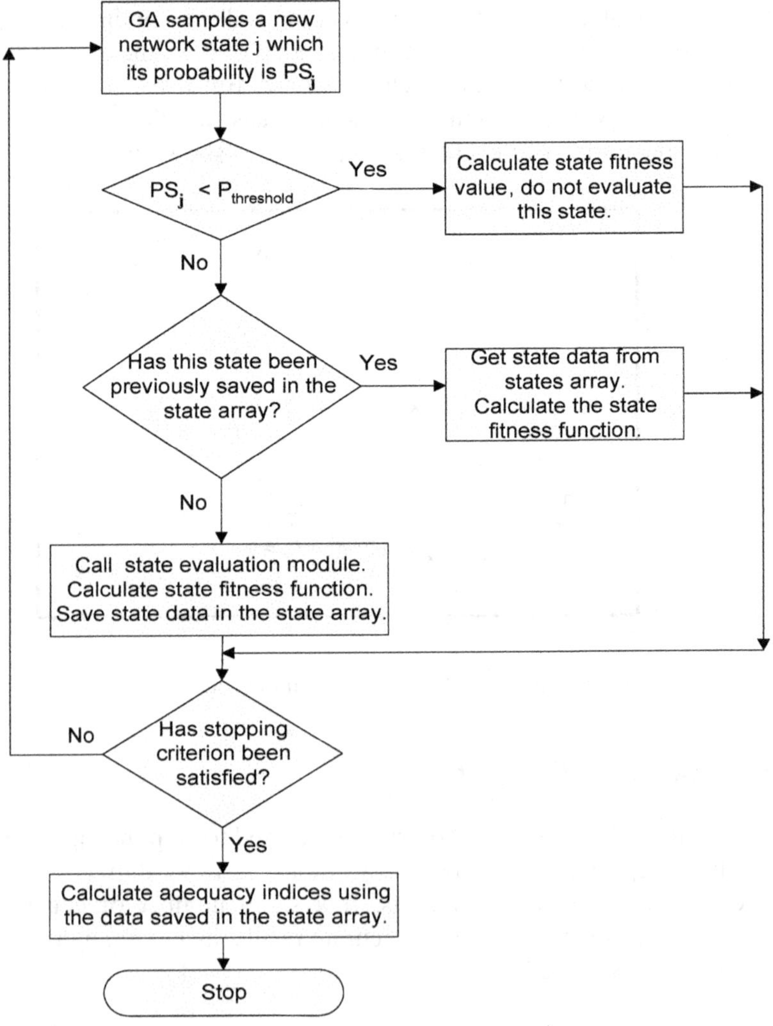

Fig. 4. GA state sampling procedures for single load level

The main role of GA is to truncate the state space by tracing states that contribute most to system failure. After the search process stops, data saved in the state array is used to calculate the full set of annualized adequacy indices for the whole system and for each load bus.

8.3.4 State Evaluation Model

State evaluation is a very important stage in composite power system reliability assessment. Through this stage the current system state is classified as a failure or success state. If it is a failure state the amount of load curtailment for the whole system and the share of each load bus in this amount is determined. These values are needed to calculate this state contribution in adequacy indices for the whole system and for load buses. Each state is evaluated using a linear programming optimization model based on dc load flow equations [8], [9]. For the current state to be evaluated, the elements of the power system susceptance matrix B are modified according to transmission line outages. The amount of available real power generation at each PV bus is also updated according to the status of generation units installed at such a bus. The objective of this optimization problem is to minimize the total load curtailment for the whole system which is equivalent to maximizing the load value at each load bus. This objective is subject to the following constraints:

1. Real power balance at each system bus.
2. Real power flow limits on each transmission line.
3. Maximum amount of load curtailed at each load bus.
4. Maximum and minimum available real power at each PV bus.

For the same optimal solution it is possible to have many scenarios of load curtailment at each bus. A load curtailment philosophy should be used; otherwise adequacy indices of load buses may be meaningless. In this work, importance of load is taken into consideration as a load curtailment philosophy as given in [9]. Each load is divided into three parts, i.e., three variables in the objective function. Weights are given for each part in the objective function according to the relative importance for each bus in comparison with the remaining buses. Weights are also adjusted so that the first part of each load is the least important and the third part is the most important. In this manner load is curtailed from the first part at each load bus in the order of importance, then from second and third parts sequentially, if possible without violating any constraint. The linear programming maximization problem is formulated as follows:

$$\max \sum_{i=1}^{nl} \sum_{p=1}^{3} W_{ip} X_{ip} \tag{9}$$

Subject to:

$$PG_i - \sum_{p=1}^{3} X_{ip} = \sum_{j=2}^{n} B_{ij} \theta_j \qquad \forall\ i=1,2,\ldots\ldots,n \tag{10}$$

$$-B_{ij}(\theta_i - \theta_j) \le PT_k \qquad \forall\ k=1,2,\ldots\ldots,nt \tag{11}$$

$$-B_{ij}(\theta_j - \theta_i) \le PT_k \qquad \forall\ \text{k}=1,2,\ldots\ldots,nt \tag{12}$$

$$0 \le X_{ip} \le C_{ip} PD_i \qquad \forall\ p=1,2,3 \quad \forall\ i=1,2,\ldots\ldots,nl \tag{13}$$

$$PG_{i\,\min} \le PG_i \le PG_{i\,\max} \qquad \forall\ i=1,2,\ldots\ldots,nv \tag{14}$$

where:

- n is the total number of system buses,
- nt is the total number of the system transmission lines,
- nl is the total number of buses that have installed load,
- nv is the total number of buses that have installed generation,
- B_{ij} is the element at the ith row and j[th] column in the system susceptance matrix,
- θ_i is the voltage angle at bus i (bus 1 is assumed the reference bus with $\theta_1 = 0$),
- PD_i is the yearly maximum load demand at bus i,
- X_{ip} is the value of part p of load installed at bus i,
- W_{ip} is the relative weight of part p of load installed at bus i, these weights are chosen so that $W_{1i} \le W_{2i} \le W_{3i}$,
- C_{ip} is the percentage of part p of load installed at bus i to total load demand at the same bus,
- PG_i is the real power generation at bus i,
- $PG_{i\,\max}$ is the maximum available generation at bus i, and
- $PG_{i\,\min}$ is the minimum available generation at bus i.
- The variables vector that is calculated by the linear programming solver is $\{X_{ip}, PG_j, \theta_k\}$
- $\forall p = 1,2,3$, $\forall i = 1,2,\ldots\ nl, \forall j = 1,2,\ldots.nv$ and $\forall k = 2,3,\ldots,$n

The optimization problem is solved using the dual simplex method. The total amount of system load curtailment "LC_s" is:

$$LC_s = \sum_{i=1}^{nl} PD_i - \sum_{i=1}^{nl} \sum_{p=1}^{3} X_{ip} \tag{15}$$

The load curtailment at load bus i "LC_i" is

$$LC_i = PD_i - \sum_{p=1}^{3} X_{ip} \tag{16}$$

8.3.5 Assessment of Composite System Adequacy Indices

Annualized adequacy indices for the whole system and for each load bus are calculated using the data saved in the state array. These indices are, Loss of Load Probability (*LOLP*), Loss of Load Expectation (*LOLE*), Expected Power Not Supplied (*EPNS*), Expected Energy Not Supplied (*EENS*), Loss of Load Frequency (*LOLF*) and Loss of Load Duration (*LOLD*). These indices are calculated considering only saved failure states and ignoring success ones. Let the total number of saved failure states to be "*nf*", then the adequacy indices for the whole system are calculated as follows:

$$LOLP = \sum_{j=1}^{nf} SP_j \tag{17}$$

$$LOLF = \sum_{j=1}^{nf} FS_j \tag{18}$$

$$EPNS = \sum_{j=1}^{nf} EPNS_j \tag{19}$$

$$LOLE = LOLP \cdot T \tag{20}$$

$$LOLD = \mathrm{LOLE} / \mathrm{LOLF} \tag{21}$$

$$EENS = EPNS \cdot T \tag{22}$$

where T is the number of hours in the period of study.

The same set of indices can be calculated for each load bus considering only failure states resulting in load curtailment at this bus and ignoring all other states.

8.3.6 Case Studies for the Assessment of Annualized Indices

The proposed algorithm has been implemented through C++ programming language. A C++ library of GA objects called GAlib developed by [19] has been integrated into the implementation. The proposed method has been tested on the RBTS [16] test system. Total number of hours in one year is considered to be 8736 instead of 8760 as only these numbers of hours are given in the RBTS load curve. The input parameters of the GA are taken as follows: *pop_size* = 40, $P_c = 0.7$, and $P_m = 0.05$. The stopping criterion used is 1000 GA generations. Linear scaling, tournament selection, one point crossover, uniform mutation, and the first evaluation function given in Eq. (1.6) are used. Calculated system annualized indices with threshold probability value of 1e-8 compared with results reported in [20] using different Monte Carlo methods techniques are given in Table 1.

It can be seen from the comparison of results that the proposed method gives similar results to those obtained using different Monte Carlo techniques. The best match is with sequential Monte Carlo Method. The slight differences between the results are due to the fact that all these methods are approximation methods. The accuracy of Monte Carlo methods depends on how low the variance has been reached. The accuracy of the proposed method will depend on the fixed threshold failure probability value and the total number of sampled and saved failure states. The total number of states that GA has sampled and has saved in the state array is 2198 states from which 1449 states result in load curtailment, i.e., 66% of saved states are failure states. It can be seen that GA truncated the huge state space (larger than a million) of the 20 component system into small fraction of it.

The failure state with the highest probability is represented by the chromosome shown in Fig. 5, in which only line 9 is down and all other components are up. This failure state probability is equal to 0.000906. If the severity of a certain contingency is considered by *EPNS*, the second evaluation function given in Eq. (7) can be used to construct state array and find the most severe state. The most severe state is represented by the chromosome given in Fig. 6, in which two generation units of 40 MW capacity installed at bus number one are in the down state and the remaining components are in the up state. The total load curtailment for this state is 25 MW. The state failure probability is 0.00075914. Hence, the *EPNS* for this state is 25*0.00075914 = 0.0189785.

Table 1. Annualized adequacy indices comparison between GA sampling and different Monte Carlo sampling techniques

Adequacy Indices	GA sampling	Sequential Sampling [20]	State Transition Sampling [20]	State Sampling [20]
LOLP	0.009753	0.00989	0.00985	0.01014
EENS(MWh/Yr)	1047.78	1081.01	1091.46	1082.63
LOLF (occ./Yr)	4.15097	4.13	4.14	5.21
LOLE (hr/Yr)	85.198	86.399	86.0496	88.58
PNS (MW/Yr)	0.119938	0.12374	0.12494	0.12393
LOLD (hr)	20.5249	20.9198	20.7849	17.0019

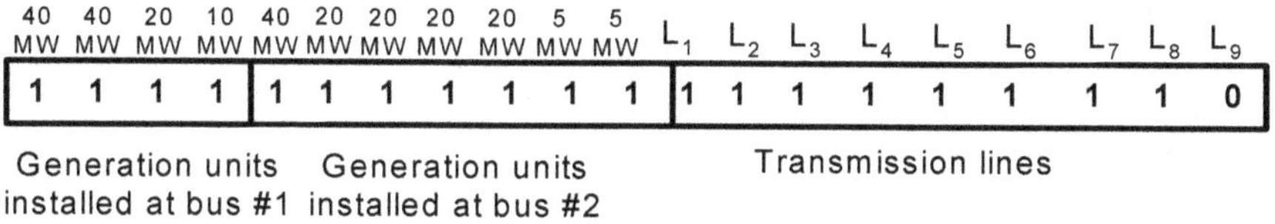

40 MW	40 MW	20 MW	10 MW	40 MW	20 MW	20 MW	20 MW	20 MW	5 MW	5 MW	L_1	L_2	L_3	L_4	L_5	L_6	L_7	L_8	L_9
1	1	1	1	1	1	1	1	1	1	1	1	1	1	1	1	1	1	1	0

Generation units installed at bus #1 — Generation units installed at bus #2 — Transmission lines

Fig. 5. Chromosome with the highest failure probability

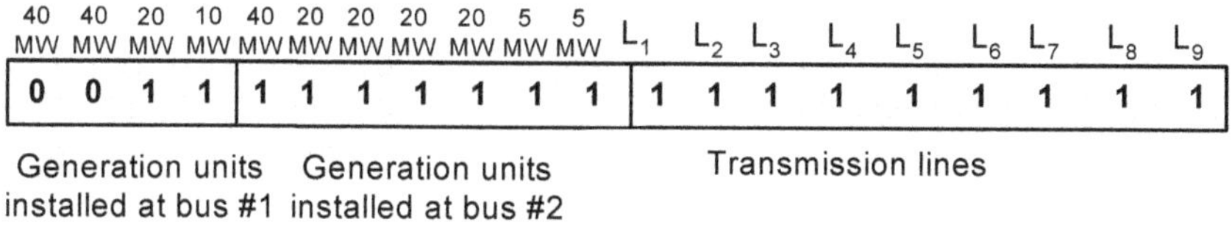

40 MW	40 MW	20 MW	10 MW	40 MW	20 MW	20 MW	20 MW	20 MW	5 MW	5 MW	L_1	L_2	L_3	L_4	L_5	L_6	L_7	L_8	L_9
0	0	1	1	1	1	1	1	1	1	1	1	1	1	1	1	1	1	1	1

Generation units installed at bus #1 — Generation units installed at bus #2 — Transmission lines

Fig. 6. Chromosome representing the most severe state

Annualized bus indices obtained using the load curtailment philosophy explained previously are given in Table 2. These indices have been obtained by dividing each load into three parts. The first and second parts range between 0 and 20% of the maximum load at the corresponding bus. Meanwhile, third part ranges from 0 to 60% of the same value. Hence curtailed load, if necessary, should be first obtained from the first part of all loads, then the second part and finally the third part. Weighting factors are used to represent importance of each part. Load at bus 2 is considered to be the most important and load at bus 6 is considered to be the least important. In this manner the weighting factor for the first part of load at bus 2 is 5 and weighting factor for the first part of load at bus 6 is 1. The biggest weighting factor is 15 which is associated with the third part of load at bus 2.

It is possible to obtain totally different bus indices if bus importance order is changed, e.g., bus 6 is the most important and bus 2 is the least important. Results in such a case are given in Table 3. Bus indices can also be varied if the maximum limit of each load part has changed, e.g., if the ranges are 0.1, 0.4 and 0.5 instead of 0.2, 0.2 and 0.6.

Table 2. Annualized adequacy indices for load buses, loads importance from the most important to the least one are 2,3,4,5,6

Adequacy Indices	*LOLP*	*EENS* (MWh/Yr)	*LOLF* (occ./year)	*LOLD* (hr)
Bus#2	0.000229	7.373	0.1204	16.616
Bus#3	0.002382	202.133	0.9845	21.137
Bus#4	0.002624	177.847	1.1145	20.568
Bus#5	0.008614	153.547	3.1537	23.861
Bus#6	0.009753	506.707	4.1509	20.526

Table 3. Annualized adequacy indices for load buses, loads importance from the most important to the least one are 6,5,4,3,2

Adequacy Indices	*LOLP*	*EENS* (MWh/Yr)	*LOLF* (occ./year)	*LOLD* (hr)
Bus#2	0.008605	306.324	3.1372	23.963
Bus#3	0.008614	437.757	3.1549	23.854
Bus #4	0.002283	90.989	0.8626	23.1245
Bus#5	0.000275	8.776	0.1505	15.999
Bus#6	0.001371	206.580	1.1208	10.684

8.4 Reliability Indices Considering Chronological Load Curves

This section presents a new technique in which the preceding approach has been extended to consider the chronological load curve at each load bus. There are many methods in the literature for representing the chronological load curve. The clustering method using k-means technique is the most developed one and is used with the proposed methods [21]-[24]. Two different approaches based on GA are presented to calculate annual adequacy indices. In the first approach, GA samples failure states for each cluster load vector separately and consequently adequacy indices for this load level are calculated. Composite system annual indices are then obtained by adding adequacy indices for each load level weighted by the probability of occurrence of its cluster load vector. In the second approach, GA samples only failure states with load buses assigned the values of maximum cluster load vector. Failure states are then reevaluated with lower cluster load vectors until a success state is obtained or all load levels have been evaluated.

Chronological loads at different load buses usually have a certain degree of correlation. Degree of correlation depends on the type of installed loads, i.e., residential, commercial, or industrial loads. It also depends on the regional time difference between load buses due to their geographical location. The two developed approaches have been applied to the RBTS test system [16]. A comparison between results of the two different approaches is given. Both fully and partially correlated chronological load curves have been considered.

8.4.1 Modeling of Chronological Load Curve

System annual load is usually represented by system load at each hour in a year. Many techniques have been used to represent system load in composite system reliability. The most common one is to approximate load curve into certain number of steps of load levels. Each load step has its probability of occurrence. A more efficient model is based on clustering techniques [21]. This model has shown good results when used for both generation system reliability [22] and multi-area reliability [23]. In this chapter, clustering has been used to represent the system load curve. Load at each bus has certain degree of correlation with load at other buses. When in a group of load buses, each bus always has an hourly load with the same percentage of group maximum load at this hour; these loads are called fully correlated. Usually in real life there is certain level of correlation between each group of fully correlated load buses. Consider that load buses are divided into n groups, each group containing a set of fully correlated buses. The vector of loads at certain hour i is:

$$\underline{L}^i = (L_1^i, L_2^i, L_3^i, \ldots\ldots\ldots, L_r^i, \ldots\ldots\ldots\ldots, L_n^i) \tag{23}$$

where L_r^i is the maximum load of group r at hour i and n is the number of load groups. The 8760 load vectors are represented by m cluster vectors. Each cluster vector j is represented as:

$$\underline{C}^j = (C_1^j, C_2^j, C_3^j, \ldots\ldots\ldots\ldots\ldots, C_r^j, \ldots\ldots\ldots\ldots\ldots, C_n^j) \tag{24}$$

where C_r^j is the cluster mean load value of group r in cluster j. Steps for applying the k-means clustering technique to obtain m clusters with their associated probability are as follows:

1. Choose initial values of cluster means. The following initial values are suggested to be used: initial cluster mean for group r for first cluster vector as $C_r^1 = 0.98L_r^{\max}$ and for the second cluster vector $C_r^2 = 0.96L_r^{\max}$. This process is repeated for all cluster vectors so that the last cluster vector m has cluster mean:

$$C_r^m = (1-0.02m).L_r^{\max} \qquad \forall\, r = 1,2,\ldots\ldots,n \tag{25}$$

   where $L_r^{\max}$ is the annual maximum load of group r. The 0.02 step is allowing maximum number of 50 clusters.

2. For each hour i calculate the Euclidean distance $DIST_{i-j}$ between its load vector and cluster j load mean values vector

$$DIST_{i-j} = \sqrt{\sum_{r=1}^{n}(C_r^j - L_r^i)^2} \tag{26}$$

   Repeat this process with all other cluster vectors. Load vector at hour i belongs to the cluster with the least Euclidean distance from it.

3. In this manner load vector at each hour belongs to a certain cluster after repeating step 2 for each of them.
4. For each cluster vector j calculate the new mean for each group r.

$$C_r^j new = \sum_{i=1}^{8760} b * L_r^i / T_j. \tag{27}$$

   where

$$- b = \begin{cases} 1 & \text{if } \underline{L}^i \in \underline{C}^j \\ 0 & \text{otherwise} \end{cases} \tag{28}$$

   -T_j is the total number of load vectors belonging to cluster j.

5. For each cluster vector calculate the Euclidean distance between old and new means.

$$change_j = \sqrt{\sum_{r=1}^{n} (C_r^j new - C_r^j)^2} \tag{29}$$

6. Repeat steps from 2 to 5 until "*change$_j$*" is less than a prespecified limit for all clusters.
7. Calculate the probability of occurrence of each cluster vector:

$$P(\underline{C}^j) = T_j / 8760 \tag{30}$$

8.4.2 Genetic Algorithms Sampling with *m* Cluster Load Vectors

When considering the annual load curve, the total number of system states increases dramatically. Considering that the annual load curve is represented by m cluster load vectors the total number of system states is:

$$N_{states} = m.2^{ng+nt} \tag{31}$$

Two different approaches have been developed to deal with the multiple load vector levels. GA parallel sampling and GA sampling for maximum cluster load vector with series state reevaluation. These two techniques are explained in the next sections.

8.4.2.1 Genetic Algorithms Parallel Sampling

In this approach GA samples system failure states with load at each bus fixed and equal to one of the cluster load vectors. Adequacy indices are then calculated for this fixed load level. This process is repeated for all cluster load vectors. The system annual adequacy indices are calculated as follows:

$$LOLP = \sum_{i=1}^{m} LOLP_i P(\underline{C}^i) \tag{32}$$

$$EENS = \sum_{i=1}^{m} EENS_i P(\underline{C}^i) \tag{33}$$

where $LOLP_i$ and $EENS_i$ are loss of load probability and expected energy not supplied calculated with cluster load vector i. This approach has the advantage of giving more accurate results but has the disadvantage of the high computational effort required as the search process is repeated *m* times. Parallel computation can be used with this approach. Failure states for each load level are sampled separately on different machines and in the final step different load level indices are added together to obtain the annual adequacy indices. An illustration for this method is shown in Fig. 7.

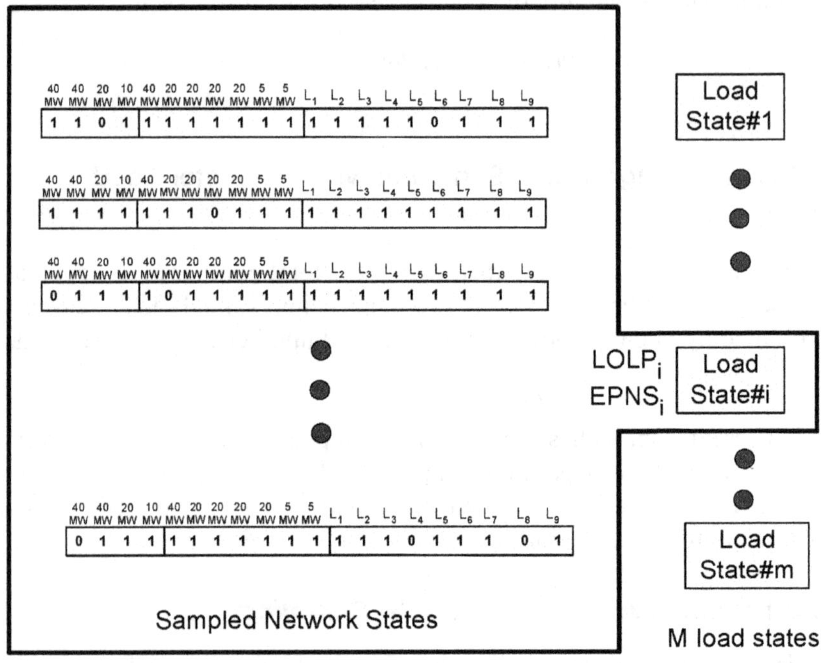

Fig. 7. GA parallel sampling for each load state

8.4.2.2 Genetic Algorithm Sampling for Maximum Cluster Load Vector with Series State Revaluation

In this approach GA searches for states which result in system failure while load buses are assigned the maximum cluster load vector. These failure states are then reevaluated while assigning load buses the values of other cluster load vectors in a descending order from the highest to the lowest cluster load vector. This series state revaluation process stops when there is no load curtailment at a certain cluster load vector, or it has been reevaluated with all cluster load vectors. Adequacy indices are updated with each state evaluation process. An illustration for this method is shown in Fig. 8.

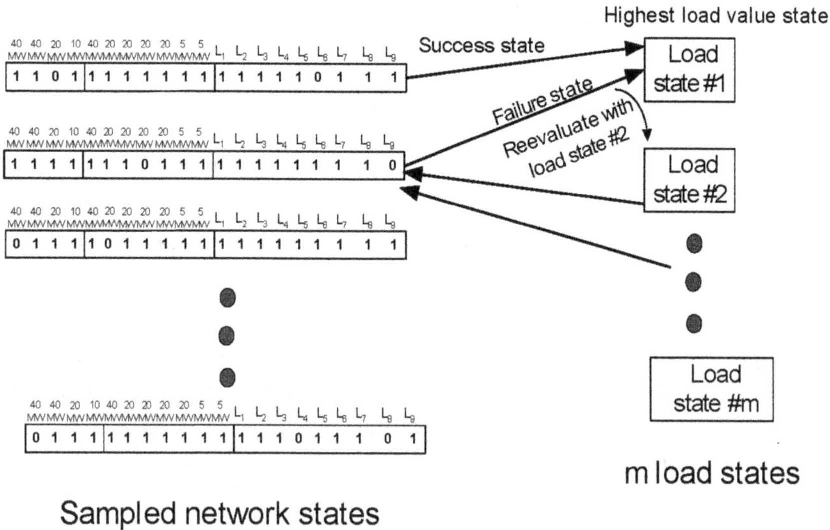

Fig. 8. GA sampling for maximum cluster load vector with series state reevaluation

The main steps for this approach are:

1. Order cluster load vectors in a descending order according to the value of their total load. Consider cluster 1 has the highest rank and cluster m has the lowest rank. It is assumed that:

$$C_r^j \le C_r^{j-1} \quad \forall\, \mathrm{r} = 1,2,.....,n,\ \forall\, \mathrm{j} = 2,.......,m \tag{34}$$

2. Assign bus loads the maximum cluster load vector $\underline{C}^1$.
3. GA samples a new network state k (states of generators and transmission lines), this state is evaluated using the assigned load values in step 2.
4. If the evaluated state represents a success state, i.e., there is no load curtailment, ignore all the remaining cluster load vectors as it is guaranteed there is no load curtailment with lower load values and jump to step 7. Otherwise, proceed to step 5.
5. If the evaluated state represents a failure state, i.e., there is load curtailment, update the annual adequacy indices.

$$LOLP_{new} = LOLP_{old} + SP_k P(\underline{C}^1) \tag{35}$$

$$EPNS_{new} = EPNS_{old} + SP_k P(\underline{C}^1) LC_k^1 \tag{36}$$

where SP_k is the probability of network state k, LC_k^1 is the amount of load curtailment for the whole system with network state k and system loads assigned the values of cluster load vector 1, and *EPNS* is the expected power not supplied.

6. Assign bus loads the lower cluster load vector $\underline{C}^2$. Hence, a new system state has been created that is formed from network state k and the new cluster load vector. This new system state is evaluated. If it represents a success state the remaining cluster load vectors are ignored and hence jump to step 7. Otherwise, it is a failure state, adequacy indices are updated using Eqs. (1.35) and (1.36) substituting cluster 1 date with cluster 2 data. A new system state is formed from state k and the next cluster load vector 3. This process for network state k is repeated until encountering a system success state or network state k has been evaluated considering all the m cluster load levels.
7. If stopping criterion for GA sampling has been satisfied proceed to step 8. Otherwise, return to step 3 for GA to sample a new network state.
8. Composite system annual adequacy indices are calculated:

$$LOLP = LOLP_{new} \tag{37}$$

$$EENS = EPNS_{new} T \tag{38}$$

8.4.3 State Evaluation Model

State evaluation depends on the power flow model used for this purpose. Linearized state evaluation model is based on dc load flow equations. In each sampled state one or more generators and/or transmission lines are in the down state. For the current state to be evaluated, elements of the power system susceptance matrix *B* are modified according to transmission line outages. The amount of available real power generation at each PV bus is also updated according to the status of generating units installed at such a bus. Load values equal the corresponding cluster load vector. State evaluation is represented as an optimization problem with the objective of minimizing the total load curtailment for the whole system, which is equivalent to maximizing the load value at each load bus. The linearized optimization model is formulated as follows (the subscript that refers to the number of the current network state is omitted from all equations):

$$\max \sum_{i=1}^{nl} X_i^z \tag{39}$$

Subject to:

$$PG_i - X_i^z = \sum_{j=2}^{n} B_{ij}.\theta_j \qquad \forall\ i=1,2,\ldots\ldots\ldots n \tag{40}$$

$$y_k.(\theta_i - \theta_j) \le PT_k \qquad \forall\ k=1,2,\ldots\ldots\ldots nt \tag{41}$$

$$y_k.(\theta_j - \theta_i) \le PT_k \qquad \forall\ k=1,2,\ldots\ldots\ldots nt \tag{42}$$

$$0 \le X_i^k \le PD_i^k \qquad \forall\ i=1,2,\ldots\ldots\ldots nl \tag{43}$$

$$PG_{i\,\min} \le PG_i \le PG_{i\,\max} \qquad \forall\ i=1,2,\ldots\ldots\ldots nv \tag{44}$$

where:

- n is the total number of system buses,
- nt is the total number of the transmission lines,
- nl is the total number of load buses,
- nv is the total number of buses that has installed generation,
- B_{ij} is the element at the ith row and jth column in the system susceptance matrix,
- θ_i is the voltage angle at bus i (bus 1 is assumed the reference bus with $\theta_1 = 0$),
- PD_i^z is the load demand at bus i corresponding to cluster z load vector,
- X_i^z is the amount of load that could be supplied at bus i while demand at load buses assigned cluster z load vector,
- PT_k, y_k are the maximum flow capacity and susceptance of transmission line k connecting between bus i and bus j,
- PG_i is the real power generation at bus i,
- $PG_{i\ \max}$ is the maximum available generation at bus i and
- $PG_{i\ \min}$ is the minimum available generation at bus i.

This model can be solved using linear programming methods like the dual simplex or interior point method. The variables vector to be calculated by the linear programming solver is $\{X_i^z, PG_j, \theta_r\}$ $\forall i$=1,2….,*nl*, $\forall j$=1,2,…..*nv* and $\forall r$= 2,3,……*n*
The total amount of system load curtailment "LC_s" is:

$$LC^z = \sum_{i=1}^{nl} PD_i^z - \sum_{i=1}^{nl} X_i^z \tag{45}$$

The load curtailment at load bus *i* "LC_i" is:

$$LC_i = PD_i^z - X_i^z \tag{46}$$

8.4.4 Case Studies for the Assessment of Annual Indices

The proposed algorithm has been implemented through C++ programming language. A C++ library of GA objects called GAlib developed by [19] has been integrated into the implementation. The proposed method has been tested on the RBTS test system [16]. Studies have been made considering partially and fully correlated load buses.

8.4.4.1 Fully Correlated Load Buses

Yearly load curve data in per unit of RBTS system maximum load (185 MW) are given in [25]. The full correlation assumption means that all the system load buses construct one load group, i.e., percentage of any load value at any load bus to system maximum load is fixed throughout the year. Hence, each cluster load vector consists of one element corresponding to system maximum load. Results of clustering the chronological load curve into 8 and 15 points are given in Table 4. Comparison of results when using different number of clusters while using GA sampling for maximum cluster load vector with series state revaluation are given in Table 5.

It can be seen from Table 5 that results obtained with 8 points are approximately equal those obtained using 30 points. Total number of evaluated system states using 8 points is about 31% of those for 30 points. These results indicate that clustering is an efficient way of representing the chronological load curve.

Table 4. Results of clustering the system chronological load curve considering all load buses belong to the same load group

No. of Clusters: 8 points		No. of Clusters: 15 points	
Cluster mean value MW	Cluster probability	Cluster mean value MW	Cluster probability
164.147	0.048191	174.294	0.007669
150.438	0.109661	164.213	0.025298
137.868	0.112523	156.763	0.041209
125.056	0.140682	150.531	0.053800
113.546	0.153159	144.263	0.059867
100.443	0.141369	137.356	0.061126
88.852	0.171932	129.915	0.074176
75.792	0.122482	122.864	0.086195
		116.193	0.089286
		109.506	0.077953
		101.880	0.081273
		94.599	0.100618
		87.824	0.099359
		79.961	0.089400
		71.461	0.052770

Table 5. Comparison of annual adequacy indices and other factors with different number of clusters

No. of Clusters	8 points	15 points	30 points
LOLP	0.00127786	0.00125581	0.00125708
EENS (MWH/Yr)	132.0736	132.6160	132.6658
no. of failure states	2691	5670	13204
no. of sampled network states by GA	2206	2175	2195

Comparison of results when using the two different GA sampling approaches, explained previously, is given in Table 6. In the first approach, GA samples each of the 8 cluster load values separately. In the second approach, GA samples failure states for the maximum load value of 164.147 MW only with failure states reevaluated for other load points in descending order until encountering a success state or considering all load levels. It can be seen from Table 6 that when GA is used for calculating adequacy indices for each load separately and then combined, the results are more accurate but the computational burden is increased. This method is equivalent to Monte Carlo simulation with random sampling in which states for each load level are sampled and evaluated separately. When parallel operation is available it is possible to calculate adequacy indices for each load level on a separate machine. When GA samples failure state for maximum load value and reevaluate failure states only with other load levels in descending order the total number of evaluated states is reduced significantly, about 27% of those obtained when evaluating each load level separately.

Table 6. Annual adequacy indices comparison using two different GA sampling Approaches with Fully Correlated Load Buses

Sampling approach		GA samples each load level separately	GA samples maximum load only
LOLP		0.00127768	0.00127786
EENS (MWH/Yr)		132.0568	132.0736
no. of failure states		2680	2691
no. of network states sampled by GA		17210	2206
no. of evaluated states		17210	4699
Bus 2	*LOLP*	0.00000416	0.00000416
	EENS	0.1216	0.1216
Bus 3	*LOLP*	0.00001543	0.00001551
	EENS	1.3944	1.3984
Bus 4	*LOLP*	0.00003732	0.00003741
	EENS	1.738606	1.7431
Bus 5	*LOLP*	0.00013831	0.00013847
	EENS	1.8409	1.8448
Bus 6	*LOLP*	0.00127768	0.00127786
	EENS	126.9613	126.9660

A comparison between annualized adequacy indices obtained previously (*EENS*≅1048 MWh/Yr) and annual adequacy indices (*EENS*≅132 MWh/Yr) shows that annual indices are much smaller than annualized indices. This is because annualized indices are basically calculated assuming system hourly load values equal to system yearly maximum load.

8.4.4.2 Partially Correlated Load Buses

System buses are assumed to be located in three different time zones. They are divided into three groups with load buses in each group fully correlated. Bus 2 belongs to the first group, bus 3 belongs to the second group and buses 4,5,6 belong to the third group. It is assumed that the bus loads of the third group have the load curve given in [25] as per unit of the group maximum load of 80 MW. Bus loads of the first group have the same load curve as per unit of the group maximum load of 20 MW but shifted earlier by one hour. Bus loads of the third group have the same load curve as per unit of the group maximum load of 85 MW but shifted later by one hour. Load vector at each hour consists of three elements. Using k-means clustering technique the 8736 load vectors have been represented by 8 cluster load vectors given in Table 7. Calculated annual adequacy indices are given in Table 8.

Comparison between the results in Table 6 and Table 8 shows that when bus load correlation is considered, annual adequacy indices are decreased. This is expected as each group peak load occurs at a different time and not simultaneously.

Table 7. Results of clustering the system chronological load curves considering load buses belong to three different load groups

Cluster Load Vectors			Cluster
Group I	Group II	Group III	probability
17.3652	74.8161	70.3534	0.0562042
15.9181	68.4577	64.3613	0.115614
14.4572	62.6707	58.9248	0.115614
13.1868	57.0087	53.5125	0.149954
12.2187	51.7321	48.6647	0.14549
10.9559	45.7813	43.2648	0.150298
9.77761	40.5867	38.2292	0.162775
8.42486	34.7781	32.8293	0.115614

Table 8. Annual adequacy indices comparison using two different GA sampling approaches with partially correlated load buses

GA sampling approach		GA samples each load level separately	GA samples maximum load only
LOLP		0.001296928	0.001297081
EENS (MWH/Yr)		130.9428	130.9637
no. of failure states		2661	2672
no. of network states by sampled GA		17161	2209
no. of evaluated states		17161	4684
Bus 2	*LOLP*	0.00000449	0.00000450
	EENS	0.1046	0.1049
Bus 3	*LOLP*	0.00001682	0.00001695
	EENS	1.3724	1.3799
Bus 4	*LOLP*	0.00004039	0.00004052
	EENS	1.5402	1.5473
Bus 5	*LOLP*	0.00004530	0.00004542
	EENS	1.3868	1.3897
Bus 6	*LOLP*	0.00129693	0.00129708
	EENS	126.5388	126.5417

8.5 Calculation of Frequency and Duration Indices

Many vital industries can suffer serious losses as a result of a few minutes of power interruption. In the current competitive environment where power customers are free to choose their power supplier, it is expected that failure frequency will be an important factor in their decision to select such a supplier. This should be a motivation for the utilities in the restructured power environment to consider failure frequency in their planning for system expansion and to improve the failure frequency and duration of existing systems. Such calculations require the development of faster and reliable methods for state sampling and evaluation.

Sequential Monte Carlo simulation is perhaps the most suitable method to calculate frequency and duration indices because of its ability to represent chronological load of buses on an hourly basis. System behavior is simulated from one year to another and the number of system transitions from success states to failure states is calculated for each year. After enough simulation years, the average value of this number represents the expected value of system failure frequency. However, this technique suffers from the extensive computational effort it needs.

Meanwhile, the assessment of composite system frequency and duration indices is more complex than the assessment of other adequacy indices when using analytical methods or non-sequential Monte Carlo simulation for state sampling. This is due to the fact that calculation of failure frequency for a single sampled state is not straightforward like other adequacy indices. The state transition for each system component in the current sampled state needs to be considered to determine if this transition results in a success state, i.e., system state crosses the boundary between failure and success states. Such an operation is computationally burdensome for large systems. To solve this problem a conditional probability based approach has been introduced in [18] and [26]. This approach is based on the forced frequency balance approach introduced in [17].

This section presents a new approach to calculate the annual frequency and duration indices. In calculating the annual indices, the system yearly chronological load curve is considered rather than considering only system maximum load in case of annualized indices. The k-means clustering technique is used to represent the system yearly load curve as a multi-state component. Transition rates between different load states are calculated. The GA is used to sample failure states while the system load is assigned its maximum value. Failure states are then reevaluated with lower load states until a success state is obtained or all load states have been evaluated. The developed methodology has been applied to a sample test system. Results are compared with those obtained by non-sequential Monte Carlo simulation. The results are analyzed to validate the efficiency of the developed method.

8.5.1 Modeling of the Chronological Load

The following procedure is used to represent the system yearly load curve by m clusters. The objective of clustering is to obtain the mean value $L(C^j)$ and its probability of occurrence $P(C^j)$ for each load cluster C^j.

1. The first step is to choose initial values of cluster means. Consider that the system load at hour i is LH^i and the system yearly maximum load is L^{max}. The following initial values are suggested to be used: initial cluster mean for first cluster is chosen as $L(C^1) = 0.98L^{\max}$ and for the second cluster $L(C^2) = 0.96L^{\max}$. This process is repeated for all clusters so that the last cluster m has cluster mean:

$$L(C^m) = (1 - 0.02m).L^{\max} \tag{47}$$

   The 0.02 step size allows maximum number of 50 clusters and can be decreased to obtain more clusters.
2. For each hour i calculate the distances between the system load value at hour i and every cluster mean value:

$$DIST_{ij} = \left| L(C^j) - LH_i \right| \qquad \forall j=1,2,\ldots\ldots,m \tag{48}$$

3. Load value at hour i belongs to the cluster with the least distance, i.e.,

$$LH_i \in C^k \text{ if } \min(DIST_{i1}, DIST_{i2}, \ldots\ldots\ldots, DIST_{im}) = DIST_{ik} \tag{49}$$

   In this manner load value at each hour is assigned to a certain cluster after repeating step 3 for each of them.
4. Calculate the new mean load value for each cluster.

$$L_{new}(C^j) = \sum_{i=1}^{8760} b * LH_i / T_j \qquad \forall j=1,2,\ldots\ldots,m \tag{50}$$

   where $b = \begin{cases} 1 & \text{if } LH_i \in C^j \\ 0 & \text{otherwise} \end{cases}$ and T_j is the total number of hourly load values belonging to cluster C^j.

5. For each cluster calculate the absolute difference between old and new means.

$$change_j = \left| L_{new}(C^j) - L_{old}(C^j) \right| \qquad \forall j = 1,2,\ldots\ldots,m \tag{51}$$

6. Repeat steps from 2 to 5 until "*change*$_j$" is less than a prespecified limit for all clusters.
7. Calculate the probability of occurrence of each cluster mean load value.

$$P(C^j) = T_j / 8760 \tag{52}$$

8. Using such initial values as given in step 1 ensures that final clusters mean values are in descending order where cluster C^1 has the highest mean value and cluster C^m has the lowest mean value, i.e.,

$$L(C^1) > L(C^2) > \ldots\ldots\ldots\ldots > L(C^m) \tag{53}$$

8.5.1.1 Calculating Transition Rates between Load Clusters

An important issue in calculating frequency and duration indices is to preserve the chronological transition of load levels from one hour to another. Load transition contribution to system failure frequency is usually higher than the combined contribution of generation and transmission systems. Using k-means clustering technique the chronological load curve is represented as a multi-state component. Each cluster represents a single state associated with its probability and capacity. It is necessary to calculate transition rates between different load states to be used later for calculation of failure frequency for each sampled failure state. The following procedure is used to calculate transition rates between load clusters:

1. Each cluster consists of hourly load values at different hours during one year. The cluster number, to which each hourly load value belongs, is saved.
2. Initialize transition frequencies between different clusters.

$$f_{xy} = 0 \qquad \forall\, x = 1,2,\ldots\ldots,m;\ \forall y = 1,2,\ldots\ldots\ldots,m;\ x \neq y \tag{54}$$

3. Transition frequencies between clusters are calculated by repeating the following process for each hourly load value:

$$f_{xy}^{new} = \begin{cases} f_{xy}^{old} + 1 & \text{if } LH_i \in C^x ,\ LH_{i+1} \in C^y ,\ x \neq y \\ f_{xy}^{old} & \text{otherwise} \end{cases} \tag{55}$$

4. Transition rates between different clusters are calculated:

$$\lambda_{xy} = \begin{cases} \dfrac{f_{xy}}{P(C^x)} & \forall x = 1,2,.......,m; \forall y = 1,2,.....,m; x \neq y \\ 0 & x = y \end{cases} \tag{56}$$

where λ_{xy} is the transition rate of system load from state x to state y.

8.5.2 Calculating Failure State Contribution to System Failure Frequency

Each sampled state represents a system contingency where one or more generation units and/or transmission lines are in the down state. Load level can also be sampled when using non-sequential Monte Carlo simulation. A sampled stated "i" is identified as a failure state if a load curtailment "LC_i" is needed for reasons of generation deficiency to meet load demand or/and transmission line overloading. Consider the load is in state r in the current sampled state i, the state probability "SP_i" is:

$$SP_i = P(C^r) . \prod_{j \in gs} (1 - FOR_j) . \prod_{j \in gf} FOR_j . \prod_{j \in ts} (1 - PT_j) . \prod_{j \in tf} PT_j \tag{57}$$

where gs is the set of generation units in the up state, gf is the set of generation units in the down state, ts is the set of transmission lines in the up state, tf is the set of transmission lines in the down state, FOR_j is the forced outage rate of generator unit j and PT_j is the failure probability of transmission line j.

Power not supplied for the current state weighted by its probability is:

$$PNS_i = SP_i \cdot LC_i \tag{58}$$

The contribution of a failure state to system failure frequency consists of three components. The first component "FG" is due to transitions of generation units, the second component "FT" is due to transition of transmission lines and the third component "FL" is due to load level transition from its current state to another load state. The failure state contribution to system failure frequency "$LOLF_i$" is calculated:

$$LOLF_i = FG_i + FT_i + FL_i \tag{59}$$

Each frequency component is calculated using the conditional probability approach described in [18] and [26]. This approach is applicable under two assumptions:

The first assumption is that system is coherent which implies that:

1. System remains in its success state if a component makes transition from its current state to a higher state. In case of generation unit, higher state means a state with higher generation capacity. In case of transmission lines, higher state means the line is restored to service. In case of load state it means load level is decreased.
2. System remains in its failure state if a component makes transition from its current state to a lower state. In case of a generation unit, lower state means a state with lower generation capacity. In case of transmission lines, lower state means the line goes out of service. In case of load state it means load level is increased.

Theoretically speaking, the first assumption is always valid for generation but not for transmission. However, if one considers the operating practices, the transmission assumption is also valid [26].

The second assumption is that system components are frequency balanced, i.e., transition frequency between two states is the same in both directions. This assumption is satisfied in case of two state components. It is artificially enforced in case of multi-state components, as is the case with load states.

Generating units are represented by two states, up state and down state. In the up state, generating unit is able to deliver power up to its rated capacity and deliver no power in the down state. Each transmission line is represented by two states, up state, i.e., in service and down state, i.e., out of service. The contributions of generating units and transmission lines transition to the sate failure frequency are:

$$FG_i = SP_i.(\sum_{k \in gf} \mu_k - \sum_{k \in gs} \lambda_k) \tag{60}$$

$$FT_i = SP_i.(\sum_{k \in tf} \mu_k - \sum_{k \in ts} \lambda_k) \tag{61}$$

where μ_k is the repair rate and λ_k is the failure rate of component k.

System load is represented as a multi-state component. The first state has the highest load value and the mth state has the lowest. Consider load in the rth state within the current sampled failure state, the contribution of load transition to the sate frequency is:

$$FL_i = SP_i.\left[\sum_{j=r+1}^{m} \lambda_{rj} - \sum_{j=1}^{r-1} \lambda_{jr}.\frac{P(C^j)}{P(C^r)}\right] \tag{62}$$

In the second term of Eq. (62) fictitious transition rates from state r to higher load levels "λ'_{rj}" have been used instead of the actual transition rates λ_{rj} to satisfy the frequency balance assumption.

$$\lambda'_{rj} = \lambda_{jr} P(C^j)/P(C^r) \qquad \text{where}\, j < r \tag{63}$$

8.5.3 Non-Sequential Monte Carlo Sampling

When using non-sequential Monte Carlo simulation for state sampling, a random number in the range [0,1] is picked for each system component. In case of two-state components, if this number is less than the component failure probability the component is considered to be in the down state, otherwise it is in the up state. In case of multi-state load model the range [0,1] is divided into m parts, load is in the rth state if the picked random number z falls in the rth part, i.e.,

$$L_i = L(C^r) \quad \text{where} \sum_{j=1}^{r-1} P(C^j) < z \le \sum_{j=1}^{r} P(C^j) \tag{64}$$

where L_i is the system load value at state i.

Each sampled state is evaluated using the minimum load curtailment linear optimization module. Composite system adequacy indices are calculated after N samples as follows:

$$LOLP = N_f / N \tag{65}$$

where $LOLP$ is the system loss of load probability and N_f is the total number of failure state in the N samples.

$$LOLF = \frac{1}{N}.\sum_{j \in fs} \frac{(FG_j + FT_j + FL_j)}{PS_j} \tag{66}$$

where $LOLF$ is the system loss of load frequency and fs is the set of sampled failure states.

$$EPNS = \frac{1}{N} \sum_{j \in fs} LC_j \tag{67}$$

where *EPNS* is the system expected power not supplied.

Expected energy not supplied is calculated from *EPNS*:

$$EENS = 8760 \cdot EPNS \tag{68}$$

Loss of load duration in hours per year can be calculated once *LOLP* and *LOLF* are known.

$$LOLD = LOLP \cdot 8760 / LOLF \tag{69}$$

Coefficient of variance for *EPNS* is usually used as a convergence indicator to stop sampling. It is calculated as follows:

$$COV(EPNS) = \sqrt{\frac{\sum_{j=1}^{N} (LC_j - EPNS)^2}{N \cdot (N-1)}} \cdot \frac{1}{EPNS} \tag{70}$$

8.5.4 Genetic Algorithm Sampling for Maximum Load State with Series State Reevaluation

In the proposed approach, GA searches for states which result in system failure while system load equals the maximum load state value as explained previously. These failure states are then reevaluated while assigning system load the values of other load states in a descending order. This series state evaluation process stops when there is no load curtailment in a certain load state or the current network sampled state i.e. states of generating units and transmission lines, has been reevaluated with all load states. Adequacy indices are updated with each state evaluation process. The main steps for this approach are:

1. Each chromosome in the current GA population represents a sampled network state, i.e., states of generators and transmission lines. Each chromosome with probability higher than the threshold value is checked wither it has been previously saved in the state array, i.e., represents old network state, or not, i.e., represents a new network state. Steps from 2 to 5 are repeated for each new network state k in the current population.
2. Evaluate the new system state "*i*" which is formed from the new network state and the system maximum load state $L(C^1)$.

3. If the evaluated state represents a success state, i.e., there is no load curtailment, ignore all the remaining load states as it is guaranteed there is no load curtailment with lower load states and return to step 2 for considering the next new network state.
4. If the evaluated state represents a failure state i.e. there is load curtailment, update the system adequacy indices.

$$LOLP_{new} = LOLP_{old} + PS_i \tag{71}$$

$$EPNS_{new} = EPNS_{old} + PNS_i \tag{72}$$

$$LOLF_{new} = LOLF_{old} + FS_i \tag{73}$$

   PS_i, PNS_i and FS_i are calculated for state i using (57), (58) and (59) respectively.
5. Assign system load the lower load state $L(C^2)$. Now a new system state "i+1" has been created which is formed from network state k and the new load state. This new system state is evaluated. If it represents a success state the remaining load states are ignored and hence jump to step 2 for considering a new network state. Otherwise, it is a failure state and adequacy indices are updated using (1.71), (1.72) and (1.73). A new system state "i+2" is formed from network state k and the next lower load state $L(C^3)$. This process for network state k is repeated until encountering a system success state or network state k has been evaluated considering all the m load states.
6. If GA sampling stopping criterion has been satisfied proceed to step 7. Otherwise produce a new population and return to step 1.
7. After GA stops the searching process, the final updated indices represent the composite system adequacy indices. *EENS* and *LOLD* can be calculated using (1.68) and (1.69).

8.5.5 Case Studies for the Assessment of Frequency and Duration Indices

Both non-sequential Monte Carlo simulation and the proposed GA based method have been applied to the RBTS test system [16] to calculate its annual frequency and duration indices. Sampled states in both methods are evaluated using linearized minimum load curtailment model based on dc load flow equations, which is explained previously.

Yearly load curve data for the RBTS system in per unit of its maximum load (185 MW) is given in [25]. The full correlation assumption means that each load bus hourly load values have a fixed percentage of the system total load throughout the year. The chronological load curve is represented by eight clusters. Load value and probability of occurrence for each load state were given previously in Table 4. Transition rates between load states are given in Table 9. The annual adequacy indices for RBTS system using both non-sequential Monte Carlo simulation method and the proposed GA based method are given in Table 10. In the GA method, states are evaluated with the highest load state and failure states are reevaluated for lower load states in descending order until encountering a success state or if all load states are considered. The percentage contributions of generation units, transmission lines and load state transitions to the system LOLF using both methods are also given in Table 10. A comparison of the number of different types of sampled states by both methods is given in Table 11.

Table 9. Transition rates per year between the eight load state

From	To state							
state	1	2	3	4	5	6	7	8
1	0	1536	0	0	0	0	0	0
2	629	0	1441	82	0	0	0	0
3	44	1271	0	1662	780	0	0	0
4	0	135	1251	0	2047	313	0	0
5	0	0	359	1684	0	0	1965	183
6	0	0	0	216	1507	0	2440	7
7	0	0	0	0	506	1605	0	1221
8	0	0	0	0	0	90	1633	0

Monte Carlo simulation is stopped after 50,000 samples as the coefficient of variance of *EPNS* reaches 13%. GA is stopped after producing 1000 generations. GA parameters are: population size = 40, crossover probability = 0.7, mutation probability = 0.05 and threshold network probability = 1e-8.

The relationship between number of samples, computational time and adequacy indices when using non-sequential Monte Carlo simulation is shown in Table 12. The relationship between GA generations, computational time and adequacy indices when using the proposed GA based method is shown in Table 13.

Table 10. Comparison of annual adequacy indices and failure frequency components with the two assessment methods

Assessment method	Non-Sequential Monte Carlo	GA Sampling	Percentage difference
LOLP	0.00130	0.00128248	1.3%
EENS (MWH/Y)	133.98	132.09	1.4%
LOLF (occ./Y)	1.3398	1.2538	6.4%
FG/*LOLF*	2.8%	4.6%	-----
FT/*LOLF*	76.5%	79.8%	-----
FL/*LOLF*	20.7%	15.6%	-----

Table 11. Comparison of the number of sampled states with the two assessment methods

Assessment method	non-sequential Monte Carlo	GA sampling
no. of sampled failure states	65	2747
no. of sampled network states by GA	N/A	2189
no. of system sampled states by Monte Carlo	50000	N/A
no. of evaluated system states	50000	4738

Table 12. Relationship between number of samples, computation time and adequacy indices when using non-sequential Monte Carlo simulation

No. of samples	Comp. time in sec[1]	EENS MWh/Y	LOLF occ/y	Coefficient of variance COV(EPNS)	No. of failure states
10 000	334	206.05	2.1431	23.3%	20
20 000	662	146.50	1.5040	19.2%	30
30 000	990	155.74	1.6537	15.4%	46
40 000	1318	136.12	1.3837	14.2%	54
50 000	1646	133.98	1.3398	12.9%	65

[1]On AMD K6-II 450 MHz processor based PC.

The following observations can be made from Tables 10, 11, 12 and 13:

1. Transitions of transmission system contribute about 80% to system failure frequency while load state transitions contribute 15.4%. Usually load state transitions have much more contribution to system failure frequency. The reason for these results for the RBTS system is that bus number 6 is connected by only one transmission line to the remaining network, hence, transition of this line from up state to down state results in system failure.

2. After about 20,000 samples, results obtained by Monte Carlo simulation fluctuate around the values obtained by GA.
3. GA was able to reach *EENS* value that is less than the final value by only 2% after 200 generations. This result is obtained after sampling 990 failure states. It took Monte Carlo simulation 40,000 samples to reach such accuracy.
4. Computational effort of the proposed GA based method is about 12% of that of non-sequential Monte Carlo simulation to reach same accuracy level.
5. As GA samples more failure states, *EENS* increases which means value obtained by GA is sure less than the actual value. However, when using Monte Carlo simulation the obtained *EENS* cannot be guaranteed to be lower or higher than the actual value.
6. In case of Monte Carlo simulation, even with coefficient of variance 15% after 30,000 samples the obtained *EENS* is higher than actual value by 17%.

Table 13. Relationship between GA generations, computation time and adequacy indices when using the proposed GA based method

No. of GA generations	Comp. time in sec[2]	*EENS* MWH/Y	*LOLF* occ/y	No. of network sampled states	No. of evaluated system states	No. of system failure states
50	25	13.40	0.1710	334	656	346
100	58	122.68	1.0995	678	1586	986
200	110	129.91	1.2245	1325	2984	1803
400	162	131.92	1.2506	1804	4077	2457
800	218	132.06	1.2535	2123	4645	2720
1000	240	132.09	1.2538	2189	4738	2747

[2]On AMD K6-II 450 MHz processor based PC.

8.6 Consideration of Multi-State Components

8.6.1 Representation of Generating Unit Derated States

It is common for generating units to operate in other states between "up" and "down", these states are called derated states. In this case, generating unit models are more detailed than the two-state model. Generating units are often modeled as three-state components. These states are "up" with full capacity, "down" with zero capacity and "derated" with a certain percentage of the full capacity. Each state has its probability of occurrence. The state transition diagram for a three-state model is shown in Fig. 9. The two state model is a special case of this model where there is no derated state 2.

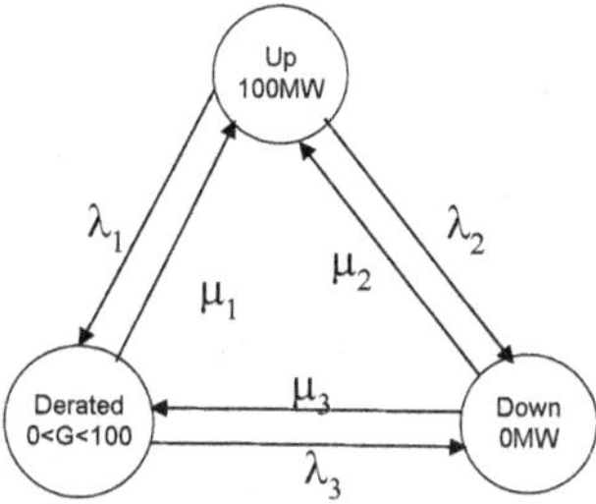

Fig. 9. Three-state model of a 100MW generating unit

The GA sampling method can be modified to consider multi-state components such as generating units with derated states and transmission line states when considering weather effect. Instead of using one gene to represent one component, n genes can be used to represent up to 2^n-state component, e.g., a three-state generating unit is represented by two genes as shown in Fig. 10.

Unit State	Gene 1	Gene 2	Capacity	Probability
Up State	1	1	100 MW	P_{up}
Derated State	0	1	0 $<G<$ 100MW	$P_{derated}$
Down State	0	0	0 MW	P_{down}
Unused State	1	0	----	0

Fig. 10. GA representation of three-state unit

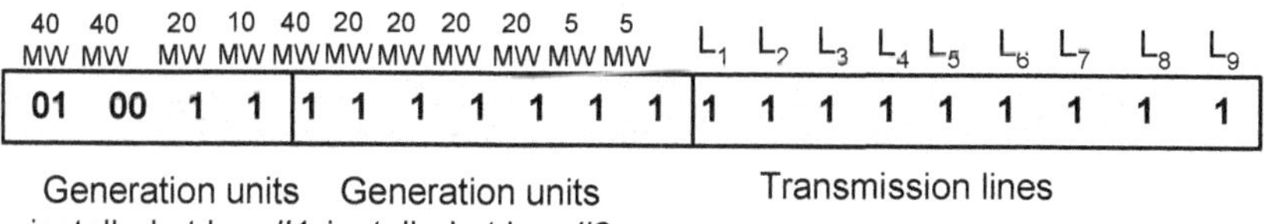

Fig. 11. Chromosome representation considering multi-state component

Assuming that each of the two 40 MW thermal units installed at bus 1 in the RBTS system is modeled as three-state component. Each of these two units is presented by two genes. Consider the state when one of these units is in the down state and the other is in derated state and all other system components are up; the chromosome representing this state is shown in Fig. 11.

8.6.2 Consideration of Common Mode Failures in Transmission Lines

The common mode failure is an event when multiple outages occur because of one common external cause. A typical example of common mode failure is the lightning stroke into a tower causing a back-flashover to two or more circuits supported by this tower. Other reasons such as the failure of a transmission tower supporting two circuits. A simple common mode failure model for two components is shown in Fig. 12.

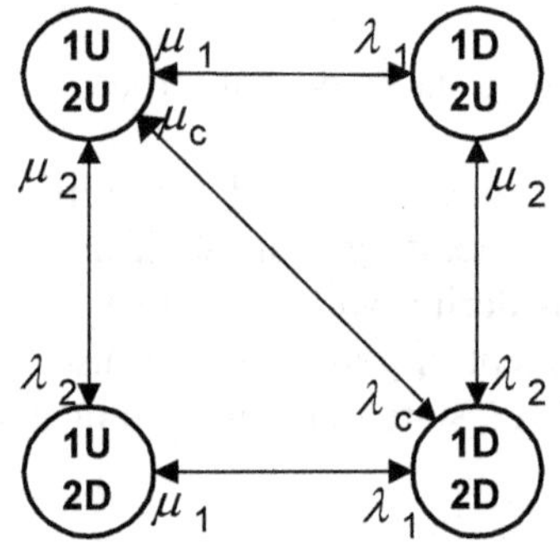

Fig. 12. State transition diagram for two transmission lines subjected to common mode failure

When two transmission lines are subjected to common mode failure, they must be treated as two dependent components. Using frequency balance equations [27] for each state, and assuming $\lambda_1=\lambda_2=\lambda$, $\mu_1=\mu_2=\mu$; the probability of each state is calculated as:

$$P(1U,2U)=\left(\frac{\mu_c\lambda+2\mu^2+\mu_c\mu}{3\mu\lambda_c+2\mu^2+4\mu\lambda+\lambda\lambda_c+3\mu_c\lambda+\mu\mu_c+2\lambda^2}\right) \tag{74}$$

$$P(1D,2D)=\left(\frac{\mu\lambda_c+\lambda\lambda_c+2\lambda^2}{3\mu\lambda_c+2\mu^2+4\mu\lambda+\lambda\lambda_c+3\mu_c\lambda+\mu\mu_c+2\lambda^2}\right) \tag{75}$$

$$P(1D,2U)=P(1U,2D)=\left(\frac{\mu\lambda_c+2\mu\lambda+\mu_c\lambda}{3\mu\lambda_c+2\mu^2+4\mu\lambda+\lambda\lambda_c+3\mu_c\lambda+\mu\mu_c+2\lambda^2}\right) \tag{76}$$

In the GA sampling approch, each transmition line in a group of lines that is subjected to common mode failures is still represented by one gene.

The only difference will be the using of combined state probaility instead of using the indpendent state probability for each transmitiom line.

8.6.3 Case Studies with Multi-State Components

The proposed algorithm has been implemented through C++ programming language. A C++ library of GA objects called GAlib developed by [19] has been integrated into the implementation. The proposed method has been tested on the RBTS test system. Studies have been made considering generation unit derated states and transmission lines common outage failure.

8.6.3.1 Generating Unit Derated States

The two 40 MW thermal units installed at bus 1 are assumed to have derated states. Four different models given in [7] are considered. These models are shown in Fig. 13. In model (a) each 40 MW is represented by four states with two derated states of 20 MW and 32 MW. In Models (b) and (c) each 40 MW is represented by three-state model. In model (d) each 40 MW unit is represent by two-state model. Data for all other components are given in Appendix.

Annualized adequacy indices are calculated by considering the system load fixed and equal to 185 MW. A comparison between annualized indices obtained using random sampling and GA based method is given in Table 14. Annual adequacy indices are calculated by considering the system chronological load curve. Yearly load curve data in per unit of RBTS system maximum load (185 MW) are given in [25]. Using k-means clustering techniques, the yearly load curve is represented by 8 states which are given in Table 4. A comparison of annual indices when using random sampling Monte Carlo simulation and the GA based method is given in Table 15.

It can be seen from Table 14 and Table 15 that consideration of derated states has larger effect on annualized indices than on annual indices.

Indices obtained by the GA based method are higher than those obtained by Monte Carlo simulation. If GA samples more failures states the value of *LOLP* and *EENS* will be higher, thus the proposed GA based method provides more accurate results than Monte Carlo simulation. When calculating annual indices the computational effort of the GA based method was about 10% of random sampling Monte Carlo simulation.

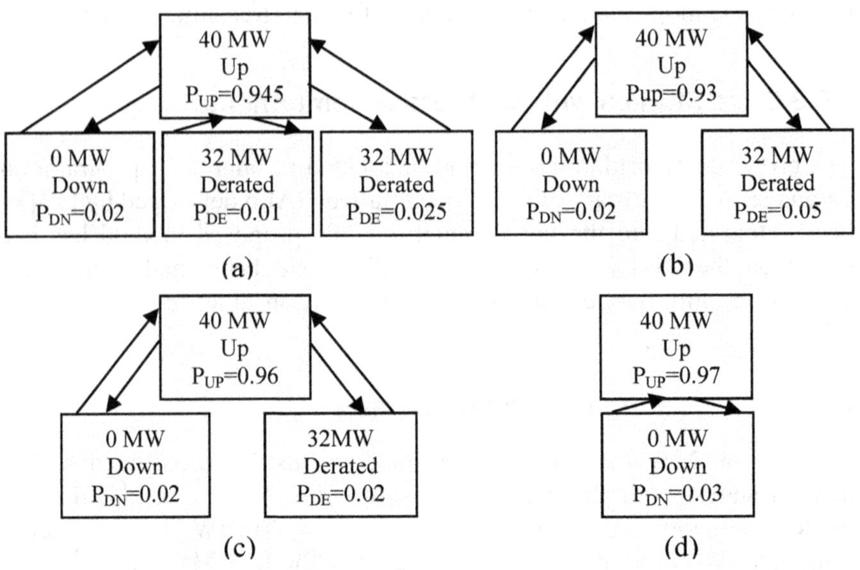

Fig. 13. 40 MW Generating unit derated state models [7]

Table 14. Comparison of annualized adequacy indices considering different derated state models

		Model (a)	Model (b)	Model (c)	Model (d)
	GA	0.008169	0.0068417	0.007692	0.009759
LOLP	Monte Carlo	0.007849	0.0067099	0.007509	0.009540
	diff. %	3.92%	1.93%	2.38%	2.24%
	GA	815.08	651.75	698.84	1052.23
EENS	Monte Carlo	776.66	638.02	678.65	1001.55
	diff. %	4.71%	2.11%	1.60%	4.8%

Table 15. Comparison of annual adequacy indices considering different derated state models

		Model (a)	Model (b)	Model (c)	Model (d)
	GA	0.001226	0.001191	0.001224	0.001284
LOLP	Monte Carlo	0.001160	0.001160	0.001169	0.001210
	diff. %	5.38%	2.60%	4.49%	5.76%
	GA	128.20	124.725	127.893	132.31
EENS	Monte Carlo	122.08	122.088	122.450	124.60
MWh/Yr	diff. %	4.77%	2.11%	4.26%	5.83%

8.6.3.2 Common Mode Outage

Two pairs of transmission lines, lines 1&6 and lines 2&7, are assumed to be installed on a common tower for their entire length. The common mode failure data for these lines as given in [16], is shown in Table 16. Each pair of transmission lines is represented by a four-state model as shown in Fig. 11. The probability of each state is calculated using Eqs. (1.70), (1.71) and (1.72). These values are given in Table 17. Annualized and annual adequacy indices have been calculated using the proposed GA based method, and results are given in Table 18. It can be seen from these results that consideration of common mode failure slightly increases the values for *LOLP* and *EENS* of the RBTS system.

Table 16. Common mode outage data for transmission lines on common tower

	First Line Pair		Second Line Pair	
Line no.	1	6	2	7
Common length km	75	75	250	250
Outage rate per year λ_c	0.150		0.500	
Outage duration (hours) r_c	16.0		16.0	

Table 17. State probability for transmission lines on common tower

	First Line Pair	Second Line Pair
Probability of the two lines being up	0.996391844	0.98806663
Probability of the two lines being down	0.001770425	0.00584654
Probability of first line down and the second is up	6.73059e-5	2.40269e-4
Probability of first line up and the second is down	6.73059e-5	2.40269e-4

Table 18. Adequacy indices with and without considering common mode outage

Indices	Annualized Indices		Annual Indices	
	Base Case	With common mode outage	Base Case	With common mode outage
LOLP	0.009753	0.009854	0.00128248	0.0012902321
EENS	1047.78	1071.251	132.090	132.965

8.7 Summary and Conclusions

Reliability studies play an important role in ensuring the quality of power delivery to customers. Developing more efficient and intelligent power system reliability assessment techniques plays a key role in improving reliability studies. This chapter has presented innovative methods based on genetic algorithms (GAs) for reliability assessment of power systems. The GA has been introduced as a state sampling tool for the first time in power system reliability assessment literature.

A GA based method for state sampling of composite generation-transmission power systems has been introduced. Binary encoded GA is used as a state sampling tool for the composite power system network states. Populations of GA generations are constructed from chromosomes, each chromosome representing a network state, i.e., the states of generation units and transmission lines. A linearized optimization load flow model is used for evaluation of sampled states. The model takes into consideration importance of load in calculating load curtailment at different buses in order to obtain a unique solution for each state.

The preceding method has been extended to evaluate adequacy indices of composite power systems while considering chronological load at buses. Hourly load is represented by cluster load vectors using the k-means clustering technique. Two different approaches have been developed. In the first approach, GA samples failure states for each load level separately. Thus adequacy indices are calculated for each load level and then combined to obtain the annual adequacy indices. In the second approach, GA samples failure states only with load buses assigned the maximum cluster load vector. Failure states are then reevaluated with lower cluster load vectors until a success state is obtained or all cluster load levels have been evaluated.

The developed GA based method is used for the assessment of annual frequency and duration indices of composite systems. Transition rates between the load states are calculated. The conditional probability based method is used to calculate the frequency of sampled failure states using different component transition rates.

The developed GA approach is generalized to recognize multi-state components such as generation units with derated states. It also considers common mode failure for transmission lines. Each two-state component is represented by one gene. Meanwhile, every multi-state component is represented by two or more genes, e.g., two genes are able to represent up to four-state component.

Case studies on the IEEE-RBTS test system were presented. It has been shown that the developed methods have several advantages over other conventional methods such as the basic Monte Carlo simulation. These advantages can be summed up as follows:

1. The effectiveness of the developed methods as compared with other conventional methods comes from the ability of GA to trace failure states in an intelligent, controlled and prespecified manner through the selection of a suitable fitness function.
2. Through its fitness function, GA can be guided to acquire certain part of the state space. This can be done by giving more credit to the states belonging to the part of the state space that is of interest.
3. The computational effort of the developed algorithms is only 10% to 20% of the computational effort when using Monte Carlo simulation to calculate annual adequacy indices for composite power systems.
4. In case of very reliable systems, Monte Carlo simulation needs much more time to converge, which is not the case with GA as it depends on fitness value comparison.
5. Parallel operation of GA sampling can be easily applied providing computational time reduction.
6. The obtained state array, after the GA states sampling stops, can be analyzed to acquire valuable information about the sensitivity of system failure to different components in the power system under study.

References

[1] R. Billinton and R. N. Allan, *Reliability Assessment of Large Electric Power System*. Boston: Kluwer Academic Publishers, 1988.

[2] R. Billinton, M. Fotuhi-Firuzabad and L. Bertling, "Bibliography on the application of probability methods in power system reliability evaluation, 1996-1999," *IEEE Trans. Power Syst.*, vol. 16, pp. 595-602, Nov. 2001.

[3] R. N. Allan, R. Billinton, A. M. Breipohl and C. H. Grigg, "Bibliography on the application of probability methods in power system reliability evaluation," *IEEE Trans. Power Syst.*, vol. 14, pp. 51-57, Feb. 1999.

[4] C. Singh, M. Schwan and W. H. Wellssow, "Reliability in liberalized electric power markets-from analysis to risk management-survey paper," presented at the *Proc. 14th Power Syst. Comput. Conf. (PSCC 2002)*, Sevilla, Spain, June 2002.

[5] N. Samaan and C. Singh, "Adequacy assessment of power system generation using a modified simple genetic algorithm," *IEEE Trans. Power Syst.*, vol. 17, pp. 974-981, Nov. 2002

[6] EPRI, "Composite system reliability evaluation methods," Final Report on Research Project 2473-10, *EPRI EL-5178*, Jun 1987.

[7] R. Billinton and W. Li, *Reliability Assessment of Electric Power Systems Using* Monte *Carlo Methods.* New York: Plenum Press, 1994.
[8] M.V.F. Pereira and L.M.V.G. Pintoand, "A new computational tool for composite reliability evaluation," *IEEE Trans. Power Syst.*, vol. 7, pp. 258-264, Feb. 1992.
[9] R. Billinton and W. Li, "Hybrid approach for reliability evaluation of composite generation and transmission systems using Monte Carlo simulation and enumeration technique," *Proc. Inst. Elect. Eng.-Gen. Transm. Dist.*, vol. 138, no. 3, pp. 233-241, May 1991.
[10] R. Billinton and A. Jonnavithula, "Composite system adequacy assessment using sequential Monte Carlo simulation with variance reduction techniques," *Proc. Inst. Elect. Eng.-Gen. Transm. Dist.*, vol. 144, no. 1, pp. 1-6, Jan. 1997.
[11] X. Luo, C. Singh and A. D. Patton, "Power system reliability evaluation using learning vector quantization and Monte Carlo simulation," *Electric Power Syst. Res.*, vol. 66, no. 2, Aug. 2003, pp. 163-169.
[12] K. Nara, "States of the arts of the modern heuristics application to power systems," presented at the *Proc. IEEE PES Winter Meeting*, Singapore, vol. 4, pp. 1238-1244, Jan. 2000.
[13] D. E. Goldberg, *Genetic Algorithms in Search, Optimization, and Machine Learning.* Reading, MA: Addison-Wesley, 1989.
[14] Z. Michalewicz, *Genetic Algorithms + Data Structures = Evolution* Programs. New York: Springer-Verlag, Third edition,1996.
[15] M. Gen and R. Cheng, *Genetic Algorithms & Engineering Optimization.* New York: John Wiley & Sons, 2000.
[16] R. Billinton, S. Kumar, N. Chowdhury, K. Chu, K. Debnath, L. Goel, E. Khan, P. Kos, G. Nourbakhsh and J. Oteng-Adjei, "A reliability test system for educational purposes-basic data," *IEEE Trans. Power Syst.*, vol. 4, pp. 1238-1244, Aug. 1989.
[17] C. Singh, " Rules for calculating the time-specific frequency of system failure," *IEEE Trans. Reliability*, vol. R-30, no. 4, pp. 364-366, Oct. 1981.
[18] A. C. G. Melo, M. V. F. Pereira and A. M. Leite da Silva, "A conditional probability approach to the calculation of frequency and duration in composite reliability evaluation," *IEEE Trans. Power Syst.*, vol. 8, pp. 1118-1125, Aug 1993.
[19] Galib genetic algorithm package, written by Matthew Wall at MIT, web sit http://lancet.mit.edu/ga/ [accessed at July 2001]
[20] R Billinton and A. Sankarakrishnan, "A comparison of Monte Carlo simulation techniques for composite power system reliability assessment," presented at the *Proc. IEEE Comm., Power and Comp. Conf.* (WESCANEX 95), Winnipeg, Canada, vol. 1, pp. 145-150, May 1995.
[21] C. Singh and Y. Kim, "An efficient technique for reliability analysis of power systems including time dependent sources," *IEEE Trans. Power Syst.*, vol. 3, pp. 1090-1096, Aug. 1988.

[22] Q. Chen and C. Singh, "Generation system reliability evaluation using a cluster based load model," *IEEE Trans. Power Syst.*, vol. 4, pp. 102-107, Feb. 1989.
[23] C. Singh and A. Lago-Gonzalez, "Improved algorithms for multi-area reliability evaluation using the decomposition-simulation approach," *IEEE Trans. Power Syst.*, vol. 4, pp. 321- 328, Feb. 1989.
[24] J. A. Hartigan, *Clustering Algorithms.* New York: John Wiley & Sons, 1975.
[25] IEEE RTS Task Force of APM Subcommittee, "IEEE reliability test system," *IEEE PAS*, vol. 98, no. 6, pp. 2047-2054, Nov/Dec. 1979.
[26] J. Mitra *and* C. Singh, "Pruning and simulation for determination of frequency and duration indices of composite power systems," *IEEE Trans. Power Syst.*, vol. 14, pp. 899-905, Aug. 1999.
[27] C. Singh and R. *Billinton, System Reliability Modeling and Evaluation.* London: Hutchinson Educational, 1977.

APPENDIX: The RBTS Test System Data

The RBTS system has been developed by the power system research group at the University of Saskatchewan [16]. The basic RBTS system data necessary for adequacy evaluation of the composite generation and transmission system is given in this appendix. The single line diagram of the RBTS test system is shown in Fig. 1.

The system has 2 generator (PV) buses, 4 load (PQ) buses, 9 transmission lines and 11 generating units. The minimum and the maximum ratings of the generating units are 5 MW and 40 MW respectively. The voltage level of the transmission system is 230 kV and the voltage limits for the system buses are assumed to be 1.05 p.u. and 0.97 p.u. The system peak load is 185 MW and the total installed generating capacity is 240 MW.

The annual peak load for the system is 185 MW. The peak load at each bus is as given in Table 19. It has been assumed that the power factor at each bus is unity.

The generating units rating and reliability data for the RBTS are given in Table 20.

The transmission network consists of 6 buses and 9 transmission lines. The transmission voltage level is 230 kV. Table 21 gives the basic transmission lines reliability data. The permanent outage rate of a given transmission line is obtained using a value of 0.02 outages per year per kilometer. The current rating is assumed on 100 MVA and 230 kV base.

Table 19. RBTS system load data

Bus number	Maximum load (MW)	User type
1	0	-----
2	20	Small users
3	85	Large users & small users
4	40	Small users
5	20	Small users
6	20	Small users

Table 20. RBTS generating system data

Unit size (MW)	Type	No. of units	Installed at bus no.	Forced Outage Rate
5	Hydro	2	2	0.01
10	Thermal	1	1	0.02
20	Hydro	4	2	0.015
20	Thermal	1	1	0.025
40	Hydro	1	1	0.02
40	Thermal	2	2	0.03

Table 21. Transmission lines lengths and outage data

Line no.	Buses		Length (km)	Permanent outage rate λ(per year)	Outage duration (hours)	Current rating (p.u.)
	From	To				
1	1	3	75	1.5	10	0.85
2	2	4	250	5	10	0.71
3	1	2	200	4	10	0.71
4	3	4	50	1	10	0.71
5	3	5	50	1	10	0.71
6	1	3	75	1.5	10	0.85
7	2	4	250	5	10	0.71
8	4	5	50	1	10	0.71
9	5	6	50	1	10	0.71

Genetic Optimization of Multidimensional Technological Process Reliability

Alexander Rotshtein

Department of Management and Industrial Engineering,
Jerusalem College of Technology – Machon Lev, Jerusalem, Israel

Serhiy Shtovba

Department of Computer–Based Information and Management Systems,
Vinnitsa National Technical University, Vinnitsa, Ukraine

9.1 Introduction

A technological process (TP) is considered multidimensional if a number of defects of diverse types occur, and they are detected and corrected simultaneously within the process execution [4, 5]. Quality of the TP is estimated by the probability of output zero-defects as well as by the probabilities of zero-defects for each of the defect types. The tasks of TP-optimization involve the choice of such a process structure that will provide the necessary output level of product quality given some certain cost limits [5]. The typical example of such optimization tasks is optimal choice of multiplicity of control-retrofit procedures in a TP. This particular optimization problem is studied in this article.

An initial perspective would view this problem as one that may be solved by using known mathematical programming methods. However, taking into account that defects of many diverse types increase the dimensionality of the state space. It means that using classical mathematical programming techniques becomes impractical. Therefore, in this article, the task of TP optimization is solved by using genetic algorithms (GA) [2, 3], which allow to find nearly global optimal solution, and additionally do not require much mathematical backgrounds about optimization. Principal

A. Rotshtein and S. Shtovba: *Genetic Optimization of Multidimensional Technological Process Reliability*, Computational Intelligence in Reliability Engineering (SCI) **39**, 287–300 (2007)
www.springerlink.com

distinction of GA methods from classical ones is in the fact that they do not use the notion of a derivative while choosing search direction and they are based on crossover, mutation and selection operations.

The first attempt of using a GA in order to solve this nonlinear optimization problem has been proposed in [6]. In that work a simple GA manages to find out the optimal solutions but the whole process proves to be very time-consuming. In this paper we discuss how to improve that GA for catching the optima quickly.

9.2 Statements of the Problems

Let's consider a discrete technological process (Fig. 1). The process consists of sequential performing of n working operation $A_1, A_2, ..., A_n$. Checking procedure ω_i for x_i-times inspects quality of working operation A_i, $i = \overline{1,n}$. Retrofit procedure U_i corrects defects detected by checking procedure ω_i.

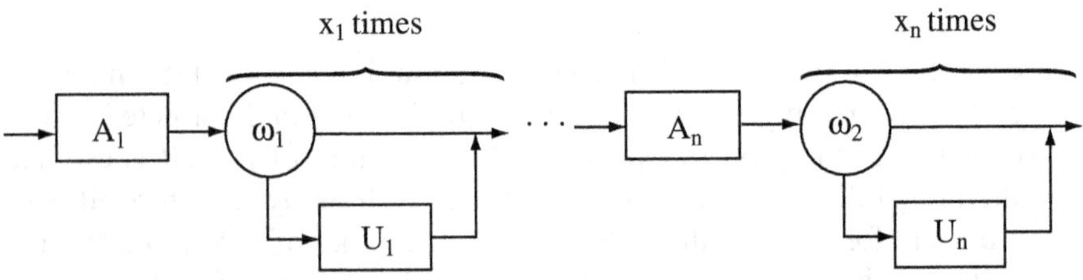

Fig. 1. A discrete technological process with multiple checking-retrofit procedures.

Let us introduce the following notations:

- m is the number of diverse types of defects;
- n is the number of working technological operations in TP;
- $x_i \in \{0, 1, 2, ...\}$ is a number of checking-retrofit procedures after i-th working operation, $i = \overline{1,n}$;
- $\mathbf{X} = (x_1, x_2, ..., x_n)$ is a vector denoting controlled variables, used to determinate the TP structure;
- $p^1(\mathbf{X})$ is the probability of executing process $\mathbf{X}$ without any defect;
- $p_j^0(\mathbf{X})$ is the probability of executing process $\mathbf{X}$ with defect of type j;
- $C(\mathbf{X})$ is cost (or other resource) required for process $\mathbf{X}$ execution.

According to [6], the problem consists of finding that $\mathbf{X}$, for which

$$\left.\begin{array}{c} p^1(\mathbf{X}) \to \max \\ \text{subject to } \; C(\mathbf{X}) \le C^* \text{ and } p_j^0(\mathbf{X}) \le q_j, \; j=\overline{1,m} \end{array}\right\}, \tag{1}$$

or

$$\left.\begin{array}{c} C(\mathbf{X}) \to \min \\ \text{subject to } \; p^1(\mathbf{X}) \ge P^* \text{ and } p_j^0(\mathbf{X}) \le q_j, \; j=\overline{1,m} \end{array}\right\}, \tag{2}$$

where q_j is the admissible probability threshold of the *j*-th type defect on process output, $j=\overline{1,m}$;

C^* is the admissible cost threshold of the TP;

P^* is the admissible threshold for zero-defect execution of the TP.

Tasks (1) and (2) look like as a nonlinear multidimensional knapsack problem. The TP is similar to the knapsack, checking-retrofit procedures correspond to the items, probabilities $p_j^0(\mathbf{X})$ are analogous to geometrical constrains of the knapsack. The linear knapsack problem is NP-hard [1], hence problems (1) and (2) are also NP-hard.

9.3 Models of Multidimensional Technological Process Reliability

The relations connecting reliability figures with TP parameters [4, 5] are described below. The reliability figures of working operation A and retrofit U are listed as follows:

$$\mathbf{P}_A = \begin{pmatrix} p_A^1 & p_A^{0_1} & \cdots & p_A^{0_m} \\ 0 & 1 & \cdots & 0 \\ \cdots & & & \\ 0 & 0 & \cdots & 1 \end{pmatrix}, \quad \mathbf{P}_U = \begin{pmatrix} 1 & 0 & \cdots & 0 \\ v_U^{1_1} & v_U^{0_1} & \cdots & 0 \\ \cdots & & & \\ v_U^{1_m} & 0 & \cdots & v_U^{0_m} \end{pmatrix},$$

where p_A^1 is the probability of zero-defect execution of operation A;

$p_A^{0_j}$ is the probability of execution of operation A with *j*-th type of defect, $j=\overline{1,m}$;

v_U^{1j} $\left(v_U^{0j}\right)$ is the probability of correcting the j-th type of defect during the execution of retrofit U, $j=\overline{1,m}$.

The probabilities of false alarm errors and defect loss errors when a checking procedure ω is carried out, are represented by the following matrices:

$$\mathbf{K}_\omega^1 = \begin{pmatrix} k_\omega^{11} & 0 & \dots & 0 \\ 0 & k_\omega^{01_1} & \dots & 0 \\ \dots & & & \\ 0 & 0 & \dots & k_\omega^{01_m} \end{pmatrix}, \quad \mathbf{K}_\omega^0 = \begin{pmatrix} k_\omega^{10} & 0 & \dots & 0 \\ 0 & k_\omega^{00_1} & \dots & 0 \\ \dots & & & \\ 0 & 0 & \dots & k_\omega^{00_m} \end{pmatrix},$$

where $k_\omega^{11}\left(k_\omega^{10}\right)$ is the probability of the right (wrong) decision made about the absence of defects during the checking procedure ω;

$k_\omega^{01_j}$ $\left(k_\omega^{00_j}\right)$ is the probability of detecting defects of type j during the checking procedure ω, $j=\overline{1,m}$.

The costs (or other resources) required for the execution of the working operation A, the checking procedure ω, and the retrofit U, are denoted by c_A, c_ω, and c_U, respectively.

A model of a working operation with x-multiple checking is employed in this paper. This operation is defined as follows [5]: a working operation with x-multiple checking is a working operation A in which, it is carried out an x times checking ω and retrofit U (if defects are detected). The output reliability figures on this operation are calculated by the following iterative scheme:

$$\left.\begin{aligned} \mathbf{P}_A^{<x>} &= \mathbf{P}_A^{<x-1>} \cdot \left(\mathbf{K}_\omega^1 + \mathbf{K}_\omega^0 \cdot \mathbf{P}_U\right) \\ \mathbf{P}_A^{<0>} &= \mathbf{P}_A \end{aligned}\right\}, \tag{3}$$

$$\left.\begin{aligned} c_A^{<x>} &= c_A^{<x-1>} + c_\omega + \sigma^{<x>} \cdot c_U \\ c_A^{<0>} &= c_A \end{aligned}\right\}, \tag{4}$$

where $\sigma^{<x>} = \sum_{j=1}^{m} \left(k_\omega^{01_j} p_A^{1^{<x-1>}} + k_\omega^{00_j} p_A^{0_j^{<x-1>}} \right)$ is probability of going to retrofit U after the x-th checking.

The reliability figures of the whole TP denoted as $\mathbf{X}$ are calculated as follows:

$$\mathbf{P}(\mathbf{X}) = \prod_{i=\overline{1,n}} \mathbf{P}_{A_i}(x_i), \quad C(\mathbf{X}) = \sum_{i=\overline{1,n}} c_{A_i}(x_i). \tag{5}$$

9.4 Basic Notions of Genetic Algorithms

One of the universal approaches for solving NP-hard combinatorial optimization tasks is Genetic Algorithms (GA). GA represents a stochastic method of optimization based on the mechanisms of natural selection acting in living nature [2, 3]. The notions of chromosome, gene, allele, and population constitute the base of GA; and classical optimization theory terms of decision variables vector, decision variable, value of decision variable, and decision set can be brought into correspondence with them.

The basic operations of GA are crossover, mutation and selection:

Crossover represents an operation on two parents-chromosomes yielding two offspring-chromosomes, each of which inherits some genes from parents-chromosomes.

Mutation is a random gene modification.

Selection represents itself as some procedure of population formation from the most adapted chromosomes according to its fitness function values.

Optimization using genetic algorithms consists of performing such a sequence of steps.

1^0. Carrying out genetic coding of decisions variants.

2^0. Generation of an initial population.

3^0. Random way choosing parents and provision of crossover.

4^0. Random way choosing chromosome and provision of mutation.

5^0. Evaluate the values of fitness function for new chromosomes.

6^0. Fixing the best decision.

7^0. Making selection (new population formation) taking into account the fitness function values.

8^0. Repeat steps 3^0-6^0 as many times as necessary.

9^0. Decoding of the decision.

9.5 Genetic Algorithm for Multidimensional Technological Process Optimization

To speed up typical GA we used:

- a smart procedure for the generation of the proper initial population;

- a fast algorithm calculating the reliability figures of the whole TP;
- a specific adaptive fitness function;
- appropriate selection scheme.

The features of the proposed GA are listed below.

9.5.1 Genetic Coding of Variants

A TP-variant can be represented by a chromosome that contains n-genes: $\mathbf{X} = (x_1, x_2, ..., x_n)$, where genes correspond to the controlled variables for optimizing tasks (1) or (2).

9.5.2 Initial Population

Typical initialization generates chromosomes with random uniform distributed alleles. In proposed initialization the allele distribution is depends upon efficiency of checking-retrofit procedures. It allows the creation of a population with high quality chromosomes. They are located into the feasible solution area or in its vicinity. Furthermore, the chromosomes have a low value of C.

For generating good chromosomes we use the so-called gradient of the checking-retrofit procedure. The gradient $\mathbf{g}_i$ of the checking-retrofit procedure with number i indicates a relatively efficient factor of impacting this procedure into the TP [5]. The gradient is calculated as follows [7]:

$$\mathbf{g}_i = \frac{\mathbf{P}(\mathbf{X}, x_i = x_i + 1) - \mathbf{P}(\mathbf{X})}{C(\mathbf{X}, x_i = x_i + 1) - C(\mathbf{X})}.$$

Our smart initialization consists of the following 4 steps.

1⁰. Predict the approximate number (N) of checking-retrofit procedures in the TP by the following formulae:

for problem (1):

$$N = \frac{C^* - C(\mathbf{X}_0)}{\tilde{c}_\omega}, \tag{6}$$

for problem (2):

$$N = \left(\max\left(\frac{p^* - p^1(\mathbf{X}_0)}{\tilde{g}_{11} \cdot \tilde{c}_\omega}, \frac{q_1 - p_1^0(\mathbf{X}_0)}{\tilde{g}_{12} \cdot \tilde{c}_\omega}, ..., \frac{q_m - p_m^0(\mathbf{X}_0)}{\tilde{g}_{1m+1} \cdot \tilde{c}_\omega} \right) \right)^{1 + \frac{m}{10}}, \tag{7}$$

where $\mathbf{X}_0 = (0, 0, 0,...)$ is the initial TP-variant without a checking-retrofit procedure; $\tilde{c}_\omega = \frac{1}{n} \sum_{i=\overline{1,n}} c_{\omega i}$ is the mean checking cost; $\tilde{\mathbf{g}} = \frac{1}{n} \sum_{i=\overline{1,n}} \mathbf{g}_i$ is the mean gradient;

Exponential factor $\left(1+\frac{m}{10}\right)$ in (7) is roughly takes into account the nonlinear dependence upon the probability of zero-defect and the cost.

2^0. Calculate the statistical expectation of multiplicity of the *i*-th checking-retrofit procedure:

$$M_i = \frac{g_{11i}}{\sum_{j=\overline{1,n}} g_{11j}} N, \quad i = \overline{1,n}.$$

3^0. Find the probabilistic distribution for each checking-retrofit multiplicity from the following systems:

$$\begin{cases} 0 \cdot prob_0(i) + 1 \cdot prob_1(i) + 2 \cdot prob_2(i) + ... = M_i \\ \dfrac{prob_1(i)}{prob_2(i)} = \dfrac{prob_2(i)}{prob_3(i)} = \dfrac{prob_3(i)}{prob_4(i)} = ... = const, \ i = \overline{1,n}, \\ prob_0(i) + prob_1(i) + prob_2(i) + ... = 1 \end{cases}$$

where $prob_k(i)$ denotes probability that the multiplicity of *i*-th checking-retrofit procedure equals $k \in \{0, 1, 2,..., \overline{x}_i\}$;

$\overline{x}_i$ is an a priori defined upper bound of the multiplicity of the *i*-th checking-retrofit procedure;

$const > 1$ is a heuristically defined constant. In our case it equals 10.

The second equation in the system describes the fact that a double check occurs more seldom than a single check in the industrial TP, and a triple check happens more rarely than a double check etc.

4^0. Generate pop_size chromosomes according to the found probabilistic distributions of the checking-retrofit multiplicities.

9.5.3 Crossover and Mutation

A uniform crossover with one cutting point and single-gene mutation [2, 3] are employed. After the mutation the value of gene x_i should be put into interval $[0, \overline{x}_i]$.

9.5.4 Fitness Function

The following adaptive fitness function is used for problem (1):

$$F(\mathbf{X})=\begin{cases} p^1(\mathbf{X}), & \text{if } \mathbf{X} \text{ is a feasible solution} \\ p^1(\mathbf{X})\cdot(1-D(\mathbf{X})), & \text{otherwise} \end{cases},$$

where $D(\mathbf{X})$ is the following penalty function:

$$D(\mathbf{X})=\frac{1}{m+1}\cdot\left(\max\left(0,\ \frac{C(\mathbf{X})-C^*}{C^*}\right)+\sum_{j=1}^{m}\left(\frac{\Delta b_j(\mathbf{X})}{\Delta b_j^{\max}}\right)^{\alpha}\right),$$

where $\alpha>0$ is a factor of importance of avoidance the defects; $\Delta b_j(\mathbf{X})=\max\left(0,\ p_j^0(\mathbf{X})-q_j\right)$ means violation of j-th constraint by chromosome $\mathbf{X}$, $j=\overline{1,m}$; $\Delta b_j^{\max} = \max\limits_{p=1,\ pop\_size}\left(\Delta b_j(\mathbf{X}_p)\right)$ is max-violation of constraint on q_j in the current population. One can associate $\Delta b_j^{\max}$ with population inferiority.

The following adaptive fitness function is used for problem (2):

$$F(\mathbf{X})=\begin{cases} \dfrac{1}{C(\mathbf{X})}, & \text{if } \mathbf{X} \text{ is a feasible solution} \\ \dfrac{1-D(\mathbf{X})}{C(\mathbf{X})}, & \text{otherwise} \end{cases},$$

where $D(\mathbf{X})$ is the following penalty function:

$$D(\mathbf{X})=\frac{1}{m+1}\cdot\left(\max\left(0,\frac{P^*-p^1(\mathbf{X})}{P^*}\right)+\sum_{j=1}^{m}\left(\frac{\Delta b_j(\mathbf{X})}{\Delta b_j^{\max}}\right)^{\alpha}\right).$$

The penalties for the same violations are different for various populations. The dependence of the penalty values upon population quality allows to separate good and poor solutions during the selection.

9.5.5 Fast Calculation of the Reliability

A profile of the GA shows that the most time-consuming part is the set of calculations by formulae (3) – (4). To speed up the process, we calculate in advance and store in memory the quantities $\mathbf{P}_{A_i}(x_i)$ and $c_{A_i}(x_i)$ for all possible values of $x_i \in \{0, 1, 2, ..., \overline{x_i}\}$, $i = \overline{1,n}$. Accordingly, for calculating the reliability figures of the whole TP we simply apply formula (5) for the corresponding (already computed) values of $\mathbf{P}_{A_i}(x_i)$ and $c_{A_i}(x_i)$.

9.5.6 Selecting Schemes

We examined 3 types of selection: 1) roulette wheel with elitism strategy on whole population, 2) roulette wheel with elitism strategy on truncated population, and 3) tournament selection. The first selection acts by the following scheme: 1) find the chromosome with the highest fitness and the chromosome with the highest fitness among the feasible ones and include them into the new population; 2) add the remaining chromosomes via the roulette wheel process [2]. In the second selection only fraction of best chromosomes can be picked up for a new population [3]. After this population truncation the selection acts as the roulette wheel with elitism strategy. In the third selection each individual in new population is defined as a winner of competition among some number of randomly chosen chromosomes [3].

9.6 Computational Experiments

The GA for problem (1) was tested on two data sets: A_4_1 and A_4_2. The GA for problem (2) was tested on two data sets: B_4_1 and B_4_2. The data sets are available at www.ksu.vstu.vinnica.ua/shtovba/benchmark . The data sets correspond to a TP with 4 diverse types of defects. The number of potential checking operations varies from 20 to 120. In data sets A_4_1 and B_4_1 the avoidance of one type of defects is the most important task, according to the very low level of the admissible probability threshold for it. All the types of defects had approximately the same importance in data sets A_4_2 and B_4_2. We restricted upper bound of checking multiplicity as follows: $\overline{x_i} = 4$ for tasks A_4_1 and A_4_2, and $\overline{x_i} = 5$ for tasks B_4_1 and B_4_2, $i = \overline{1,n}$.

Test results are shown in Table 1 and Table 2. As an alternative optimization routine we used the greedy method [7] with fast gradient computing. Its basic idea is similar to the knapsack problem greedy solving. The greedy method is an iterative algorithm with two epochs. On the first epoch, it increases the number of checking-retrofit procedures with maximal gradient iteratively. On the second epoch, the greedy algorithm tries to backtrack the solution into the feasible area.

Table 1. Probabilities $p^1(\mathbf{X})$ for optimal solutions (boldface font points to unfeasible solutions).

Dimension (n)	Tasks A_4_1		Tasks A_4_2	
	GA	Greedy search	GA	Greedy search
20	0.9693	**0.9652**	0.9546	0.9539
40	0.9425	0.9423	0.9233	0.9225
60	0.9309	**0.9254**	0.9040	0.9037
80	0.8697	0.8689	0.8649	0.8642
100	0.8548	**0.8524**	0.8372	0.8363
120	0.8482	0.8479	0.8091	0.8080

Table 2. Costs $C(\mathbf{X})$ for optimal solutions.

Dimension (n)	Tasks B_4_1		Tasks B_4_2	
	GA	Greedy search	GA	Greedy search
20	199.67	203.02	153.27	154.37
40	451.97	460.99	322.35	325.25
60	531.88	540.25	478.45	482.24
80	708.40	719.90	697.51	700.97
100	975.04	992.01	859.29	860.59
120	1105.15	1118.06	958.92	962.69

The GA provides better solutions than the greedy method, especially for tasks A_4_1 and B_4_1. Fig. 2 and Fig. 3 show a satisfaction of the constraints by solutions found by the GA and by the greedy method. In these figures y-axes correspond to the following factor of relative satisfaction:

$$\text{for } C(\mathbf{X}): \quad \psi(\boldsymbol{X}) = 100\% \cdot (C^* - C(\mathrm{X}))/C^*,$$

$$\text{for } p_j^0(\boldsymbol{X}): \quad \psi(\boldsymbol{X}) = 100\% \cdot (q_j - p_j^0(\boldsymbol{X}))/q_j, \quad j = \overline{1,m}.$$

A negative value of $\psi(\mathbf{X})$ means, that solution $\mathbf{X}$ does not satisfy a constraint. There are 3 such solutions, found by the greedy search for task B_4_1 with n=20, n=60, and n=100.

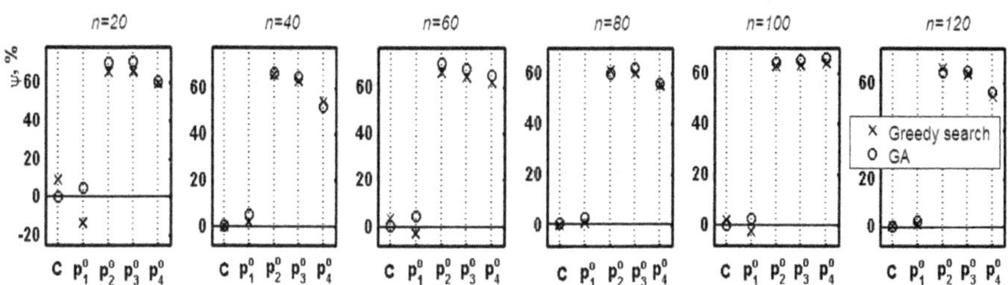

Fig. 2. A satisfaction of constrains for tasks B_4_1 by optimal solutions, found by the GA and by the greedy search.

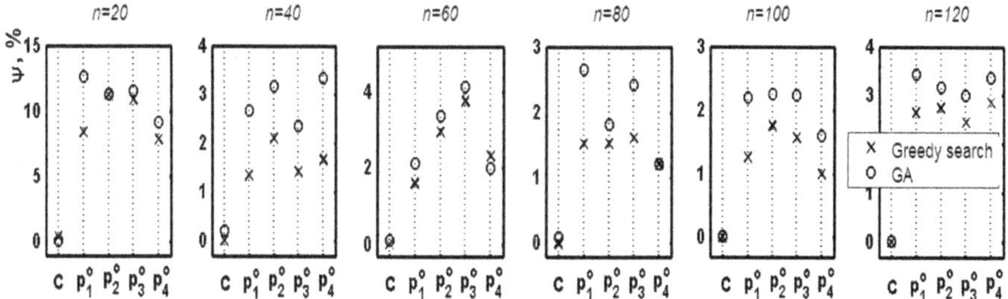

Fig. 3. A satisfaction of constrains for tasks B_4_2 by optimal solutions, found by the GA and by the greedy search.

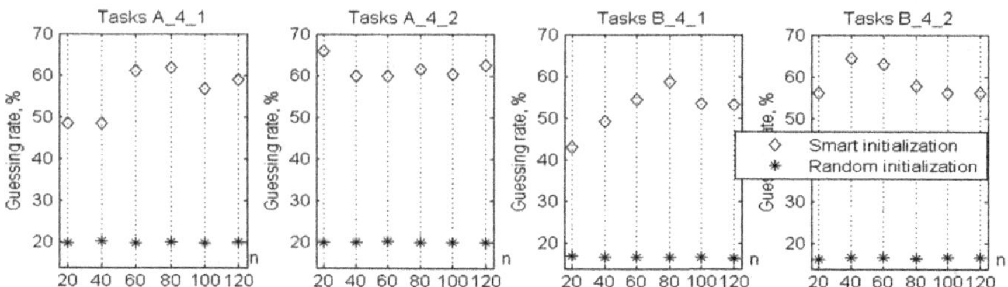

Fig. 4. Matching optimal gene values by random and proposed initializations (average means out of 1000 generations).

Fig. 4 compares random and proposed smart chromosome initializations by the criterion of guessing the optimal gene values. The random initialization produces chromosomes that have about 15-20% genes with the optimal values. The guessing rate of the smart initialization is about 50-60%. Such improvement of the initial population more then doubles the speed of optimization. Fig. 5 shows that the proposed models (6) – (7) predict satisfactorily the number of checking-retrofit procedures.

Profiles of GA with various procedures of calculations of reliability figures are shown on Fig. 6. It is clear that the proposed procedure of calculations of reliability figures accelerates genetic search several fold.

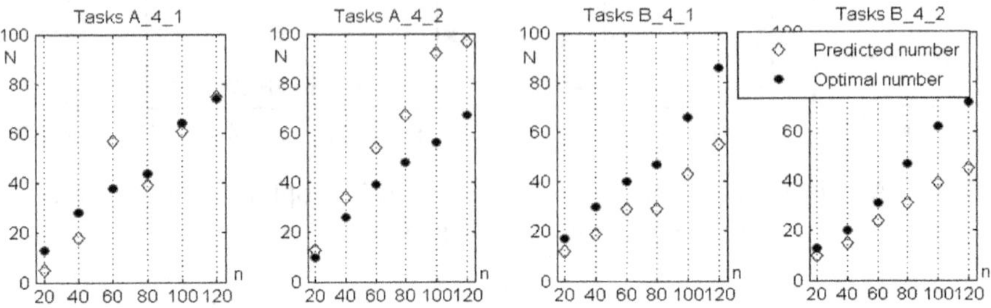

Fig. 5. Predicted and optimal numbers of checking-retrofit procedures.

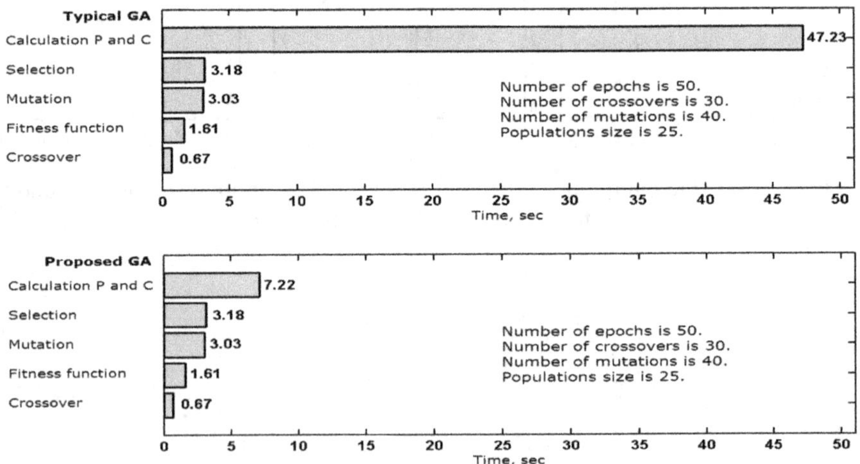

Fig. 6. Profiles of the GA with typical and proposed calculations of reliability figures (task A_4_1, n=20).

Fig. 7 compares average dynamics of the GAs with different selection schemes for task B_4_2 with n=20. All the selections were tested on 10 various initial populations. For every initial population the GAs run 10 times. The left part of Fig. 7 shows dynamics the optimization in cases of various thresholds τ of the truncation selection. The threshold means a percentage of chromosomes, rejected from the population before a roulette wheel spinning. Good dynamics takes place when $\tau \approx 50\%$. The right part of Fig. 7 shows dynamics the optimization in cases of various number participants t in the competition. Good dynamics takes place when $t \approx 6$.

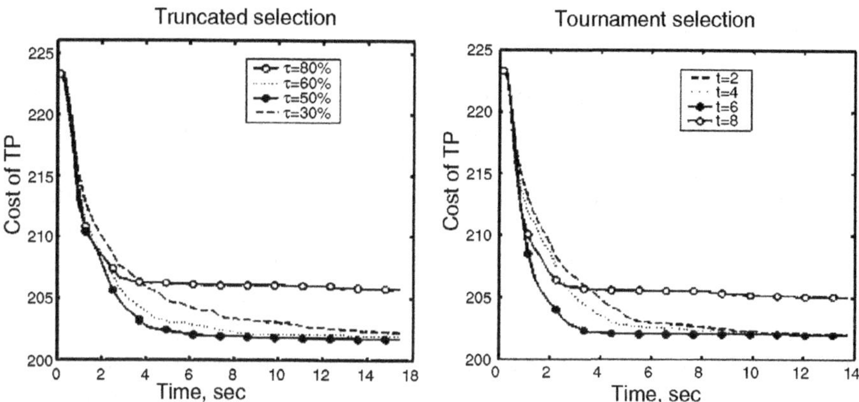

Fig. 7. Average dynamics of the GA with truncation selection and tournament selection (task B_4_1, dimension n=20).

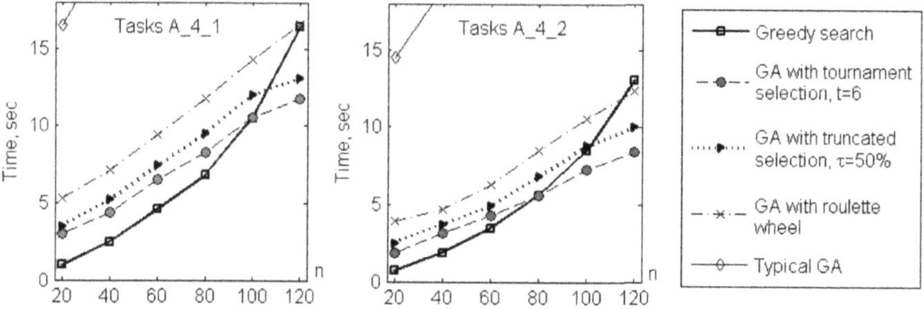

Fig. 8. Average time for solving testing tasks by the greedy method and by various GAs (the GAs run 100 times).

Fig. 8 shows that roulette wheel is out of competition among 3 selections. Timing of truncation selection and timing of tournament selection are comparative, but GA with tournament selection produces solutions near to the optima faster. Fig. 8 also compares the proposed fast GAs, typical GA and greedy method timing on tasks A_4_1 and A_4_2. This figure shows that the proposed GAs are faster than the typical GA. For large-scale tasks the GAs find out the optimal solutions even quicker the greedy method. Note that the GA with the tournament selection provides the best performance level.

9.7 Conclusions

A fast GA of optimization for the multiplicity of checking – retrofit procedures in multidimensional TP is proposed. The acceleration of the GA is achieved through: (a) a smart procedure for the generation of a high-quality initial population, (b) a fast algorithm-calculation of the reliability figures of the whole TP, (c) a specific adaptive fitness function proposed, and (d) appropriate tournament selection. Computational experiments carried out, show that the GA finds out better solutions, and proceeds faster than greedy method for large-scale instances. Proposed schema of GA may be useful for solving hard optimization problems in design of reliable algorithms, checking point allocation in technological processes, diagnosis of defects in complex systems etc.

Acknowledgement

The authors thank MS Olexiy Kozachko for conducting the computational experiments.

References

[1] Garey, M., Johnson, D. (2000) Computers and Intractability. A Guide to the Theory of NP–Completeness, *W.H. Freeman and Company*, New York.

[2] Gen, M., Cheng, R. (1997) Genetic Algorithms and Engineering Design, *John Wiley & Sons.*

[3] Mitchell, M. (1996) An Introduction to Genetic Algorithms, *The MIT Press*, Cambridge.

[4] Rotshtein, A. (1987) Probabilistic-Algorithmic Models of Man-Machine Systems, *Soviet Journal Automation and Information Science,* No 5: 81-86.

[5] Rotshtein, A., Kuznetcsov, P. (1992) Design of Zero-Defect Man-Machine Technologies, Technika, Kiev, [In Russian].

[6] Rotshtein, A., Shtovba, S. (1999) Optimization of Multidimensional Technological Processes by Genetic Algorithms, *Measuring and Computing Devices in Technological Processes,* No. 2: 7-13, [In Ukrainian].

[7] Rotshtein, A., Shtovba, S., Dubinenko, S., Kozachko O. (2004) A Heuristic Optimization of Checking Procedure Allocation in Technological Processes for Multidimensional Space of Defects, *Journal of Vinnitsa State Technical University,* No. 1: 54-62, [In Ukrainian].

Scheduling Multiple-version Programs on Multiple Processors

Piotr Jędrzejowicz, Ewa Ratajczak

Departament of Information Systems,
Gdynia Maritime University, Gdynia, Poland

10.1 Introduction

Complex computer systems are often facing the conflicting requirements. One of the frequently encountered conflicts involves high dependability and safety standards required versus system performance and cost. In this paper the focus is on managing the trade-off between requirements on software dependability and time-performance in the real-time computer systems with hard and soft time constraints.

The demand for highly dependable software can be met by applying fault avoidance, fault removal and fault-tolerance techniques. Due to high costs and complexity, the later has been little used. It is, however, nowadays widely accepted that highly dependable software in many applications is possible only through software fault-tolerance [24].

A unit of software is fault-tolerant (f-t) if it can continue delivering the required service after dormant imperfections called software faults have become active by producing errors. When the errors disrupt the expected service one says that software unit has failed for the duration of service disruption. To make simplex software units fault-tolerant, the corresponding solution is to add one, two or more program variants to form a set of N ≥ 2 units. The redundant units are intended to compensate for, or mask a failed software unit. The evolution of techniques for building f-t software out of simplex units has taken several directions.

Basic models of f-t software are N-version programming (NVP), recovery blocks, and N-version self-checking programming (NSCP). A hybrid solution integrating NVP and recovery block is known as the consensus recovery block. Yet another approach to software f-t is known as a t/(n-1) variant programming. In all the above listed techniques, the required fault tolerance is achieved by increasing the number of independently developed program variants, which in turn leads to a higher dependability at a cost of the additional resources used.

P. Jędrzejowicz and E. Ratajczak: *Scheduling Multiple-version Programs on Multiple Processors*, Computational Intelligence in Reliability Engineering (SCI) **39**, 301–328 (2007)
www.springerlink.com

Self-configuring optimal programming i another attempt to combine some techniques used in RB and NVP in order to exploit a trade-off between software dependability and efficiency. This scheme organises the execution of software variants in stages, configuring at each stage a currently active variant set, so as to produce the acceptable fault tolerance with relatively small effort and efficient use of resources. In fact, the approach does attempt to dynamically optimize a number of program variants run but it offers local optimization capabilities rather then global ones. All known f-t techniques require additional effort during design, testing and implementation. However, once additional variants or checkers, correctors and trail generators have been developed they can be run indefinite number of times at a cost of additional processing time only.

In this chapter we propose to manage complexity and costs issues of the fault-tolerant programs not at a single program level, as traditionally done, but rather from the point of view of the whole set of such programs, which are to be run under time constraints. To solve the emerging problems of the fault-tolerant computing under time-constraints we will use models and techniques of the multiprocessor (m-p) task scheduling. It should be noted that the m-p task concept could be used to model a variety of the fault-tolerant structures, since all require processing of redundant variants, internal communication and execution of an adjudication algorithm. In such a model, the size of a task would correspond to the number of its redundant variants.

The algorithms introduced in this chapter paper can be applied to scheduling computer programs constructed of redundant versions in safety-critical, multiple-processor environment, under criteria of the number of versions executed or schedule reliability.

In recent years an intensive increase of computation speed has enabled development of computer systems and their complexity. Parallel and distributed systems are more and more widely applied in practical use. Consequently, appropriate scheduling techniques have been proposed. The idea of multiprocessor task extends the classical scheduling theory assumption that a task is executed on one processor at a time only. A multiprocessor task requires more than one processor at the same time [8]. A multiple-version task may be treated as a special case of the multiprocessor task. The multiple-version task idea originates from the theory of fault-tolerant computing [23, 5], where so called N-version programs are a common fault tolerance technique in highly reliable systems [2, 3].

Scheduling m-p tasks is understood as assigning processing elements to tasks and, at the same time, deciding on task size and structure, in such a way that all problem constraints are satisfied and some global performance goal is optimized. Scheduling m-p tasks differs from traditional, single-variant task scheduling problems, by the extended solution space. It includes not only an assignment of tasks to processors but also a decision as to which combination of redundant variants, from a set of the available ones, should be used to construct each task (at least one variant of each task has to be included within a schedule). The chapter summarizes and reviews several results of the authors presented earlier in [6, 7, 17, 18, 19, 20, 21].

In this paper the global optimization criterion is either a number of versions executed or schedule reliability understood as a probability that all scheduled tasks will be executed without failures. In both cases a criterion is maximized under the requirement that all scheduled tasks meet their respective deadlines. Two basic classes of multiple program variants models are considered. The first class involves models based on the assumption that failures of multiple program variants are statistically independent. The second class includes models for which the earlier assumption is not needed any longer.

Unfortunately, the discussed scheduling problems belong to the NP-hard class To enable obtaining solutions within reasonable time the following approximation algorithms, based on soft-computing methods, have been designed and tested:

- Tabu search metaheuristic.
- Population learning algorithm.
- Neural network algorithm.
- Island-based evolutionary algorithm.
- Hybrid algorithms.

The paper is organized as follows. Section 2 deals with the case of statistically independent program variant failures. It includes the multiple-version task scheduling problem formulation, main results on the complexity of multiple-version task scheduling problems, description of the algorithms used to solve problem instances and the results of the computational experiment carried out with a view to evaluating effectiveness of the proposed algorithms. Section 3 deals with the case of scheduling m-p tasks in presence of correlated failures. It includes problem formulation and the corresponding reliability model, a description of the proposed approximation algorithms, numerical example and the results of computational experiment carried out. Finally, in Section 4 suggestions for future work and conclusions are made.

10.2 Scheduling Multiprocessor Tasks – Case of the Statistically Independent Failures

10.2.1 Problem Formulation

In general, scheduling problems are characterized by three sets: the set of m processors $P=\{P_1, P_2,\ldots,P_m\}$, the set of n tasks $T=\{T_1, T_2, \ldots,T_n\}$ and the set of s types of other resources $X=\{X_1, X_2,\ldots,X_s\}$. Scheduling means assigning processors from P (and other resources from R) to tasks from T in order to complete all tasks under the imposed constraints.

A multiprocessor task is built from a finite set of possibly independently developed program variants. It is assumed that scheduling such a task requires two types of resources – processors and time. The decision problem consists of selecting the set of variants to be executed and scheduling these tasks in such a way that each task gets access to the required number of processors. In the multiprocessor task scheduling problems each task requires more then one processor at the same time. The number of processors needed is equal to the number of program variants used to construct an m-p task.

A schedule is an assignment in time processors from P to tasks from T satisfying the following requirements:

- Each task (program) $T_j, j=1,\ldots,n$, consists of $NV_j \in Z^+$ versions $T_j=\{t_{j1},\ldots, t_{jk_j}\}$. The version t_{ji}, $i=1,\ldots,NV_j$, is an independent, one-processor, nonpreemtable program or unit performing the task T_j.
- Processors $P=\{P_1,P_2,\ldots,P_m\}$ are identical. Each processor executes at most one version at a time.
- The task T_j is executed in time interval $[a_j, \infty)$, where a_j is an arrival (ready) time for task $T_j, j=1,\ldots,n$.
- For each task a set of versions to be executed $\overline{T}_j \subseteq T_j$ is selected.
- The task T_j is executed if at least one version of the task T_j is executed, $j=1,\ldots,n$.
- All tasks are executed.

The multiprocessor tasks may be executed in three different modes [19]:

- Basic multiprocessor tasks mode (m-p mode) – the selected versions of task $T_j \in T$ are executed in parallel by available processors, the processors are available for the next task when all the selected versions of task T_j are completed.

- Multiprocessor tasks with processors release mode (m-r mode) – the selected versions of task $T_j \in T$ are executed in parallel by the available processors, each of the processors becomes available for the next task when the version assigned to it is completed.
- Multiple-version tasks mode (m-v mode) – the selected versions of task $T_j \in T$ are executed in parallel or in sequence by the available processors. There is no requirement to execute all or even some of the m-v tasks in parallel.

In this chapter we investigate the above defined problems, assuming each task $T_j \in T$ is characterized by the following data:

- Arrival time (ready time) a_j – the time at which task T_j is ready for processing.
- Deadline d_j – the time limit by which T_j must be completed.
- Set of versions $T_j=\{t_{j1},\ldots,t_{jNV_j}\}$ in which task T_j may be executed. Let $k = \max_{1 \le j \le n} \{k_j\}$.
- Processing time p_{ji} – the time needed to execute version t_{ji} of task T_j, $i=1,\ldots,k_j$. The time required to process an m-p task in m-p or m-r modes can be calculated from its version processing times. The task T_j processing time is calculated as follows:

$$p_j = \begin{cases} \max\limits_{\{i \mid t_{ji} \in \overline{T}_j\}} \{p_{ji}\} + h_j & \text{if } \left|\overline{T}_j\right| \ge 2, \\ p_{ji} & \text{otherwise,} \end{cases}$$

  where $\overline{T}_j$ denotes a set of task T_j variants, which are selected to be executed, and h_j (overhead) denotes the time needed to compare the results of the versions for performing an acceptance test or running a voting algorithm;
- Reliability r_{ji} – the probability that the version t_{ji} of task T_j will be executed without a failure, $i=1,\ldots,k_j$.

A schedule is feasible if the tardiness constraint is satisfied i.e. $TC = \sum_{j=1}^{n} \max_{1 \le j \le n} \left(0, c_j - d_j\right) = 0$, where c_j denotes the completion time of the task T_j, which equals the maximum completion time taken over all executed versions of task T_j.

The proposed models use optimality criteria:

– Schedule reliability $R = 1 - \prod_{j=1}^{n} \prod_{\{i|t_{ji} \in \overline{T}_j\}} (1 - r_{ji})$.

– Number of versions executed $V = \sum_{j=1}^{n} \left|\overline{T}_j\right|$.

The analyzed scheduling problems in a standard three-field notation for scheduling problems $\alpha|\beta|\gamma$ are denoted as: $P|\text{m-p},a_j,d_j|V$, $P|\text{m-p},a_j,d_j|R$, $P|\text{m-r},a_j,d_j|V$, $P|\text{m-r},a_j,d_j|R$, $P|\text{m-v},a_j,d_j|V$ and $P|\text{m-v},a_j,d_j|R$.

10.2.2 Computational Complexity of Multiprocessor Task Scheduling Problems

In this section we will show that the considered problems with arrival time and deadline constraints are NP-hard in the strong sense.

Let P|m--,a_j,d_j|- denote a general multiprocessor task scheduling problem with m-p, m-r or m-v mode and arbitrary optimality criterion. To prove strong NP-completeness of the general multiprocessor task scheduling problem it is sufficient to reduce the problem to the known sequencing within intervals problem, which was proved to be strongly NP-complete in [11].

Definition 1: Sequencing within intervals problem.
INSTANCE: A finite set $T=\{T_1, \ldots, T_n\}$ of tasks and, for each $T_j \in T$ an arrival time $a_j \geq 0$, a deadline $d_j \in Z^+$, a length $p_j \in Z^+$.
QUESTION: Does there exist a feasible schedule for T, such that for each $T_j \in T$, $c_j - p_j \geq a_j$, $c_j \leq d_j$ and for $j,l=1,\ldots,n$, $j \neq l$ either $c_j \leq c_l - p_l$ or $c_l \leq c_j + p_j$.

Definition 2: General multiprocessor task scheduling problem P|m--, a_j,d_j|-.
INSTANCE: A finite set T of tasks and, for each $T_j \in T$ an arrival time, deadline, set of versions and processin times defined in section 2.
QUESTION: Does there exist a feasible schedule for T, such that for each $T_j \in T$, $c_{ji} - p_{ji} \geq a_j$, $c_{ji} \leq d_j$ for $j=1,\ldots,n$, $i=1,\ldots,m$, and for $i=1,\ldots,m$, $j,l=1,\ldots,n$, $j \neq l$ either $c_{ji} \leq c_{li} - p_{li}$ or $c_{li} \leq c_{ji} + p_{ji}$.

Theorem 1: Problem $P|\text{m--},a_j,d_j|\text{-}$ is NP-complete in the strong sense.

Proof: Let π_s denote sequencing within intervals problem according to Definition 1 and π_v general multiprocessor task scheduling problem according to Definitions 2. To prove that $\pi_v \in$NP one needs to assure that the following inequalities hold:

1. $c_{ji} \leq d_j$, for each task $T_j \in T$;
2. $c_{ji} \leq c_{li} - p_{li}$ or $c_{li} \leq c_{ji} + p_{ji}$, for each task $T_j \in T$ and each task $T_l \in T$ such that $j \neq l$.

An easy computation shows that both conditions can be checked in $O(n^2)$ time.

To reduce the π_v problem to the π_s problem it suffices to assume that tasks are scheduled on one processor and that each task T_j consists of one version only $T_j = \{t_{j1}\}$. Therefore, it is obvious that π_v reduced in such a way is equal to the π_s.□

From the above theorem it is clear that the whole family of the considered multiprocessor task scheduling problems is NP-hard in the strong sense.

10.2.3 Approximation Algorithms for Solving Multiprocessor Task Scheduling Problems in the m-p Mode

Since the discussed problems are strongly NP-hard it is quite unlikely that there exist polynomial time algorithms solving them to optimality in a reasonable time. Therefore, approximation algorithms are proposed. In this subsection two such algorithms for solving $P|\text{m-p},a_j,d_j|V$ problem instances are presented: tabu search algorithm (TSA) [12], and population learning algorithm (PLA) [17], in which TSA is used as one of the learning procedures. Algorithms based on the same idea were also used to solve the $P|\text{m-p},a_j,d_j|R$ problem instances.

10.2.3.1 Tabu Search Algorithm Implementation

To describe the proposed TSA implementation the following notation is used:

- *lit* - Iteration number.
- Q – Schedule representation, which is a list of objects, each object representing a task and consisting of its number and a list of versions selected to be executed. The objects (tasks) are allocated to processors by the Multiprocessor_First_Fit algorithm.
- Q_1 – An initial solution, in which only the shortest version of each task is selected to be executed. Objects (tasks) are ordered by a non-decreasing arrival times a_j as the first criterion and a non-decreasing deadlines d_j as the second.
- Q_2 – An initial solution, in which only, the shortest version of each task is selected to be executed. Objects are ordered by a non-decreasing

deadlines d_j as the first criterion and a non-decreasing arrival times a_j as the second.

- $T(Q)$ – Schedule tardiness (see Section 2.1).
- $v(Q)$ – Value of the optimality function for the solution Q;
- $N(Q)$ – Neighborhood of the solution Q.
- *LTA* – Tabu active list containing tabu active moves.
- *lt* - Tabu tenure i.e. the numer of iterations for which an element (move) is tabu active.
- $N^*(Q)$ – Modified neighborhood of the solution Q.
- *PAM* – Short term memory of the tabu algorithm.
- move(j, m_s, k_s, m_d, k_d, v_m) – Move of the task T_j from position m_s to position m_d in the schedule Q. Simultaneously, the list of selected versions of task T_j may be changed from k_s to k_d; v_m denotes the value of the move and it is equal to the value of the optimality function obtained after the move is performed.
- The following aspiration criteria are enforced: if the currently performed move produces the best value obtained so far but it is on the list *LTA*, it is removed from this list; if the moves performed directly before the last one have better values than the last one, then these moves are removed from *LTA* (in the proposed TSA the three last moves were evaluated and removed if their values were better by more then 20% of the last move value).

The proposed algorithm is shown in the following pseudo code:

```
Procedure MultiProcessor_First_Fit_m-p(Q)
Begin
  For each task Tj from schedule Q do
    allocate selected versions of task Tj to |T̄j|
    processors which are first released, in such a
    way that the task execution is not started before
    time aj;
  End for
End

Procedure TSA
Begin
  LTA=∅;
  PAM=∅;
  Q=Q1 (initial solution);
  calculate the tardiness of the schedule TC(Q);
  If TC(Q)>0 then exit (no solution); endif
  vm^max=v(Q);
  For i=1 to lit do
    While the best move is not chosen do
```

```
        from the neighborhood N*(Q) choose
        move(j,m_s,k_s,m_d,k_d,v_m) with the best value, such
        that it is not on the LTA or its value is
        greater than v_m^max (aspiration criterion);
        If no move is chosen then
          remove from LTA moves with the shortest tabu
          tenure lt;
        Endif
      Endwhile
      perform move (Q=Q(move));
      update LTA:
        remove the moves, for which lt=0;
        add to LTA the following moves:
        move_1(j,m_d,0,m_s,0,0), lt=2n;
        move_2(j,0,0,0,k_s,0), lt=⌊1/2 n (k-1)⌋;
        move_3(j,0,0,m_s,0,0), lt=⌊3/4 n⌋;
        move_4(j,0,0,0,0,0), lt=⌊1/3 n⌋;
        move_5(j,m_s,k_s,m_d,0,0), lt=⌊3/4 n k⌋;
      update PAM:
        remove the first move;
        add move;
      If v_m>v_m^max then
        v_m^max=v_m;
        remember the new solution Q;
      Endif
    Endfor
  End;
```

All the parameter values showed in the pseudo-code were chosen by a trial and error procedure carried out during the algorithm fine-tuning phase. The above described code was run in sequence with the two different initial solutions Q_1 and Q_2, and the algorithm is further referred to as the TSA_2s.

10.2.3.2 Population Learning Algorithm Implementation

Population learning algorithm (PLA), was proposed in [17] as yet another population-based method, which can be applied to solve combinatorial optimization problems. PLA has been inspired by analogies to a social phenomenon rather than to evolutionary processes. Whereas evolutionary algorithms emulate basic features of natural evolution including natural selection, hereditary variations, the survival of the fittest and production of far more offspring than are necessary to replace current generation, population learning algorithms take advantage of features that are common to social education systems:

- A generation of individuals enters the system.
- Individuals learn through organized tuition, interaction, self-study and self-improvement.
- Learning process is inherently parallel with different schools, curricula, teachers, etc.
- Learning process is divided into stages, each being managed by an autonomous agent.
- More advanced and more demanding stages are entered by a diminishing number of individuals from the initial population (generation).
- At higher stages agents use more advanced education techniques.
- The final stage can be reached by only a fraction of the initial population.

In the PLA an individual represents a coded solution of the considered problem. Initially, a number of individuals, known as the initial population, is randomly generated. Once the initial population has been generated, individuals enter the first learning stage. It involves applying some, possibly basic and elementary, improvement schemes. These can be based on some simple local search procedures. The improved individuals are then evaluated and better ones pass to a subsequent stage. A strategy of selecting better or more promising individuals must be defined and duly applied. In the following stages the whole cycle is repeated. Individuals are subject to improvement and learning, either individually or through information exchange, and the selected ones are again promoted to a higher stage with the remaining ones dropped-out from the process. At the final stage the remaining individuals are reviewed and the best represents a solution to the problem at hand.

Within the proposed PLA implementation a solution is represented by the same list of objects as in the TSA algorithm case. The initial solutions Q_1 and Q_2 are individuals attached to the initial population. Additionally, the following notation is used:

- *POP* – Population of solutions (individuals), the initial population includes randomly generated solutions as well as Q_1 and Q_2; $|POP|$ denotes the population size.
- Criterion_Average – Selection procedure which removes from the population all individuals for which the optimality function value is lower than its average value for the whole current population. The condition $|POP|\geq 1$ holds.
- Improver(Q) – the procedure which, if feasible, adds to a randomly chosen task from the solution Q one randomly chosen version of this task; the action is repeated n times.

- The following evolution operators are used: Mutation(*perc*) – the procedure which adds or removes a randomly chosen version to or from randomly chosen task from the population; the *perc* argument determines a part of the population towards which the operator is applied; Crossover(*perc*) – the procedure which performs one point crossover producing an offspring from two randomly chosen individuals from the population and assuring its feasibility through removing and adding redundant and missing tasks, respectively; the *perc* argument determines a part of the population towards which the operator is applied.
- TSA(*Q*, *lit*) – the algorithm TSA with *Q* as the initial solution and *lit* as the number of the algorithm iterations.
- TSA(*Q*, *lit*, *perc*) – the algorithm TSA, the additional argument indicates a part of the population towards which TSA is applied.

The proposed PLA implementation involves three learning stages. At the end of each stage the procedure Criterion_Average is run to select and remove less promising individuals resulting in decrease of the number of individuals in the population. The pseudo-code of the proposed algorithm follows:

```
Procedure PLA
Begin
  |POP|=400;
  //first learning stage
  randomly generate POP;
  For each individual Q in POP do
    Improver(Q);
  Endfor
  Criterion_Average;
  //second learning stage
  For i=1 to n do
    Mutation(0.2); Crossover(0.4); TSA(Q,50,0.02);
    For each individual in POP do
      Improver;
    Endfor
  Endfor
  Criterion_Average;
  //third learning stage
  For each individual in POP do
    TSA(Q,100);
  Endfor
End;
```

All the parameter values showed in the pseudo-code were chosen by a trial and error procedure carried out during the algorithm fine-tuning phase.

10.2.4 Computational Experiment Results

Both algorithms described in Sections 2.3.1 and 2.3.2 were used to solve instances of the multiple-version task scheduling problem P|m-p,a_j,d_j|V. The computational experiments were carried out for 50 randomly generated data sets. Each data set is an instance of scheduling problem. Number of tasks per instance have been generated as random integers from the uniformly distributed interval U[10,28], numbers of available processors from U[2,12], number of variants per task from U[2,5], variant processing times from U[1,15]. The effectiveness of the algorithms was shown based on the mean (MRE) and maximal (MaxRE) relative errors obtained during the computation realized for all data sets. The errors were calculated by comparison to the upper bound known for each instance. Thus obtained computation errors together with some other earlier published results are presented in Table 1 and 2.

Table 1. Mean and maximal relative errors for P|m-p,a_j,d_j|V problem for the algorithms from [7, 20, 21, 18].

	Evolutionary – ANN algorithm [7]	Simulated Annealing - ANN algorithm [20]	Island-Based Evolutionary algorithm [21]	PLA-based on heuristics [18]
MRE	3.39%	2.38%	3.09%	3.55%
MaxRE	14.00%	9.00%	19.00%	11.00%

Table 2. Mean and maximal relative errors for P|m-p,a_j,d_j|V problem for the presented algorithms. The TSA argument denotes a number of iterations

	TSA(100)	TSA_2s(100)	PLA-based on TSA
MRE	5.00%	2.82%	1.49%
MaxRE	14.00%	9.00%	9.00%

Similar algorithms were used to solve multiple-version task scheduling problem P|m-p,a_j,d_j|R. The computational experiment was carried out for 20 randomly generated data sets. Each data set is an instance of scheduling problem where number of tasks have been generated as random integers from U[11,29], number of versions from U[2,5] and number of processors from U[5,8]. Additionally variant reliabilities have been generated as random real numbers from U[0.8999,0.9999]. The results are presented in Table 3.

Table 3. Mean and maximal relative errors for $P|\text{m-p}, a_j, d_j|R$ problem. The TSA argument denotes the number of iterations.

	ANN algorithm [7]	Evolutionary algorithm [21]	TSA(100)	TSA_2s(100)	PLA Based on TSA
MRE	7.44%	4.77%	4.33%	3.16%	0.91%
MaxRE	18.40%	15.86%	14.99%	13.12%	5.79%

10.3 Scheduling Multiprocessor Tasks in the Presence of Correlated Failures

10.3.1 Multiprocessor Task Reliability Model

Algorithms proposed in the earlier Section were based on the assumption that failures of multiple program variants are statistically independent. The present approach differs by allowing correlation of failures within a task. Hence the earlier assumption is not needed any longer. Now, a non-trivial aspect of our investigation is the computation of the task reliability R_j from the variant reliabilities r_{ji}, $i \in \overline{T}_j$. Models for the reliability computation of fault-tolerant structures with variants the failures of which are correlated have been developed in [10, 25] and [26]. These approaches, however, assume that the estimated reliability values of the single variants of a task are all equal, which is a too severe restriction within our context. We need an extended model where these estimates may vary. Nevertheless, in the special case of equal single reliability estimates, the model should be consistent with the framework proposed in the cited papers.

In the following, we describe an extension of the Nicola-Goyal approach serving this purpose. The formal background of this extension is presented in [14]. Here we concentrate on the basic ideas. For simplicity, let us assume that $\overline{T}_j=\{1,\ldots,k\}$ with $1 \le k \le NV_j$ (which can always be achieved by a suitable re-assignment of the variant indices).

The key for an appropriate modeling of failure dependencies between redundant program variants has been given in [10]. There is no reason why independently developed variants should not fail independently on some fixed program input y; nevertheless, on a random program input Y, the failure events for different variants get correlated simply by the fact that not all inputs y have the same probability of leading to a program failure: some of them are very easy to process, such that they are most probably not failure-causing, while others are difficult to process, such that all variants may have a high probability of failing on them. This induces a positive correlation between the failures of the variants in an execution of the task.

Conceptually, we may consider a population *POP* of programs potentially written to the given specification, and assume that the program variants 1,...,k are (randomly) selected from subsets $POP_1,...,POP_k$ of *POP*, respectively. By $p(y)$ and $p_i(y)$ ($1 \leq i \leq k$), we denote the fraction of programs in *POP,* respectively in POP_i, that fail at input y. Now let us suppose that also the input is selected randomly according to some usage distribution, i.e., the input is a random variable Y. Then, the fractions of failing programs in *POP*, respectively in POP_i, which is $p(Y)$, respectively $p_i(Y)$, are random variables as well. The random variable $p(Y)$ can be interpreted as the probability that during operational usage, a program variant selected arbitrarily from *POP* will fail, and $p_i(Y)$ has an analogous interpretation. Nicola's and Goyal's approach is based on the assumption that $p(Y)$ has a Beta distribution with parameters α and β, which are chosen as follows:

$$\alpha = \pi / \theta, \qquad \beta = (1-\pi)/\theta.$$

The density f(p) of Beta(α, β) is given by:

$$f(p) = p^{\alpha-1}(1-p)^{\beta-1}/B(\alpha,\beta),$$

where B(α,β) is the Beta function. In the proposed approach, π is the estimated unreliability of a variant drawn from *POP* (i.e., the expected value of $p(Y)$), and $\theta \geq 0$ is a correlation level, where the boundary case $\theta = 0$ represents failure independence, whereas the other boundary case $\theta = \infty$ represents identical failure behavior of all versions (one of them fails exactly if all others fail).

In our extended model, we assume that also the random variables $p_i(Y)$ ($1 \leq i \leq k$) have Beta distributions, with parameters α_i and β_i chosen as follows:

$$\alpha_i = \pi_i / \theta, \qquad \beta_i = (1-\pi_i)/\theta,$$

where π_i is the estimated unreliability (probability of failing) of variant i drawn from POP_i , and θ is the correlation level as explained above.

Furthermore, a certain functional relation between $p(y)$ and $p_i(y)$ is assumed: We suppose that the fraction $p(y)$ of programs in *POP* failing on a certain input y already determines the fractions $p_i(y)$ of programs in POP_i failing on this input y. Of course, this functional dependency (we denote it by φ_i) cannot be chosen arbitrarily. Instead, it must be established in such a way that the distribution Beta(α,β) of $p(Y)$ is transformed just to the distribution Beta(α_i,β_i) of $p_i(Y)$.

In [14] it is shown that the following function φ_i satisfies this requirement: Let F be the distribution function of Beta(α,β), and let F_i be the distribution function of Beta(α_i,β_i). We set

$$\varphi_i(p)=F_i^{-1}(F(p)).$$

The relation between $p(y)$ and $p_i(y)$ is then given by

$$p_i(y) = \varphi_i(p(y)).$$

On the assumptions above, the unreliability $unrel_j$=1–R_j of task j, i.e., the probability that all its variants i (i= 1,...,k) fail on a random input Y, can be computed by elementary probabilistic calculations: One obtains

$$unrel_j = \int_0^1 \varphi_1(p) \; ..... \; \varphi_k(p)\, f(p)\, dp,$$

where $f(p)=F'(p)$ is the first derivative of the distribution function F, i.e., the corresponding density function.

As it has been demonstrated in [14], the value $unrel_j$ computed from the formula above is *independent* of the chosen basic distribution F: another F yields another density $f(p)$, but also other transformation functions $\varphi_i(p)$, leading finally to the same computation result. The distribution function F of a Beta(α,β) distribution is given by

$$F(p) = \int_0^p u^{\alpha-1} / B(\alpha,\beta) du$$

with Beta(α,β) denoting the Beta function as before. Using this fact and a Taylor expansion of the Beta function, we obtain the following numerical procedures for the computation of $unrel_j$:

```
Function B(α,β,x)

Begin
  value=0;
  For j=0 to β-1 do
    value=value+(-1)^j (β-1)!x^(α+j)/(j!(β-1-j)!(α+j));
  Endfor
  Return value;
End

Function phi(α,β,α₀,β₀,p)
Begin
  c=B(α₀,β₀,p)*B(α,β,1)/B(α₀,β₀,1); x=0.1;
  While B(α,β,x)>c do

    x=x/10; x_old=x;
    Repeat until (|x-x_old| smaller then some pre-
    defined real precision constant)
      x_old=x; x=x-(B(α,β,x)-c)/(x^(α-1)*(1-x)^(β-1));
  Endwhile
```

```
    Return x;
  End

  Function integrand(k,α[],β[],α0,β0,p)
  Begin
    q=1;
    For i=1 to k do
      q=q*phi(α[i],β[i],α0,β0,p);
    Endfor
    Return q*p^(α0-1)*(1-p)^(β0-1)/B(α0,β0,1);
  End

  Function unrel(k,α[],β[],α0,β0)
  Begin
    return result of numerical integration of
    integrand(k,α[],β[],α0,β0,p) between p=0 and p=1,
    e.g., by Simpson's rule;
  End
```

Therein, α_0 and β_0 are the parameters chosen for the basic Beta distribution F, and α[] and β[] are the arrays of the numbers α_i and β_i . Since $\beta_i=(1-\pi_i)/\theta$, and since π_i and θ are small in practical applications, the values β_i can be rounded to integers. The procedures above are based on the assumption that this has been done.

The respective multiprocessor task scheduling problems are denoted as: $P|\text{m-p},a_j,d_j,\theta_j|V$, $P|\text{m-p},a_j,d_j,\theta_j|R$, $P|\text{m-r},a_j,d_j,\theta_j|V$, $\text{P}|\text{m-r},a_j,d_j,\theta_j|\text{R}$, $P|\text{m-v},a_j,d_j,\theta_j|V$ and $P|\text{m-v},a_j,d_j,\theta_j|R$.

10.3.2 Approximation Algorithms for Solving Multiprocessor Task Scheduling Problems in the Presence of Correlated Failures

From the problem assumptions it is clear that tasks are allocated to processors only after their size and the internal structure, i.e. combination of variants to be processed, have been chosen. It also follows from these assumptions that when constructing a task from, say, NV available variants there are $2NV$-1 possible structures to be considered.

To reduce the solution space a concept of an effective task structure is introduced. For a task j of the size equal to x ($1 \le x \le NV_j$) there are at most NV_j+1-x effective structures. A structure of a given size is considered effective if there exists no other combination of program variants with the same (or shorter) processing time and a smaller reliability. Generating effective structures for a given task size x is straightforward and requires polynomial time as shown in the following procedure ESG:

```
Procedure ESG
Begin
  Order variants according to their non-decreasing
  processing times as the first criterion and reli-
  ability as the second.
  For x=2 to NV_j do
    For i=x to NV_j do
      among all variants v_k, k<i choose x-1 variants
      of the highest reliability;
      produce a new combination consisting of the v_i
      and x-1 variants chosen in the previous step;
    Endfor;
  Endfor;
  For each size x (1≤x≤NV_j) do
    choose the effective structures among the above
    obtained;
  Endfor
End
```

10.3.2.1 Island Based Evolution Algorithm – IBEA

An island-based evolution algorithm (IBEA) belongs to the distributed algorithm class. To improve efficiency of genetic algorithms (GA) several distributed GA's were proposed in [13, 4, 1]. Their ideas included an island-based approach where a set of independent populations of individuals evolves on "islands" cooperating with each other. The island-based approach brings two benefits: a model that maps easily onto the parallel hardware and extended search area (due to multiplicity of islands) preventing from sticking in local optima. Promising results of the island-based approach motivated the authors to design the IBEA for scheduling multiple-variant tasks.

The proposed island-based evolutionary algorithm (IBEA) works on two levels with two corresponding types of individuals. To evolve individuals of the lower level a population-based evolutionary algorithm (PBEA) is proposed. On the higher level the following assumptions are made:

- An individual is an island I_k, k=1,2,...,K, where K is the number of islands.
- An island is represented by a set of the lower level individuals.
- All islands are located on the directed ring.
- Populations of lower level individuals evolve on each island independently.
- Each island I_k regularly sends its best solution to the successor $I_{(k \bmod K)+1}$ in the ring.
- The algorithm stops when an optimality criterion is satisfied or the preset number of populations on each island have been generated.

- When IBEA stops the best overall solution is the final one.

On the lower level the following assumptions are made:

- An individual is represented by a *n*-element vector $S_u=\{ c_i^j \mid i=1,2,...,n; 1 \leq j \leq n)$, used to construct a list of tasks, where *i* is an index describing a place of the task *j* on the list, and c_i^j is a code representing the combination of variants used to construct task *j*.
- All S_u representing feasible solutions are potential individuals.
- An initial population is composed in part from the potential individuals for whom combination and order of tasks on the list is random, and in part from the potential individuals for whom combination of tasks is random with order determined by a non-decreasing ready time as the first criterion and a non-decreasing deadlines as the second.
- Each individual can be transformed into a solution by applying LSG-$P|$m-p,$a_j,d_j,\theta_j|R$, which is a specially designed algorithm for list-scheduling m-p tasks.
- Each solution produced by the LSG-$P|$m-p,$a_j,d_j,\theta_j|R$ can be directly evaluated in terms of its fitness.
- New population is formed by applying several evolution operators: selection and transfer of some more "fit" individuals, random generation of individuals, crossover, and mutation.

In the following pseudo-code main stages of the IBEA-$P|$m-p,$a_j,d_j,\theta_j|R$ algorithm are shown:

```
Procedure IBEA-P|m-p,a_j,d_j,θ_j|R
 Begin
   Set number of islands K, number of populations PN
   for each island, size of the population for each
   island PS. For each island I_k, generate an initial
   population PP_0;
   While no stopping criteria is met do
     For each island I_k do
       Evolve PN populations using PBEA;
       Send the best solution to I_(k mod K)+1;
       Incorporate the best solution from
       I_((K+k-2) mod K)+1 instead of the worst one;
     Endfor
   Endwhile
   Find the best solution across all islands and save
   it as the final one;
 End
```

PBEA-$P|$m-p,$a_j,d_j,\theta_j|R$ algorithm is shown in the following pseudo-code:

```
Procedure PBEA-P|m-p,a_j,d_j,θ_j|R
Begin
```

```
    Set population size PS, generate a set of PS indi-
    viduals to form an initial population PP_0;
    Set ic=0; (ic - iteration counter);
    While no stopping criteria is met do
      Set ic=ic+1;
      Calculate fitness factor for each individual in
      PP_ic-1 using LSG-P|m-p,a_j,d_j,θ_j|R;
      Form new population PP_ic:
      Select randomly a quarter of PS individuals from
      PP_ic-1 (probability of selection depends on fitness
      of an individual);
      Produce a quarter of PS individuals by applying
      crossover operator to previously selected indi-
      viduals from PP_ic-1;
      Produce a quarter of PS individuals by applying
      mutation operators to previously selected indi-
      viduals from PP_ic-1;
      Generate half of a quarter of PS individuals from
      set of potential individuals (random size, and
      order);
      Generate half of a quarter of PS individuals from
      set of potential individuals (random size, and
      fixed order);
    Endwhile
  End
```

LSG-P|m-p,a_j,d_j,θ_j|R algorithm used within PBEA-P|m-p,a_j,d_j,θ_j|R is carried in the three following steps:

```
  Procedure LSG-P|m-p,a_j,d_j,θ_j|R
  Begin
    Construct a list of tasks from the code represent-
    ing individuals. Set loop over tasks on the list.
    Within the loop, allocate current task to multiple
    processors minimizing the beginning time of its
    processing. Continue with tasks until all have been
    allocated.
    If the resulting schedule has task delays, a fit-
    ness of the individual S_u is calculated as R_u=-(1-
    ΠR_l) where l belongs to a set of the delayed tasks.
    Otherwise, R_u=ΠR_j, j=1,...,n.
  End
```

10.3.2.2 Neural Network Algorithm – NNA

Neural networks have been used to obtain solutions of problems belonging to NP-hard class (see for example [15]). NNA have been also applied to job-shop scheduling [27, 16]. The NNA proposed in this paper is based on a dedicated neural network. To train the network an evolution algorithm is used. The NNA-P|m-p,a_j,d_j,θ_j|R architecture depends on a number and properties of m-p tasks, which are to be scheduled. Set of these tasks is partitioned to form subsets of tasks with identical ready times. The number of such subsets corresponds to the number of layers in the neural network. Each layer consists of neurons representing the tasks to be scheduled.

The described architecture is additionally extended and includes elements known as adders. These elements are situated in each layer. Number of adders at each layer corresponds to the number of available processors. The neurons of each layer send their signals to adders. Signals between layers pass through decision blocks, which are equivalent to decoders. The decision blocks take decisions with respect to correcting (or not) signals from adders. They may restructure connections between neurons and adders in a layer if signals from adders require such a correction. If the change of connections does not yield a positive result then a decision block can stimulate an action of the decision block in earlier layer if it exists.

Decision blocks adjust connection weights in learning process using genetic algorithm. Weights are adjusted after the end of the subsequent epoch. This method is known as batch training procedure [9].

During the learning process a genetic algorithm is searching for suitable weights of connections. Each neuron must send a signal to at least one adder. Chromosome for a layer is a string of bits which length depends on number of tasks and processors in this layer. In searching for weights of connections the genetic algorithm uses standard genetic operators: crossover and mutation. Crossover is based on exchanging parts of bit strings between two chromosomes. Operation of mutation is realized by random changes of weight values. In the selection process elitist method is used.

The proposed Neural Network Algorithm involves executing steps shown in the following pseudo-code:

```
Procedure NNA-P|m-p,a_j,d_j,θ_j|R
Begin
  Set number of layers k;
  While no stopping criteria is met do
    For i=1 to k do
      Train network using GA;
    Endfor
    Evaluate the objective function for the obtained
    network structure;
    If the objective function is improved then
     best network structure = obtained structure;
    Endif
  Endwhile
  Treat the best network structure as the final one
  and decode the problem solution;
End
```

10.3.2.3 Hybrid 3opt-tabu Search Algorithm – TSA

The proposed approach is an extension of the tabu-search meta-heuristic described in Section 2.3.1. The algorithm TSA-$P|\text{m-p},a_j,d_j,\theta_j|R$ is based on the following assumptions and notations:

- A solution Q is a list of objects, each representing a single variant of a task. An object has attributes including task number j, j=1,...n and variant number i, i=1,...NV_j. At least one variant of each task has to be scheduled.
- For each processor b a list of objects is created. For k processors there are k such lists. Originally lists are generated by consecutively allocating objects to the least loaded processor;
- objects on the lists are ordered using the 3-opt algorithm (i.e. searching for the best schedule for the first three objects, then 2nd, 3rd, and 4th, etc.).
- A move involves adding or erasing a variant i of the task j from the list owned by a processor b.
- Attributes of the move are denoted as $m(b, j, i, h, ea, mr)$, where b, j, i are the respective indexes, h is a position of task j on the respective list, ea is a move status (+1) for adding, and (-1) for erasing, and mr is a move value.
- $N(Q)$ denotes a set of possible moves from Q, called neighbourhood of Q.
- $N'(Q)$ denotes a modified neighbourhood of Q, that is neighbourhood without all tabu-active moves and moves of which it is easy to say that they will not improve the current solution.
- TAM denotes list of tabu-active moves.

TSA-$P|\text{m-p},a_j,d_j,\theta_j|R$ involves executing steps shown in the following pseudo-code:

```
Procedure TSA-P|m-p,a_j,d_j,θ_j|R
Begin
  SM=∅;
  TAM=∅;
  Generate two initial solutions:
    Q1- includes one shortest variant for each task
    order by their non-decreasing ready times as the
    first criterion and non-decreasing deadlines as
    the second;
    Q2 - includes one shortest variant for each task
    order by their non-decreasing deadlines as the
    first criterion and non-decreasing ready times as
    the second;
  Calculate total tardiness;
```

```
Try to improve both solutions using the 3-opt ap-
proach providing the total tardiness value remains
zero;
Set number of iterations it;
If total tardiness>0 then scheduling problem has no
solution;
else
  For both initial solutions Q1 and Q2 do
    max = reliability of the initial solution;
    For 1 to it do
      While best move is not chosen do
        Consider all moves in N'(Qx), select and re-
        member the best move (with the best move
        value) m_best(b, j, i, h, ea, mr);
        If the best move cannot be found then
          delete from TAM all moves with the least
          tabu_tenure;
        Endif
      Endwhile
      Add the new solution to the SM;
      Update TAM:
      Add m1(b_best,0,0,h_best,0,0), it/4;
      Add m2(b_best,0,i_best,0,-1,0), it/6;
      Add m3(b_best,j_best,i_best,-ea_best,0), it/2;
      Delete moves with tabu_tenure=0.
      If mr_best>max then max=mr_best;
    Endfor
    Choose the best solution from the solutions
    found for Q1 and Q2;
  Endfor
 Endif
End
```

10.3.2.4 Population Learning Scheme – PLS

To obtain solutions of the $P|\text{m-p},a_j,d_j,\theta_j|R$ problem instances we have adopted the Lin-Kerninghan algorithm [22] originally applied to solving traveling salesman problems, and used it as a population learning tool. The approach is based on the improvement procedure, which can be applied to the existing schedule, possibly not even a feasible one. The procedure aims at improving a schedule by reducing the number of the delayed tasks and, at the same time, increasing schedule reliability. Within the proposed PLS the improvement procedure is applied to a population of solutions. Constructing a list of tasks, sorting it according to non-decreasing deadlines and allocating tasks from this list to least loaded processors generates an initial population. Tasks are allocated to processors in order of their appearance on the list with some probability q and randomly with probability

1-q. For each task a number of program variants to be executed is randomly drawn.

The improvement procedure aims at maximizing schedule reliability minus the number of delayed tasks. Its idea is to consider one by one the delayed tasks. For each task z a set K of other tasks, such that inserting task $k \in K$ instead of task z best improves value of the goal function, is constructed. As the result a schedule with a double task k appears. Improvement is again applied to the second appearance of the task k. If set K can not be constructed a schedule is upgraded in such a way that each task is scheduled exactly once and the result compared with the best schedule achieved so far. At the first two levels of recursion set K contains 5 elements, later on, only 1 element.

10.3.3 Numerical Example and the Results of the Computational Experiment

As an example the $P|\text{m-p},a_j,d_j,\theta_j|R$ problem instance with 11 tasks and 5 processors has been solved using all of the proposed approaches. The respective data is shown in Table 4.

All the proposed algorithms have generated feasible schedules – that is schedules without delayed tasks, with at least one variant of each task scheduled, and all tasks having the required resources (processor and time) allocated. Thus generated schedules are shown in Fig. 1. To evaluate the proposed algorithms computational experiment has been carried. It has involved 20 randomly generated problems. Numbers of tasks per problem have been generated as random integers from the uniformly distributed interval U[10,28], numbers of available processors from U[2,12], numbers of variants per task from U[2,5], variant processing times from U[1,15]. Variant reliabilities have been generated as random real numbers from U[0.8999,0.9999], and variants failure correlation from U[0.1,0.2].

The following measures have been used to evaluate the algorithms:

- Mean relative error related to the best known solution – *MRE.*
- Maximum relative error encountered – *Max RE.*
- Minimum relative error encountered – *Min RE.*
- Best result ratio – *BRR.*

Results of the experiment are shown in Table 5. They show that the proposed algorithms generate good quality solutions. The hybrid 3opt-tabu search algorithm outperforms the remaining ones in terms of both - mean relative error and best result ratio. It also has a very low maximum relative error level. Island based evolution algorithm also offers good performance and a low maximum relative error level. However both approaches require more computational resources then the remaining two. Population learning scheme has been faster then IBEA by the factor of 2 to 3.

Table 4. Example $P|\text{m-p},a_j,d_j,\theta_j|R$ problem instance data

Task no	Var. per task	Var. nr	a_j	p_{ij}	d_j	r_{ij}	θ_j	h_j
1	5	1	0	2	9	0,9250	0,1705	2
		2	0	2	9	0,9301		
		3	0	4	9	0,9365		
		4	0	4	9	0,9873		
		5	0	5	9	0,9998		
2	5	1	0	3	16	0,9450	0,1533	2
		2	0	3	16	0,9569		
		3	0	3	16	0,9600		
		4	0	4	16	0,9723		
		5	0	4	16	0,9825		
3	4	1	1	5	21	0,9029	0,1579	1
		2	1	5	21	0,9563		
		3	1	6	21	0,9762		
		4	1	7	21	0,9835		
4	4	1	3	7	25	0,9441	0,1289	2
		2	3	8	25	0,9619		
		3	3	9	25	0,9558		
		4	3	9	25	0,9608		
5	5	1	4	3	19	0,9143	0,1301	2
		2	4	3	19	0,9230		
		3	4	4	19	0,9312		
		4	4	4	19	0,9337		
		5	4	4	19	0,9801		
6	5	1	9	8	25	0,9060	0,1774	2
		2	9	8	25	0,9092		
		3	9	10	25	0,9237		
		4	9	11	25	0,9254		
		5	9	11	25	0,9367		
7	3	1	13	5	26	0,9074	0,1014	1
		2	13	5	26	0,9497		
		3	13	6	26	0,9633		
8	4	1	14	6	30	0,9029	0,176	2
		2	14	7	30	0,9448		
		3	14	7	30	0,9840		
		4	14	8	30	0,9896		
9	2	1	18	5	27	0,9274	0,1814	1
		2	18	6	27	0,9794		
10	2	1	23	5	32	0,9187	0,1709	1
		2	23	6	32	0,9621		
11	4	1	26	2	32	0,9047	0,1045	2
		2	26	2	32	0,9749		
		3	26	2	32	0,9831		
		4	26	2	32	0,9915		

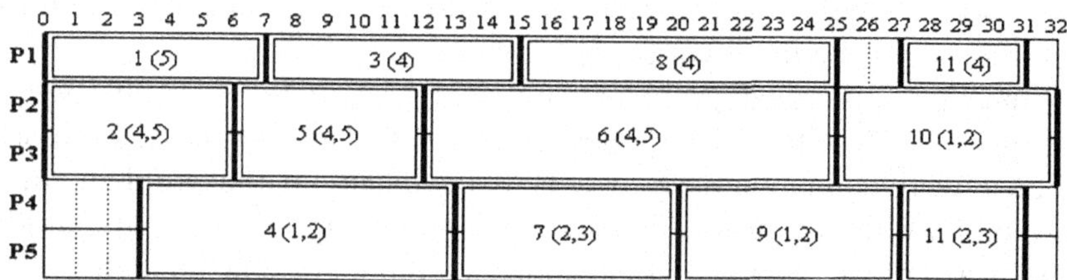

a) Island based evolution algorithm and population learning scheme (R = 0,917186).

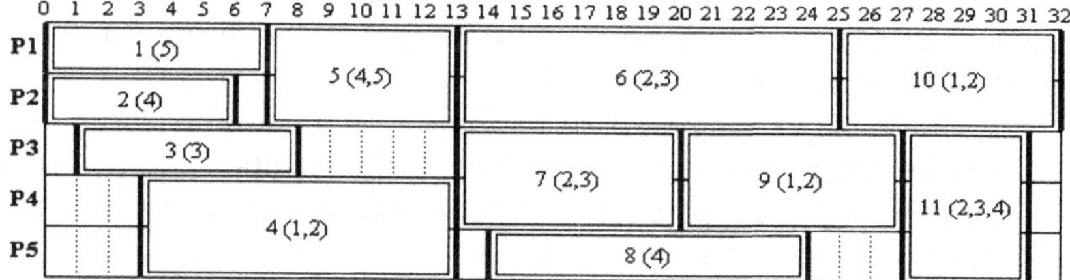

b) Neural network algorithm (R = 0,89425)

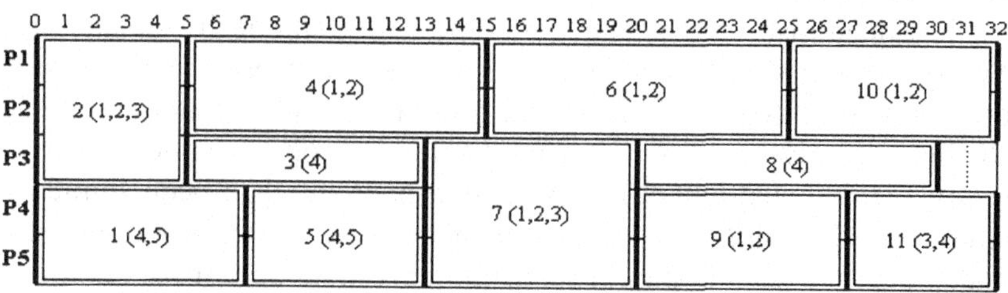

c) Hybrid 3opt-tabu search algorithm (R = 0,915393)

Fig. 1. Schedules generated by the proposed algorithms

Table 5. Computational experiment results

Measure	IBEA	NNA	TSA	PLS
MRE	0.371%	6.501%	0.243%	1.255%
Max RE	2.21%	18.01%	2.21%	3.34%
Min RE	0%	0%	0%	0%
BRR	31.6%	10.5%	47.4%	15.8%

10.4 Conclusions

The chapter deals with the multiple-version task scheduling problems. In the first part of the chapter the complexity of the problems m-p, m-r and m-v was considered. Theorem 1 showed that these problems are NP-hard in the strong sense for any processor number.

Since the discussed problems are computationally difficult, two approximation algorithms have been proposed for solving them. Both seem to be effective for randomly generated instances of the problem. In the case of TSA algorithm for $P|\text{m-p},a_j,d_j|V$ problem applying two initial solutions was crucial. Sequential execution of the algorithm for two solution results in the reduction of the mean error by 43.6%.

The population learning algorithm is even more effective. The PLA, which applies TSA as one of its learning procedures and mutation operator is also better than another population learning algorithm using heuristics as learning procedures.

To improve the effectiveness of the presented algorithms an algorithm TSA may be constructed with two or more initial solutions. Each solution might be searched for in parallel on different processors. The same idea of parallelization used in the PLA could enable the simultaneously learning for two or more populations. It could also be useful to exchange individuals between different populations.

It has been also shown that novel modeling approach overcoming the idealized assumption that redundant variants fail independently from each other lends itself to optimization and could be useful in generating realistic schedules for systems involving fault-tolerant program structures run under hard time constraints. The presented approach has been limited to a static case where information on set of tasks to be run is known beforehand. Further research should lead into investigating construction of dynamic schedulers where the suggested approach (static one) could be used as a base for an online scheduler. Another direction of research is further improvement of the soft computing algorithms and techniques resulting in improving either quality of result or reducing the demand for computational resources.

References

[1] Alba, E., J. Troya (1999) Analysis of Synchronous and Asynchronous Parallel Distributed Genetic Algorithms with Structured and Panmictic Islands, in: Jose Rolim et al. (Eds.), *Proceedings of the 10th Symposium on Parallel and Distributed Processing, 12-16 April,* San Juan, Puerto Rico, USA: 248-256

[2] Avivžienis, A. (1985) The N-version approach to fault-tolerant software, IEEE Transactions on Software Engineering SE-11 (12): 1491-1501

[3] Avivžienis, A., J. Xu (1995) The methodology of N-version programming, Software Fault Tolerance, Trends in Software 3, John Wiley & Sons, Chichester: 23-46

[4] Belding, T.C. (1995) The Distributed Genetic Algorithm Revisited, in: Eshelman, L.J. (ed.): *Proceedings of the Sixth International Conference on Genetic Algorithms,* Morgan Kaufmann, San Francisco, CA.: 114-121

[5] Bondavalli, A., L. Simoncini, J. Xu (1993) Cost-effective and flexible scheme for software fault tolerance, Computer System Science & Engineering 4: 234-244.

[6] Czarnowski, I., W.J. Gutjahr, P. Jędrzejowicz, E. Ratajczak, A. Skakowski, I. Wierzbowska (2003) Scheduling multiprocessor tasks in presence of the correlated failures, Central European Journal of Operational Research, 11:163-182

[7] Czarnowski, I., P. Jędrzejowicz (2000) Artificial neural network for multiprocessor task scheduling, Intelligent Information Systems, Proceedings of the IIS'2000 Symposium, Bystra, Poland, June 12-16 2000, Phisica-Verlag: 207-216

[8] Drozdowski, M. (1996) Scheduling multiprocessor tasks – An overview, Elsevier Science, European Journal of Operational Research 94: 215-230

[9] Duch, W., J. Korczak (1998) Optimization and Global Minimization Method Suitable for Neural Networks. N*eural Computing Surveys 2,* http://www. icsi.berceley.edu/~jagopta/NCS

[10] Eckhardt, D.E., L.D. Lee (1985) A theoretical basis for the analysis of multiversion software subject to coincident errors, *IEEE Transactions on Software Engineering,* vol. SE-11: 1511-1517

[11] Garey, M.R., D.S. Johnson (1977) Computers and intractability. A guide to the theory of NP-completeness, W.H. Freeman and Company, San Francisco

[12] Glover, F., M. Laguna (1997) Tabu Search, Kluwer, Boston

[13] Gordon, V.S., D. Whitley (1993) Serial and Parallel Genetic Algorithms as Function Optimizers, in: Forrest S. (ed.): *Proceedings of the Fifth International Conference on Genetic Algorithms,* Morgan Kaufmann, San Mateo, CA.: 177-183

[14] Gutjahr, W.J. (2000) A Reliability Model for Inhomogeneous Redundant Software Versions with Correlated Failures, *Working Paper – Department of Statistics and Decision Support Systems*, University of Vienna, Vienna

[15] Hopfield, J.J., D.W. Tank (1985) Neural Computations of Decision in Optimization Problems, *Biological Cybernetics*, vol. 52: 141-152

[16] Janiak, A. (1999) Wybrane Problemy i Algorytmy Szeregowania Zadań i Rozdziału Zasobów, *Akademicka Oficyna Wydawnicza PLJ*, Warszawa (in Polish)

[17] Jędrzejowicz, P. (1999) Social learning algorithm as a tool for solving some difficult scheduling problems, Foundation of Computing and Decision Sciences, 24(2): 51-66.

[18] Jędrzejowicz, P., I. Czarnowski, M. Forkiewicz, E. Ratajczak, A. Skakowski, I. Wierzbowska (2001) Population-based scheduling on multiple processors,

Proceedings 4th Metaheuristics International Conference, MIC'2001, Porto: 613-618

[19] Jędrzejowicz, P., I. Czarnowski, E. Ratajczak, A. Skakowski, H. Szreder (2000) Maximizing schedule reliability in presence of multiple-variant tasks, in: M.P. Cottam, D.W. Harvey, R.R. Pape, J. Tait (ed.) Foresight and Precaution, Proceedings of ESREL 2000, A.A. Balkema Publ., Rotterdam, 1: 679-687

[20] Jędrzejowicz, P., I. Czarnowski, A. Skakowski, H. Szreder (2001) Evolution-based scheduling of multiple variant and multiple processor programs, Future Generation Computer Systems, 17: 405-414

[21] Jędrzejowicz, P., A. Skakowski (2000) An island-based evolutionary algorithm for scheduling multiple-variant tasks, Proceedings of ICSC Symposium Engineering of Intelligent Systems, Academic Press, Paisley: 1-9

[22] Lin, S., B.W. Kerningham (1973) An Effective Heuristic Algorithm for the Travelling-Salesman Problem, *Operations Research,* 21: 498 - 516.

[23] Lyu, M.R. (1995) Software fault tolerance – Preface, Software Fault Tolerance, Trends in Software 3, John Wiley & Sons, Chichester

[24] Lyu, M.R. (1995) Preface, in: M.R.Lyu (ed.), *Software Fault Tolerance,* Wiley, New York: xi-xiv

[25] Nicola, V.F., A. Goyal (1990) Modeling of correlated failures and community error recovery in multiversion software, *IEEE Transactions on Software Engineering,* vol. SE-16: 350-359

[26] Tomek, L.A. J.K. Muppala, K.S. Trivedi (1993) Modeling correlation in software recovery blocks, *IEEE Transactions on Software Engineering,* vol. SE-19: 1071-1086

[27] Zohu, D.N., V. Cherkassky, T.R. Balwin, D.E. Olson (1991) A Neural Approach to job shop scheduling, *IEEE Transactions on Neural Networks*, vol 2: 175-179

Redundancy Optimization Problems with Uncertain Lifetimes

Ruiqing Zhao and Wansheng Tang

Institute of Systems Engineering
Tianjin University, Tianjin 300072, China

11.1 Introduction

Component redundancy plays a key role in engineering design and can be effectively used to increase system performances. Often two underlying component redundancy techniques are considered. One is parallel redundancy where all redundant units are in parallel and working simultaneously. This method is usually employed when the system is required to operate for a long period of time without interruption. The other is standby redundancy where one of redundant units (i.e., spares) begins to work only when the active one failed. This method is usually employed when the replacement takes a negligible amount of time and does not cause system failure. The problem of determining the number of redundant units for improving the system performances under some constraints such as cost constraint is well known as the redundancy optimization problem. It has been addressed in much research work on redundancy optimization theory such as Coit (2001), Coit and Smith (1998), Kuo and Prasad (2000), Levitin et al. (1998), Prasad et al. (1999) and Zhao and Liu (2003).

In a classical redundancy optimization model, the lifetimes of system and components have been basically assumed to be random variables, and system performances such as system reliability are evaluated using the probability measure. Although this assumption has been adopted and accorded with the facts in widespread cases, it is not reasonable in a vast range of situations. For many systems such as space shuttle system, the estimations of probability distributions of lifetimes of systems and components are very difficult due to uncertainties and imprecision of data (Cai et al. (1991)). Instead, fuzzy theory can be employed to handle this case.

R. Zhao and W. Tang: *Redundancy Optimization Problems with Uncertain Lifetimes*, Computational Intelligence in Reliability Engineering (SCI) **39**, 329–374 (2007)
www.springerlink.com

Possibility theory was introduced by Zadeh (1978) and developed by many investigators such as Dubois and Prade (1988), Yager (1993) and Liu (2004). The use of fuzzy methodology in reliability engineering can be traced back to Kaufmann (1975). The main work of fuzzy methodology in reliability engineering appeared in 1980s. Cai et al. (1991) gave an insight by introducing the possibility assumption and the fuzzy state assumption to replace the probability and binary state assumptions. Recently, fuzzy methodology has been widely applied in reliability engineering such as human reliability (Karwowski and Mital 1986), hardware reliability (Cai et al. 1991), structural reliability (Shitaishi and Furuta 1983) and so on.

The use of fuzzy theory in representing unknown parameters provides an interesting alternative to the conventional approaches using probabilistic modeling (Zhao and Liu 2005). In fact, from a practical viewpoint, the fuzziness and randomness of the element lifetimes are often mixed up with each other. For example, the element lifetimes are assumed to be exponentially distributed variables with unknown parameters. Often these parameters can be estimated from historical data. But sometimes obtaining these data by means of experiments is difficult. Instead, expert opinion is used to provide the estimations of these parameters. In this case, fuzziness and randomness of the element lifetimes are required to be considered simultaneously and the effectiveness of the classical redundancy optimization theory is lost (Zhao and Liu 2004).

The concept of fuzzy random variables was introduced by Kwakernaak (1978, 1979). Roughly speaking, a fuzzy random variable is a measurable function from a probability space to a collection of fuzzy variables. Recently, a new variable, random fuzzy variable, was presented by Liu (2000) and defined as a function from a possibility space to a collection of random variables. In addition, an expected value operator of random fuzzy variable was introduced by Liu and Liu (2003). Both fuzzy random theory and random fuzzy theory offer powerful tools for describing and analyzing the uncertainty of combining randomness and fuzziness. These two concepts play important role in redundancy optimization problems involving both fuzziness and randomness (Zhao and Liu 2004).

In this chapter, we will introduce the uncertainty theory into redundancy optimization problems and establish a theory of redundancy optimization in uncertain environments. In Section 2, we take a look at the concepts and properties of uncertain variables. In Section 3, a general redundancy system is presented and the concept of the system structure function is reviewed. In Section 4, we formulate three constructive types of the system performances and design some simulations to estimate them. We present a spectrum of redundancy uncertain programming models, including redundancy expected value model (Redundancy EVM), redundancy chance-constrained programming (Redundancy CCP) and redundancy dependent-chance programming (Redundancy DCP), for general redundancy optimization problems in Section 5. The genetic algorithm is employed to solve these models in Section 6. Finally, some numerical examples are provided to illustrate the effectiveness of the algorithm proposed in Section 7.

11.2 Uncertain Variables

In this section we will give a presentation of the basic concepts of the uncertainty theory. The principles discussed here will serve as background for the study of redundancy optimization problems.

11.2.1 Fuzzy Variable

Let Θ be a nonempty set, and $P(\Theta)$ the power set of Θ (i.e., $P(\Theta)$ is the collection of all subsets of Θ). First, we concentrate on the following four axioms provided by Nahmias (1978) and Liu (2004):

Axiom 1. $\text{Pos}\{\Theta\} = 1$.

Axiom 2. $\text{Pos}\{\phi\} = 0$.

Axiom 3. $\text{Pos}\{\bigcup_i A_i\} = \sup_i \text{Pos}\{A_i\}$ for any collection $\{A_i\}$ in $P(\Theta)$.

Axiom 4. Let Θ_i be nonempty sets on which $\text{Pos}_i\{\cdot\}$ satisfy the first three axioms, $i = 1, 2, \cdots, n$, respectively, and $\Theta = \Theta_1 \times \Theta_2 \times \cdots \times \Theta_n$. Then

$$\text{Pos}\{A\} = \sup_{(\theta_1,\theta_2,\cdots,\theta_n)\in A} \text{Pos}_1\{\theta_1\} \wedge \text{Pos}_2\{\theta_2\} \wedge \cdots \wedge \text{Pos}_n\{\theta_n\} \tag{1}$$

for each $A \in P(\Theta)$. In that case we write $\text{Pos} = \text{Pos}_1 \wedge \text{Pos}_2 \wedge \cdots \wedge \text{Pos}_n$.

Here we can see that the set function Pos assigns a number $\text{Pos}\{A\}$ to each set A in $P(\Theta)$. This number is employed to indicate the possibility that the set A occurs in practice. We call Pos a possibility measure if it satisfies the first three axioms. Further, the triplet $(\Theta, P(\Theta), \text{Pos})$ is called a possibility space.

Example 1: Let Θ =[0, 2], and $P(\Theta)$ the power set of Θ. Define

$$\text{Pos}\{\theta\} = \begin{cases} \theta, & \text{if } 0 \le \theta < 1 \\ 2-\theta, & \text{if } 1 \le \theta \le 2. \end{cases}$$

Then the set function Pos is a possibility measure and the triplet $(\Theta, P(\Theta), \text{Pos})$ is a possibility space.

Besides the possibility measure, necessity measure is another measure to assess a set in the power set $P(\Theta)$ of Θ in fuzzy theory.

Definition 1 (Zadeh 1979) Let $(\Theta, P(\Theta), \text{Pos})$ be a possibility space, and A a set in $P(\Theta)$. Then the necessity measure of A is defined by

$$\text{Nec}\{A\} = 1 - \text{Pos}\{A^c\}. \tag{2}$$

In fuzzy theory, possibility measure and necessity measure are two important measures. It is well known that the necessity measure is the dual of possibility measure.

Recently, Liu and Liu (2002) provided a new measure — credibility measure which is defined as the average of possibility measure and necessity measure.

Definition 2 (Liu and Liu 2002) Let $(\Theta, P(\Theta), \text{Pos})$ be a possibility space, and A a set in $P(\Theta)$. Then the credibility measure of A is defined by

$$\text{Cr}\{A\} = \frac{1}{2}(\text{Pos}\{A\} + \text{Nec}\{A\}). \tag{3}$$

The main difference among the above-mentioned three measures is that they consider the same question from the different angle. Actually, the possibility measure assesses the possibility of a set A from the angle of affirmation, while the necessary measure from the angle of impossibility of the opposite set A^c. This fact will lead that the possibility measure overrates the possibility of a set while the necessary measure underrates the possibility of a set. The credibility measure is just right a corrective of possibility measure and necessity measure.

Example 2: Let us continue to consider Example 1. Let A=[0,1), then we have

$$\begin{aligned}
&\text{Pos}\{A\} = 1,\\
&\text{Nec}\{A\} = 1 - \text{Pos}\{A^c\} = 0,\\
&\text{Cr}\{A\} = \frac{1}{2}(\text{Pos}\{A\} + \text{Nec}\{A\}) = 0.5.
\end{aligned}$$

It is obvious that $\text{Nec}\{A\} \le \text{Cr}\{A\} \le \text{Pos}\{A\}$.

In stochastic case, as we all know, an event must hold if its probability is 1, and fail if its probability is 0. In fuzzy case, however, a fuzzy event may fail even though its possibility is 1, and hold even though its necessity is 0. This is the greatest shortcoming for both possibility measure and necessity measure. From the point of mathematical programming, it is very hard for decision maker to adopt the decision based on possibility measure or necessity measure. In all combinations of possibility measure and necessity measure, only the average of possibility and necessity, the credibility measure, is self dual. A fuzzy event must hold if its credibility is 1, and fail if its credibility is 0. Hence, it is reasonable for us to use the credibility measure to characterize the performance of the redundancy systems in our later sections. Here we emphasize that we don't oppose to characterize the system in the context of possibility measure or necessity measure.

In addition, there is an essential difference between credibility measure and probability measure. As we know, if events A and B are disjoint in probability sense, we have $\Pr\{A \cup B\} = \Pr\{A\} + \Pr\{B\}$. However, in fuzzy sense, generally, $\mathrm{Cr}\{A \cup B\} \neq \mathrm{Cr}\{A\} + \mathrm{Cr}\{B\}$. For example, let A=(0, 0.5) and B=(1.6, 2) in Example 1. Then we have $\mathrm{Cr}\{A \cup B\} = 0.25$, $\mathrm{Cr}\{A\}$=0.25 and $\mathrm{Cr}\{B\}$=0.2. We can see that $\mathrm{Cr}\{A \cup B\} \neq \mathrm{Cr}\{A\} + \mathrm{Cr}\{B\}$. Thus the credibility measure is not a probability measure. For details of propositions of the possibility measure, necessity measure and credibility measure, see Liu (2002).

Kaufmann (1975) is the first to use the term fuzzy variable as a generalization of a Boolean variable and then Nahmias (1978) defined it as a function from a pattern space to a set of real numbers. This variable plays the same role in fuzzy theory as the random variable does in probability theory.

Definition 3 A fuzzy variable is defined as a mapping from a possibility space (Θ, $P(\Theta)$, Pos) to the set of real numbers, and its membership function is derived by

$$\mu(x) = \mathrm{Pos}\{\theta \in \Theta \mid \xi(\theta) = x\}, \quad x \in \Re . \tag{4}$$

Example 3: Let us continue to consider Example 1. Define

$$\xi(\theta) = \begin{cases} \theta, & \text{if } \theta \in [0,2] \\ 0, & \text{otherwise.} \end{cases}$$

It is clear that ξ is a fuzzy variable defined on the possibility space (Θ, $P(\Theta)$, Pos) with the following membership function

$$\mu_\xi(x) = \begin{cases} x, & \text{if } x \in [0,1] \\ 2-x, & \text{if } x \in (1,2] \\ 0, & \text{otherwise.} \end{cases}$$

Since the figure of $\mu_\xi(x)$ in coordinates is a triangle, we call ξ a triangle fuzzy variable, denoted by $\xi = (0,1,2)$. More general case is that we instead $(0,1,2)$ by (r_1, r_2, r_3) with $r_1 < r_2 < r_3$, and the membership function by

$$\mu_\xi(x) = \begin{cases} \dfrac{x - r_1}{r_2 - r_1}, & \text{if } r_1 \le \mathrm{x} \le r_2 \\ \dfrac{x - r_3}{r_2 - r_3}, & \text{if } r_2 \le \mathrm{x} \le r_3 \\ 0, & \text{otherwise.} \end{cases}$$

Example 4: Let ξ be a fuzzy variable defined on the possibility space $(\Theta, P(\Theta), \mathrm{Pos})$ with the following membership function

$$\mu_\xi(x) = \begin{cases} \dfrac{x - r_1}{r_2 - r_1}, & \text{if } r_1 \le x < r_2 \\ 1, & \text{if } r_2 \le x < r_3 \\ \dfrac{x - r_4}{r_3 - r_4}, & \text{if } r_3 \le x \le r_4 \\ 0, & \text{otherwise,} \end{cases}$$

where $\Theta = \Re$, and $r_1 < r_2 < r_3 < r_4$. Then ξ is called a trapezoidal fuzzy variable and denoted by $\xi = (r_1, r_2, r_3, r_4)$.

Expected value operator of fuzzy variable have been discussed by many authors such as Dubois and Prade (1987), Heilpern (1992), Campos and Gonzάlez (1989), Gonzάlez (1990) and Yager (1981, 2002). Following the idea of Choquet integral, Liu and Liu (2002) used the credibility measure to define a most general definition of expected value operator of fuzzy variable. This definition is not only applicable to continuous fuzzy variables but also discrete ones. Moreover, the advantage of the expected value of fuzzy variable defined by Liu and Liu (2002) is a scalar value.

Definition 4 (Liu and Liu 2002) Let ξ be a fuzzy variable. Then the expected value of ξ is defined by

$$E[\xi] = \int_0^{+\infty} \mathrm{Cr}\{\xi \ge r\}\mathrm{d}r - \int_{-\infty}^0 \mathrm{Cr}\{\xi \le r\}\mathrm{d}r \tag{5}$$

provided that at least one of the two integrals is finite.

Example 5: The expected value of a trapezoidal fuzzy variable $\xi = (r_1, r_2, r_3, r_4)$ is $E[\xi] = (r_1 + r_2 + r_3 + r_4)/4$.

Further, if $r_2 = r_3$, we have $E[\xi] = (r_1 + 2r_2 + r_4)/4$, this is just the expected value of the triangular fuzzy variable $\xi = (r_1, r_2, r_4)$.

Remark 1 (Liu 2004) The definition of expected value operator is also applicable to discrete case. Assume that ξ is a discrete fuzzy variable whose membership function is given by $\mu(a_i) = \mu_i, i = 1, 2, \cdots, n$. Without loss of generality, we also assume that $a_1 < a_2 < \cdots < a_m$. Definition 4 implies that the expected value of ξ is

$$E[\xi] = \sum_{i=1}^{n} \omega_i a_i, \tag{6}$$

where the weights ω_i, $i = 1, 2, \cdots, n$ are given by

$$\begin{aligned}
\omega_1 &= \frac{1}{2}\left(\mu_1 + \max_{1 \le j \le n} \mu_j - \max_{1 < j \le n} \mu_j \right), \\
\omega_i &= \frac{1}{2}\left(\max_{1 \le j \le i} \mu_j - \max_{1 \le j < i} \mu_j + \max_{i \le j \le n} \mu_j - \max_{i < j \le n} \mu_j \right), \quad 2 \le i \le n-1, \\
\omega_n &= \frac{1}{2}\left(\max_{1 \le j \le n} \mu_j - \max_{1 \le j < n} \mu_j + \mu_m \right).
\end{aligned}$$

The independence of fuzzy variables has been discussed by many authors from different angle, for example, Zadeh (1978), Nahmias (1978), Yager (1992), Liu (2004) and Liu and Gao (2005).

Definition 5 (Liu 2004) The fuzzy variables $\xi_1, \xi_2, \cdots, \xi_n$ are said to be independent if and only if

$$\text{Cr}\{\xi_i \in B_i, i = 1, 2, \cdots, n\} = \min_{1 \le i \le n} \text{Cr}\{\xi_i \in B_i\} \tag{7}$$

for any sets $B_1, B_2, \cdots, B_n$ of $\Re$.

Definition 6 (Liu 2004) The fuzzy variables $\xi_1, \xi_2, \cdots, \xi_n$ are said to be identically distributed if and only if

$$\text{Cr}\{\xi_i \in B\} = \text{Cr}\{\xi_j \in B\}, \quad i, j = 1, 2, \cdots, n \tag{8}$$

for any sets B of $\Re$.

Proposition 1 (Liu and Liu 2003) Let ξ and η be independent fuzzy variables with finite expected values. Then for any numbers a and b, we have

$$E[a\xi + b\eta] = aE[\xi] + bE[\eta]. \quad (9)$$

Definition 7 (Liu 2004) Let ξ be a fuzzy variable on possibility space $(\Theta, P(\Theta), \text{Pos})$, and $\alpha \in (0, 1]$. Then

$$\xi'_\alpha = \inf\{r \mid \text{Cr}\{\xi \le r\} \ge \alpha\} \text{ and } \xi''_\alpha = \sup\{r \mid \text{Cr}\{\xi \ge r\} \ge \alpha\} \quad (10)$$

are called the α-pessimistic value and the α-optimistic value to ξ, respectively.

Definition 8 (Liu 2004) Let $(\Theta_i, P(\Theta_i), \text{Pos}_i)$, $i = 1, 2, \cdots, n$ be possibility spaces. The product possibility space is defined as $(\Theta, P(\Theta), \text{Pos})$, where Θ and Pos are determined by

$$\Theta = \Theta_1 \times \Theta_2 \times \cdots \times \Theta_n$$

and

$$\text{Pos}\{A\} = \sup_{(\theta_1, \theta_2, \cdots \theta_n) \in A} \min_{1 \le i \le n} \text{Pos}_i(\theta_i)$$

for any $A \in P(\Theta)$, respectively.

Definition 9 (Liu 2002) (Fuzzy Arithmetic on Different Possibility Spaces) Let $f: \Re^n \to \Re$ be a function, and ξ_i fuzzy variables on the possibility space $(\Theta_i, P(\Theta_i), \text{Pos}_i)$, $i = 1, 2, \cdots, n$, respectively. Then $\xi = f(\xi_1, \xi_2, \cdots, \xi_n)$ is a fuzzy variable defined on the possibility space $(\Theta, P(\Theta), \text{Pos})$ as

$$\xi(\theta_1, \theta_2, \cdots, \theta_n) = f(\xi_1(\theta_1), \xi_2(\theta_2), \cdots, \xi_n(\theta_n))$$

for any $(\theta_1, \theta_2, \cdots, \theta_n) \in \Theta$.

Example 6: (Liu 2002) Let ξ_1 and ξ_2 be two fuzzy variables on $(\Theta_1, P(\Theta_1), \text{Pos}_1)$ and $(\Theta_2, P(\Theta_2), \text{Pos}_2)$. Then their sum and product, written as $\xi_1 + \xi_2$ and $\xi_1 \cdot \xi_2$, are defined by

$$(\xi_1 + \xi_2)(\theta_1, \theta_2) = \xi_1(\theta_1) + \xi_2(\theta_2), \quad (\xi_1 \cdot \xi_2)(\theta_1, \theta_2) = \xi_1(\theta_1) \cdot \xi_2(\theta_2),$$

for any $(\theta_1, \theta_2) \in \Theta_1 \times \Theta_2$.

11.2.2 Fuzzy Random Variable

The definition of fuzzy random variable has been discussed by several authors. Kwakernaak (1978, 1979) first introduced the notion of fuzzy random variable. Further concepts were then developed by Puri and Ralescu

(1985, 1986), and Kurse and Meyer (1987). Recently, Liu and Liu (2003) make a comparison among these definitions and presented a new definition of fuzzy random variable at the same time. In what follows, we will introduce the results in Liu and Liu (2003) that are related to our work.

Let $(\Omega, \mathrm{A}, \Pr)$ be a probability space and F a collection of fuzzy variables defined on the possibility space.

Definition 10 (Liu and Liu 2003) A fuzzy random variable is a function $\xi:\Omega\to F$ such that $\mathrm{Pos}\{\xi(\omega)\in B\}$ is a measurable function of ω for any Borel set B of $\Re$.

Example 7: Let $(\Omega, \mathrm{A}, \Pr)$ be a probability space.

If $\Omega=\{\omega_1,\omega_2,\cdots,\omega_n\}$, and $u_1,u_2,\cdots,u_n$ are fuzzy variables, then the function

$$\xi(\omega)=\begin{cases} u_1, & \text{if } \omega=\omega_1 \\ u_2, & \text{if } \omega=\omega_2 \\ \cdots & \\ u_n, & \text{if } \omega=\omega_n \end{cases}$$

is clearly a fuzzy random variable.

Expected value of fuzzy random variable has been defined as a fuzzy number in several ways, for example, Kwakernaak (1978), Puri and Ralescu (1986), and Kruse and Meyer (1987). However, in practice, we need a scalar expected value operator of fuzzy random variables.

Definition 11 (Liu and Liu 2003) Let ξ be a fuzzy random variable. Then its expected value is defined by

$$E[\xi]=\int_0^{+\infty}\Pr\{\omega\in\Omega \mid E[\xi(\omega)]\ge r\}\mathrm{d}r-\int_{-\infty}^{0}\Pr\{\omega\in\Omega \mid E[\xi(\omega)]\le r\}\mathrm{d}r \tag{11}$$

provided that at least one of the two integrals is finite.

Example 8: (Liu 2002): Assume that ξ is a fuzzy random variable defined as

$$\xi=(\rho,\rho+1,\rho+2),\quad \text{with } \rho\sim N(0,1).$$

Then for each $\omega\in\Omega$, we have $E[\xi(\omega)]=\frac{1}{4}[\rho(\omega)+2(\rho(\omega)+1)+(\rho(\omega)+2)]=\rho(\omega)+1$. Thus $E[\xi]=E[\rho]+1=1$.

Proposition 2 Assume that ξ and η are fuzzy random variables with finite expected values. If for each $\omega\in\Omega$, the fuzzy variables $\xi(\omega)$ and $\eta(\omega)$ are independent, then for any real numbers a and b, we have

$$E[a\xi+b\eta]=aE[\xi]+bE[\eta]. \tag{12}$$

Definition 12 (Liu and Liu 2003) The fuzzy random variables $\xi_1, \xi_2, \cdots, \xi_n$ are said to be independent and identically distributed (iid) if and only if

$(\text{Pos}\{\xi_i(\omega) \in B_1\}, \text{Pos}\{\xi_i(\omega) \in B_2\}, \cdots, \text{Pos}\{\xi_i(\omega) \in B_m\}), \quad i = 1, 2, \cdots, n$

are iid random vectors for any Borel sets $B_1, B_2, \cdots, B_m$ of $\Re$ and any positive integer m.

Proposition 3 (Liu 2004) Let $\xi_1, \xi_2, \cdots, \xi_n$ be iid fuzzy random variables. Then for any Borel set B of $\Re$, we have

(a) $\text{Pos}\{\xi_i(\omega) \in B\}, i = 1, 2, \cdots, n$ are iid random variables;

(b) $\text{Nec}\{\xi_i(\omega) \in B\}, i = 1, 2, \cdots, n$ are iid random variables;

(c) $\text{Cr}\{\xi_i(\omega) \in B\}, i = 1, 2, \cdots, n$ are iid random variables.

Definition 13 (Liu 2001; Gao and Liu 2001) Let ξ be a fuzzy random variable, and B a Borel set of $\Re$. Then the chance of fuzzy random event $\xi \in B$ is a function from $[0, 1]$ to $[0, 1]$, defined as

$$\text{Ch}\{\xi \in B\}(\alpha) = \sup_{\Pr\{A\} \geq \alpha} \inf_{\omega \in A} \text{Cr}\{\xi(\omega) \in B\}. \tag{13}$$

Definition 14 (Liu and Liu 2005) Let ξ be a fuzzy random variable. Then the mean chance of fuzzy random event $\xi \in B$ is defined as

$$\text{Ch}\{\xi \in B\} = \int_0^1 \Pr\{\omega \in \Omega \mid \text{Cr}\{\xi(\omega) \in B\} \geq \alpha\} \, \mathrm{d}\alpha. \tag{14}$$

Definition 15 (Liu 2001) Let ξ be a fuzzy random variable, and $\gamma, \delta \in (0, 1]$. Then

$$\xi_{\sup}(\gamma, \delta) = \sup\{r \mid \text{Ch}\{\xi \geq r\}(\gamma) \geq \delta\} \tag{15}$$

is called the (γ, δ)-optimistic value to ξ, and

$$\xi_{\inf}(\gamma, \delta) = \inf\{r \mid \text{Ch}\{\xi \leq r\}(\gamma) \geq \delta\} \tag{16}$$

is called the (γ, δ)-pessimistic value to ξ.

11.2.3 Random Fuzzy Variable

In this section, we shall state some basic concepts and results on random fuzzy variables. An interested reader may consult Liu (2000) and Liu and

Liu (2003) where some important properties of random fuzzy variables are recorded.

Definition 16 (Liu 2000) A random fuzzy variable is a function ξ from a possibility space $(\Theta, P(\Theta), \text{Pos})$ to a collection of random variables.

Example 9: (Liu 2002) Let $\xi \sim \text{N}(\rho, 1)$, where ρ is a fuzzy variable with membership function $\mu_\rho(x) = [1-|x-2|] \vee 0$. Then ξ is a random fuzzy variable taking "normally distributed variable $\text{N}(\rho, 1)$" values.

Remark 2 (Liu 2002) Let ξ be a random fuzzy variable from possibility space $(\Theta, P(\Theta), \text{Pos})$ to *R*, where *R* is a collection of random variables. Roughly speaking, if Θ consists of a single element, then the random fuzzy variable degenerates to a random variable. If *R* is a collection of real numbers (rather than random variables), then the random fuzzy variable degenerates to a fuzzy variable. Hence, whichever of random variable and fuzzy variable is a special case of random fuzzy variable. This fact makes us to discuss the mixture of fuzzy lifetimes and random lifetimes at the same time in a system.

Definition 17 (Liu 2002) Let ξ be a random fuzzy variable, and *B* a Borel set of $\Re$. Then the chance of random fuzzy event $\xi \in B$ is a function from $(0, 1]$ to $[0, 1]$, defined as

$$\text{Ch}\{\xi \in B\}(\alpha) = \sup_{\text{Cr}\{A\} \geq \alpha} \inf_{\theta \in A} \Pr\{\xi(\theta) \in B\} \tag{17}$$

Definition 18 (Liu and Liu 2002) Let ξ be a random fuzzy variable. Then the mean chance of random fuzzy event $\xi \in B$ is defined as

$$\text{Ch}\{\xi \in B\} = \int_0^1 Cr\{\theta \in \Theta \mid \Pr\{\xi(\theta) \in B\} \geq \alpha\} \, \text{d}\alpha \,. \tag{18}$$

Definition 19 (Liu and Liu 2003) Let ξ be a random fuzzy variable defined on the possibility space $(\Theta, P(\Theta), \text{Pos})$. The expected value $E[\xi]$ is defined by

$$E[\xi] = \int_0^{+\infty} \text{Cr}\{\theta \in \Theta \mid E[\xi(\theta)] \geq r\} \text{d}r - \int_{-\infty}^0 \text{Cr}\{\theta \in \Theta \mid E[\xi(\theta)] \leq r\} \text{d}r \tag{19}$$

provided that at least one of the two integrals if finite.

It is easy to know that the expected value of a random fuzzy variable becomes the expected value of the random variable when it degenerates to a random variable while becomes the expected value of the fuzzy variable when it degenerates to a fuzzy variable.

Example 10: (Liu 2002) Suppose that ξ is a random fuzzy variable defined as $\xi \sim \mathrm{U}(\rho, \rho+2)$, with $\rho=(0,1,2)$.
Without loss of generality, we assume that ρ is defined on the possibility space $(\Theta, P(\Theta), \mathrm{Pos})$. Then for each $\theta \in \Theta$, $\xi(\theta)$ is a random variable and $E[\xi(\theta)]=\rho(\theta)+1$. Thus the expected value of ξ is $E[\xi]=E[\rho]+1=2$.

Definition 20 (Liu 2004) The random fuzzy variables $\xi_1, \xi_2, \cdots, \xi_n$ are said to be independent and identically distributed (iid) if and only if

$$(\Pr\{\xi_i(\theta) \in B_1\}, \Pr\{\xi_i(\theta) \in B_2\}, \cdots, \Pr\{\xi_i(\theta) \in B_m\}), \quad i=1,2,\cdots,n$$

are iid fuzzy vectors for any Borel sets $B_1, B_2, \cdots, B_m$ of $\Re$ and any positive integer m.

Proposition 4 (Liu and Liu 2003) Assume that ξ and η are random fuzzy variables with finite expected values. If $E[\xi(\theta)]$ and $E[\eta(\theta)]$ are independent fuzzy variables, then for any real numbers a and b, we have

$$E[a\xi+b\eta]=aE[\xi]+bE[\eta]. \tag{20}$$

Definition 21 (Liu 2001) Let ξ be a fuzzy random variable, and $\gamma, \delta \in (0,1]$. Then

$$\xi_{\sup}(\gamma,\delta)=\sup\{r \mid \mathrm{Ch}\{\xi \ge r\}(\gamma) \ge \delta\} \tag{21}$$

is called the (γ,δ)-optimistic value to ξ, and

$$\xi_{\inf}(\gamma,\delta)=\inf\{r \mid \mathrm{Ch}\{\xi \le r\}(\gamma) \ge \delta\} \tag{22}$$

is called the (γ,δ)-pessimistic value to ξ.

11.3 Redundancy Optimization Problem

Consider a redundant system consisting of n components. For each component i, there is only one type of elements available, $i=1,2,\cdots,n$. Let x_i indicate the numbers of the ith types of elements selected as redundant elements. The redundant elements are arranged in one of two ways: parallel or standby. Let ξ_{ij} indicate the lifetimes of the jth redundant elements in components i, $T_i(\mathbf{x},\xi)$ indicate the lifetimes of components i, and $T(\mathbf{x},\xi)$ indicate the system lifetime, where $\mathbf{x}=(x_1,x_2,\cdots,x_n)$, and

$\xi=(\xi_{11},\xi_{12},\cdots,\xi_{1x_1},\xi_{21},\xi_{22},\cdots,\xi_{2x_2},\cdots,\xi_{n1},\xi_{n2},\cdots,\xi_{nx_n})$, $j=1,2,\cdots,x_i$, $i=1,2,\cdots,n$, respectively. For a parallel redundant system we have $T_i(\mathbf{x},\xi)=\max_{1\le j\le x_i}\xi_{ij}$, while for a standby redundant system we have $T_i(\mathbf{x},\xi)=\sum_{j=1}^{x_i}\xi_{ij}$, $i=1,2,\cdots,n$, respectively.

We make the following assumptions about the redundant system.

1. Lifetimes of elements ξ_{ij} are fuzzy variables or fuzzy random variables or random fuzzy variables according to the different needs of the realistic problems, $j=1,2,\cdots,x_i$, $i=1,2,\cdots,n$.
2. There is no element (or system) repair or preventive maintenance.
3. The switching device of the standby system is assumed to be perfect.
4. The system and all redundant elements are in one of two states: operating (denoted by 1) or failed (denoted by 0).

Let y_{ij} represent the state of the jth redundant element in component i, y_i represent the state of the component i, $j=1,2,\cdots,x_i$, $i=1,2,\cdots,n$, respectively, and $\Psi(\mathbf{y})$ indicate the state of the system, where $\mathbf{y}=(y_1,y_2,\cdots,y_n)$. The function $\Psi(\mathbf{y})$ is called the *system structure function*. In this chapter we assume that system structure function is known. This leads to the following assumption.

5. There exists a system structure function $\Psi:\{0,1\}^n\to\{0,1\}$ that assigns a system state $\Psi(\mathbf{y})\in\{0,1\}$ to each component state vector $\mathbf{y}\in\{0,1\}^n$.

The redundancy optimization problem is to find the optimal value of $\mathbf{x}=(x_1,x_2,\cdots,x_n)$ such that some system performances are optimized under some cost constraints.

Remark 3 *It is meaningless to study the parallel redundancy optimization problem in fuzzy environment, because there is no distinguish between multiple parallel elements and only one element.*

11.4 System Performances

If the lifetimes of the redundant elements ξ_{ij} are fuzzy variables, then both the lifetimes of the components $T_i(\mathbf{x},\xi)$ and the lifetime of the system $T(\mathbf{x},\xi)$ are fuzzy variables, $j=1,\cdots,x_i$, $i=1,2,\ldots,n$, respectively. Similar results also hold in fuzzy random or random fuzzy environments.

In practice, it is natural for the decision maker to maximize the system lifetime $T(\mathbf{x}, \xi)$. However, it is meaningless to maximize an uncertain variable unless a criterion of ranking uncertain variables is offered. An alternative is to optimize the various characteristics of the system lifetime $T(\mathbf{x}, \xi)$.

First, we give three types of system performances characterized in the context of credibility in fuzzy environment:

- expected system lifetime $E[T(\mathbf{x}, \xi)]$;
- α-system lifetime which is defined as the largest value $\overline{T}$ satisfying $\mathrm{Cr}\{T(\mathbf{x}, \xi) \geq \overline{T}\} \geq \alpha$, where α is a predetermined confidence level;
- system reliability $\mathrm{Cr}\{T(\mathbf{x}, \xi) \geq T^0\}$, where T^0 is a predetermined confidence level.

Second, we list the states in fuzzy random or random fuzzy environments:

- expected system lifetime $E[T(\mathbf{x}, \xi)]$;
- (α, β)-system lifetime which is defined as the largest value $\overline{T}$ satisfying $\mathrm{Ch}\{T(\mathbf{x},\xi) \geq \overline{T}\}(\alpha) \geq \beta$, where α and β are predetermined confidence levels;
- system reliability defined as the α-chance $\mathrm{Ch}\{T(\mathbf{x},\xi) \geq T^0\}(\alpha)$, where T^0 is a determined number and α is a predetermined confidence level.

In many cases, it is difficult to design an analytic method to calculate these system performances due to the complexity of system structure. In order to handle them, we shall introduce the simulations to estimate them (About the simulation techniques, see Liu and Iwamura (1998, 1999), Liu and Liu (2002). In addition, fuzzy simulation was also discussed by Chanas and Nowakowski (1988) and this method was used to generate a single value of a fuzzy variable).

11.4.1 Fuzzy Simulations

In this subsection, we shall introduce the fuzzy simulation designed by Zhao and Liu (2005).

11.4.1.1 Fuzzy Simulation for $E[T(x,\xi)]$

The fuzzy simulation for computing the expected value of a fuzzy variable has been provided by Liu and Liu (2002). The key for evaluating the expected value of a continuous fuzzy variable is to discretize this fuzzy variable into a discrete fuzzy variable and then calculate the expected value of the latter. The expected value of the discrete fuzzy variable will approximate the expected value of the continuous fuzzy variable provided that the selected points are large enough. For example, let ξ be a continuous fuzzy variable whose membership function is shown in Fig. 1. In order to estimate the expected value $E[\xi]$, we generate N points a_i uniformly from the δ-level set of ξ for $i = 1, 2, \ldots, N$. Thus we get a new discrete fuzzy variable ξ' and $\mu_{\xi'}(a_i) = \mu_{\xi}(a_i)$ for $i = 1, 2, \ldots, N$. We can calculate $E[\xi']$ by (6). Then we use $E[\xi']$ as the estimation of $E[\xi]$ provided that N is a sufficiently large integer.

In addition, in order to estimate $T(\mathbf{x}, \mathbf{u})$ for a fixed decision $\mathbf{x}$ and an observational point $\mathbf{u}$ of the standby unit lifetime vector $\boldsymbol{\xi}$, we have to clarify the relationship between system structure function and system lifetime. In fact, it is reasonable to assume that the system structure function $\Psi(\mathbf{y}(t))$ is a decreasing function of time t, where $\mathbf{y}(t) = (y_1(t), y_2(t), \cdots, y_n(t))$ is the state vector of the components at time t. This implies that $\text{Cr}\{T(\mathbf{x}, \boldsymbol{\xi}) \geq t\} = 1$ if and only if $\Psi(\mathbf{y}(t)) = 1$.

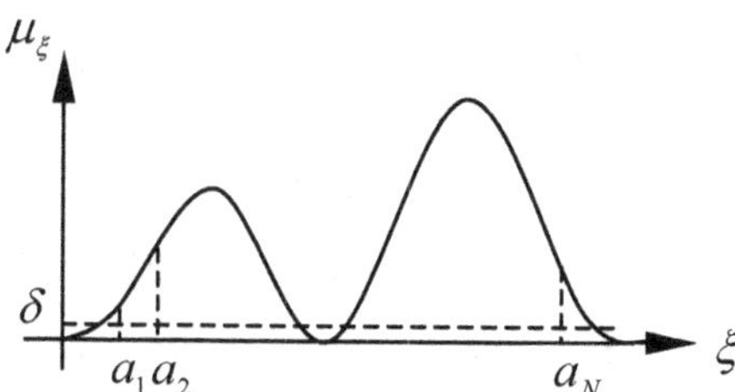

Fig. 1. Discretization of a continuous fuzzy variable

This result allows us to estimate $T(\mathbf{x}, \mathbf{u})$ with a bisection search by computing system structure function $\Psi(\mathbf{y}(t))$. Here we summarize the fuzzy simulation for $E[T(\mathbf{x}, \boldsymbol{\xi})]$ as follows.

Step 1. Randomly generate N points $\mathbf{u}_i$ uniformly from the δ-level set of $\boldsymbol{\xi}$ for $i = 1, 2, \ldots, N$, where δ is a sufficiently small positive number.

Step 2. Give two bounds t_{1i} and t_{2i}, such that $\Psi(\mathbf{y}(t_{1i})) = 1$ and $\Psi(\mathbf{y}(t_{2i})) = 0$ for $\mathbf{u}_i$, $i = 1, 2, \ldots, N$, respectively (For example, we can take t_{1i} as 0 and t_{2i} as a sufficiently large number).

Step 3. Set $t_{0i} = (t_{1i} + t_{2i})/2, i = 1, 2, \cdots, N$. If $\Psi(\mathbf{y}(t_{0i})) = 1$, then we set $t_{1i} = t_{0i}$, otherwise $t_{2i} = t_{0i}$.

Step 4. Repeat Step 3 until $|t_{1i} - t_{2i}| < \varepsilon, i = 1, 2, \cdots, N$, where ε is a pre-determined level of accuracy.

Step 5. $T(\mathbf{x}, \mathrm{u}_i) = (t_{1i} + t_{2i})/2, i = 1, 2, \cdots, N$.

Step 6. Rearrange $\mathbf{u}_1, \mathbf{u}_2, \cdots, \mathbf{u}_N$ such that $T(\mathbf{x}, \mathbf{u}_i) \le T(\mathbf{x}, \mathbf{u}_{i+1})$ for $i = 1, 2, \cdots, N-1$.

Step 7. Set $E[T(\mathbf{x}, \xi)] = \sum_{i=1}^{N} \omega_i T(\mathbf{x}, \mathbf{u}_i)$, where the weights ω_i are given by (6) and $\mu_i = \mu(\mathbf{u}_i)$ for $i = 1, 2, \ldots, N$.

11.4.1.2 Fuzzy Simulation for $\mathrm{Cr}\{T(x,\xi) \ge T^0\}$

Note that

$$\begin{aligned}\mathrm{Cr}\{T(\mathbf{x}, \xi) \ge T^0\} &= [\mathrm{Pos}\{T(\mathbf{x}, \xi) \ge T^0\} + \mathrm{Nec}\{T(\mathbf{x}, \xi) \ge T^0\}]/2 \\ &= [\mathrm{Pos}\{T(\mathbf{x}, \xi) \ge T^0\} + 1 - \mathrm{Pos}\{T(\mathbf{x}, \xi) < T^0\}]/2 .\end{aligned}$$

In order to estimate system reliability $\mathrm{Cr}\{T(\mathbf{x}, \xi) \ge T^0\}$ for a given decision $\mathbf{x}$, we first set $\mathrm{Pos}\{T(\mathbf{x}, \xi) \ge T^0\} = 0$ and $\mathrm{Pos}\{T(\mathbf{x}, \xi) < T^0\} = 0$. Then we generate two points $\mathbf{u}_1$ and $\mathbf{u}_2$ from the δ-level set of ξ such that $T(\mathbf{x}, \mathbf{u}_1) \ge T_0$ and $T(\mathbf{x}, \mathbf{u}_2) < T_0$ respectively (the technique for estimating $T(\mathbf{x}, \mathbf{u}_i)$ is the same in fuzzy simulation for $E[T(\mathbf{x}, \xi)], i = 1,2$), where δ is a sufficiently small positive number. If $\mathrm{Pos}\{T(\mathbf{x}, \xi) \ge T^0\} < \mu(\mathbf{u}_1)$, we set $\mathrm{Pos}\{T(\mathbf{x}, \xi) \ge T^0\} = \mu(\mathbf{u}_1)$. Similarly, if $\mathrm{Pos}\{T(\mathbf{x}, \xi) < T^0\} < \mu(\mathbf{u}_2)$, we set $\mathrm{Pos}\{T(\mathbf{x}, \xi) < T^0\} = \mu(\mathbf{u}_2)$. Repeat this process for N times. Now we summarize the fuzzy simulation for $\mathrm{Cr}\{T(\mathbf{x}, \xi) \ge T^0\}$ as follows.

Step 1. Set $\mathrm{Pos}\{T(\mathbf{x}, \xi) \ge T^0\} = 0$, and $\mathrm{Pos}\{T(\mathbf{x}, \xi) < T^0\} = 0$.

Step 2. Generate two points u_1 and u_2 from the δ-level set of ξ such that $T(\mathbf{x}, \mathbf{u}_1) \ge T_0$ and $T(\mathbf{x}, \mathbf{u}_2) < T_0$.

Step 3. Set $\mathrm{Pos}\{T(\mathbf{x}, \xi) \ge T^0\} = \mu(\mathbf{u}_1)$ provided that $\mathrm{Pos}\{T(\mathbf{x}, \xi) \ge T^0\} < \mu(\mathbf{u}_1)$; and set $\mathrm{Pos}\{T(\mathbf{x}, \xi) < T^0\} = \mu(\mathbf{u}_2)$ provided that $\mathrm{Pos}\{T(\mathbf{x}, \xi) < T^0\} < \mu(\mathbf{u}_2)$.

Step 4. Repeat Step 2 and Step 3 for N times.

Step 5. $\text{Cr}\{T(\mathbf{x},\xi)\ge T^0\}=\frac{1}{2}(1+\text{Pos}\{T(\mathbf{x},\xi)\ge T^0\}-\text{Pos}\{T(\mathbf{x},\xi)<T^0\})$.

11.4.1.3 Fuzzy Simulation for $\overline{T}$

In order to find the largest value $\overline{T}$ such that $\text{Cr}\{T(\mathbf{x},\xi)\ge \overline{T}\}\ge\alpha$ for a given decision vector $\mathbf{x}$, we first set $\overline{T}=0$. Then we generate a point $\mathbf{u}$ uniformly from the δ-level set of ξ, where δ is a sufficiently positive small number. We set $\overline{T}=T(\mathbf{x},\mathbf{u})$ if $\overline{T}<T(\mathbf{x},\mathbf{u})$ and $\text{Cr}\{T(\mathbf{x},\xi)\ge T(\mathbf{x},\mathbf{u})\}\ge\alpha$. Repeat this process for N times. The value $\overline{T}$ is regarded as the α-system lifetime at the decision $\mathbf{x}$.

Step 1. Randomly generate $\mathbf{u}_k$ from the δ-level set of the fuzzy vector ξ for $k=1,2,\cdots,N$.

Step 2. Calculate $T(\mathbf{x},\mathbf{u}_k)$ for $k=1,2,\cdots,N$.

Step 3. Define a function of r as $L(r)=\max\limits_{1\le k\le N}\{\mu(\mathbf{u}_k)\,|\,T(\mathbf{x},\mathbf{u}_k)\ge r\}+\min\limits_{1\le k\le N}\{1-\mu(\mathbf{u}_k)\,|\,T(\mathbf{x},\mathbf{u}_k)<r\}$, where $\mu(\mathbf{u}_k)$ is the membership of the vector $\mathbf{u}_k$.

Step 4. Employ the bisection search to find the maximal value $\overline{T}$ of r such that $L(\overline{T})\ge\alpha$.

Step 5. Return $\overline{T}$.

Remark 4 *In the above system performance, the symbol Cr may be replaced with Pos and Nec for the different purposes of management.*

We will provide an example to show the effectiveness of the technique of fuzzy simulations.

Example 11: Let us consider a system with only one unit. The lifetime of the unit is assumed to be a trapezoidal fuzzy variable ξ=(100, 108, 112, 120). Therefore, the system lifetime $T(\mathbf{x},\xi)=\xi$.

First we use the fuzzy simulation to estimate the expected system lifetime $E[T(\mathbf{x},\xi)]=E[\xi]$. The results of the simulation are shown in Fig. 2. The straight line represents the known expected system lifetime 110.0 and the curve represents the expected system lifetime obtained by different numbers of cycles in simulation. From Fig. 2, we can see that the relative error of the results obtained by performing over 2500 cycles is less than 0.5%.

We now use fuzzy simulation to obtain the maximal $\overline{T}$ such that

$$\mathrm{Cr}\{T(\mathbf{x}, \xi) \geq \overline{T}\} \geq 0.90, \quad \text{i.e., } \mathrm{Cr}\{\xi \geq \overline{T}\} \geq 0.90.$$

By Definition 7, we can gain the precise value of $\overline{T}$ is 101.60. The results of fuzzy simulation are shown in Fig. 3, from which it is easy to know that the value by simulation is almost equal to the real value by performing over 250 cycles.

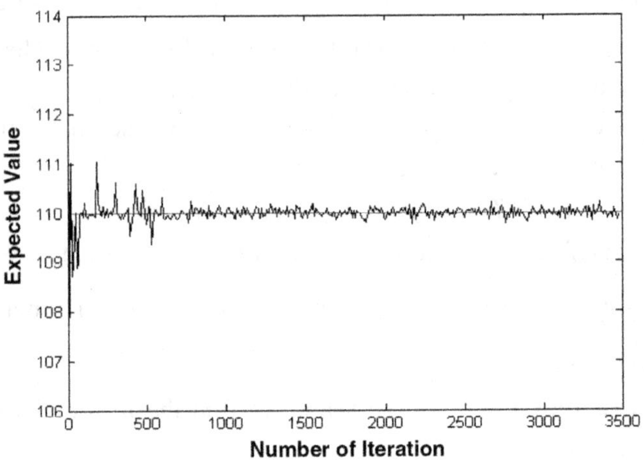

Fig. 2. A fuzzy simulation process of the expected value

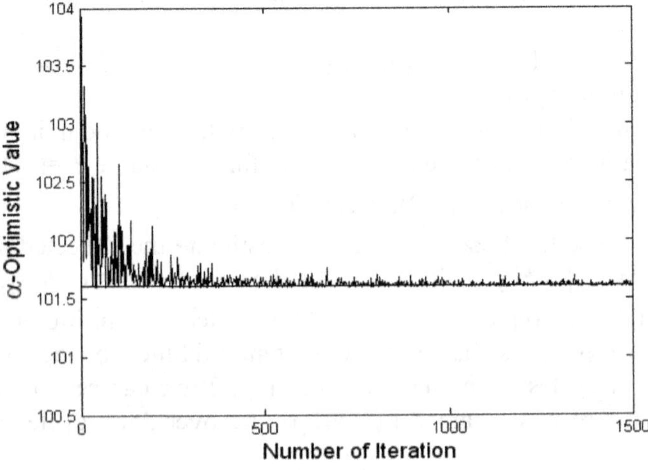

Fig. 3. A fuzzy simulation process of α-system lifetime

Finally, we employ the fuzzy simulation to estimate $\mathrm{Cr}\{\xi \geq 101.60\}$. From the above, it is apparent that the theoretical value is 0.90. The results of fuzzy simulation are shown in Fig. 4, which implies the value of simulation and the real one are almost the same by performing over 280 cycles.

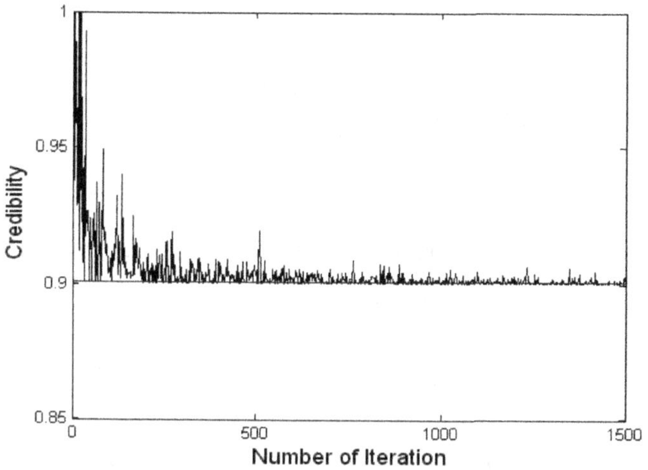

Fig. 4. A fuzzy simulation process of $\mathrm{Cr}\{\xi \geq 101.60\}$

11.4.2 Fuzzy Random Simulation

The following fuzzy random simulations were designed by Zhao et al. (2001)

11.4.2.1 Fuzzy Random Simulation for $E[T(\mathbf{x}, \xi)]$

For any given $\mathbf{x}$, in order to estimate the expected system lifetime $E[T(\mathbf{x}, \xi)]$, we first sample $\boldsymbol{\omega}_1, \boldsymbol{\omega}_2, \cdots, \boldsymbol{\omega}_N$ from Ω according to probability measure Pr. For each $\boldsymbol{\omega}_i$, we can use the fuzzy simulation (see Zhao and Liu 2005) to estimate $E[\xi(\boldsymbol{\omega}_i)]$ since that $\xi(\boldsymbol{\omega}_i)$ is a fuzzy variables, $i = 1, 2, \cdots, N$. Then we use $\frac{1}{N}\sum_{i=1}^{N} E[\xi(\boldsymbol{\omega}_i)]$ as the estimation of $E[T(\mathbf{x}, \xi)]$.

Step 1. Randomly generate $\boldsymbol{\omega}_1, \boldsymbol{\omega}_2, \cdots, \boldsymbol{\omega}_N$ from Ω according to probability measure Pr.

Step 2. Compute $E[\xi(\boldsymbol{\omega}_i)]$ by the fuzzy simulation for $i = 1, 2, \cdots, N$.

Step 3. $E[T(\mathbf{x},\xi)]=\frac{1}{N}\sum_{i=1}^{N}E[\xi(\omega_i)]$.

11.4.2.2 Fuzzy Random Simulation for $\overline{T}$

For any given decision $\mathbf{x}$ and confidence level α and β, in order to find the maximal $\overline{T}$ such that $\mathrm{Ch}\{T(\mathbf{x},\xi)\geq\overline{T}\}(\alpha)\geq\beta$, we sample $\omega_1,\omega_2,\cdots,\omega_N$ from Ω according to probability measure Pr. We can find the maximal values $\overline{T}_i$ such that $\mathrm{Ch}\{T(\mathbf{x},\xi(\omega_i))\geq\overline{T}_i\}\geq\beta$ by the fuzzy simulation (see Zhao and Liu (2005)) for $i=1,2,\cdots,N$ since that $T(\mathbf{x},\xi(\omega_i)$ are fuzzy variables. By the law of large numbers, the N'th largest element in the sequence $\{\overline{T}_1,\overline{T}_2,\cdots,\overline{T}_N\}$ is the estimation of $\overline{T}$, where N' is the integer part of αN.

Step 1. Randomly generate $\omega_1,\omega_2,\cdots,\omega_N$ from Ω according to probability measure Pr.

Step 2. Find the maximal values $\overline{T}_i$ such that $\mathrm{Ch}\{T(\mathbf{x},\xi(\omega_i))\geq\overline{T}_i\}\geq\beta$ by the fuzzy simulation for $i=1,2,\cdots,N$.

Step 3. Set N' as the integer part of αN.

Step 4. Return the N'th largest element in $\{\overline{T}_1,\overline{T}_2,\cdots,\overline{T}_N\}$.

11.4.2.3 Fuzzy Random Simulation for System Reliability

Similarly, for any given decision $\mathbf{x}$ and confidence levels α, in order to find the maximal $\overline{\beta}$ such that $\mathrm{Ch}\{T(\mathbf{x},\xi)\geq T^0\}(\alpha)\geq\overline{\beta}$, we sample $\omega_1,\omega_2,\cdots,\omega_N$ from Ω according to probability measure Pr. We can calculate the credibility $\overline{\beta}_i=\mathrm{Cr}\{T(\mathbf{x},\xi(\omega_i))\geq T_i^0\}$ by the fuzzy simulation (see Zhao and Liu 2005) for $i=1,2,\cdots,N$ since that $T(\mathbf{x},\xi(\omega_i)$ are fuzzy variables. By the law of large numbers, the N'th largest element in the sequence $\{\overline{\beta}_1,\overline{\beta}_2,\cdots,\overline{\beta}_N\}$ is the estimation $\overline{\beta}$, where N' is the integer part of αN.

Step 1. Randomly generate $\omega_1,\omega_2,\cdots,\omega_N$ from Ω according to probability measure Pr.

Step 2. Compute the credibility $\overline{\beta}_i=\mathrm{Cr}\{T(\mathbf{x},\xi(\omega_i))\geq T_i^0\}$ by the fuzzy simulation for $i=1,2,\cdots,N$.

Step 3. Set N' as the integer part of αN.

Step 4. Return the N'th largest element in $\{\overline{\beta}_1, \overline{\beta}_2, \cdots, \overline{\beta}_N\}$.

We will provide an example to show the effectiveness of the technique of fuzzy random simulations.

Example 12: Let us consider a system with only one unit. The lifetime of the unit is considered to be a fuzzy random variable defined as $\xi \sim (\rho-10, \rho-5, \rho+5, \rho+10)$, with $\rho \sim \mathrm{N}(100, 10^2)$ .Correspondingly, the system lifetime $T(\mathbf{x}, \xi) = \xi$.

First we use the fuzzy random simulation to estimate the expected system lifetime $E[T(\mathbf{x}, \xi)] = E[\xi]$. The results of fuzzy random simulation are shown in Fig. 5, in which the straight line represents the known expected system lifetime 100.0 and the curve represents the expected system lifetime obtained by different numbers of cycles in simulation. From Fig. 5, we can see that the relative error of the results obtained by performing over 3000 cycles is less than 0.5%.

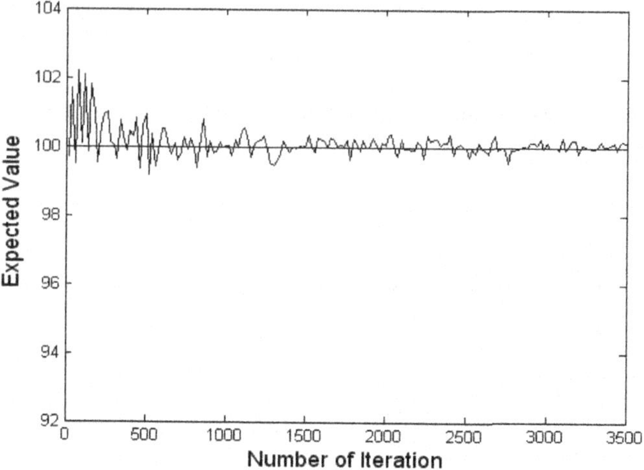

Fig. 5. A fuzzy random simulation process of the expected system lifetime

We now use fuzzy random simulation to obtain the maximal $\overline{T}$ such that $\mathrm{Ch}\{T(\mathbf{x}, \xi) \geq \overline{T}\}(0.9) \geq 0.90$, i.e., $\mathrm{Ch}\{\xi \geq \overline{T}\}(0.9) \geq 0.90$. The results of fuzzy random simulation are shown in Fig. 6.

From Fig. 6, it is easy to know that the relative error of the results obtained by performing over 3500 cycles tends to 0.

Finally, we employ the fuzzy random simulation to estimate the value of $\mathrm{Ch}\{T(\mathbf{x}, \xi) \geq \overline{T}\}(0.9) = \mathrm{Ch}\{\xi \geq \overline{T}\}(0.9)$. From the above, it is apparent that its theoretical value is 0.90. The results of fuzzy random simulation are shown in Fig. 7, which implies that the value of simulation and the real one are almost the same by performing over 3500 cycles.

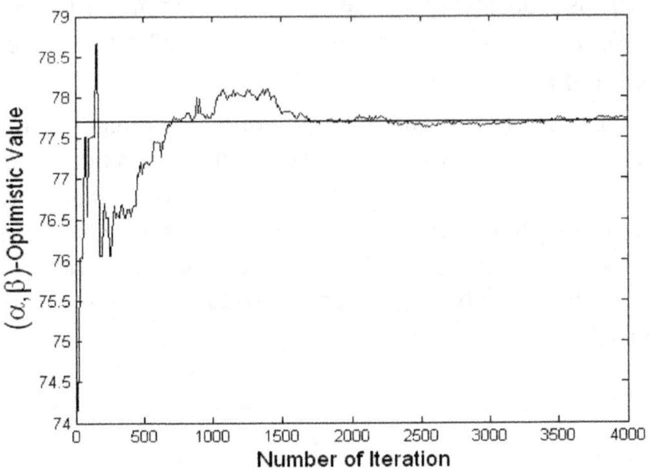

Fig. 6. A fuzzy random simulation process of (α, β)-system lifetime

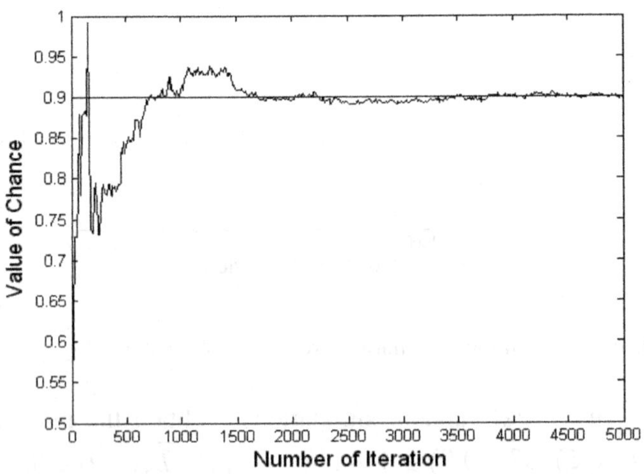

Fig. 7. A fuzzy random simulation process of system reliability

11.4.3 Random Fuzzy Simulation

Finally, we introduce the random fuzzy simulations designed by Zhao and Liu (2004).

11.4.3.1 Random Fuzzy simulation for $E[T(\mathbf{x}, \xi)]$

For any given decision $\mathbf{x}$, in order to calculate the expected system lifetime $E[T(\mathbf{x}, \xi)]$, we sample a sequence $\boldsymbol{\theta}_1, \boldsymbol{\theta}_2, \cdots, \boldsymbol{\theta}_N$ from the universe Θ such that $\text{Pos}\{\boldsymbol{\theta}_i\} \geq \varepsilon$ for $i = 1, 2, \cdots, N$, where ε is a sufficiently small positive number. We can estimate $E[T(\mathbf{x}, \xi(\boldsymbol{\theta}_i))]$ by the stochastic simulation (see Zhao and Liu 2003) since $T(\mathbf{x}, \xi(\boldsymbol{\theta}_i))$ are random variables for $\boldsymbol{\theta}_i$, $i = 1, 2, \cdots, N$. Rearrange $\boldsymbol{\theta}_1, \boldsymbol{\theta}_2, \cdots, \boldsymbol{\theta}_N$ such that $E[T(\mathbf{x}, \xi(\boldsymbol{\theta}_i))] \leq E[T(\mathbf{x}, \xi(\boldsymbol{\theta}_{i+1}))]$ for $i = 1, 2, \cdots, N-1$. We can use $\sum_{i=1}^{N} \omega_i E[T(\mathbf{x}, \xi(\boldsymbol{\theta}_i))]$ as the estimation of $E[T(\mathbf{x}, \xi)]$ by Liu and Liu (2003), where the weights are given by

$$\omega_1 = \frac{1}{2}\left(\mu_1 + \max_{1 \leq j \leq N} \mu_j - \max_{1 < j \leq N} \mu_j\right),$$

$$\omega_i = \frac{1}{2}\left(\max_{1 \leq j \leq i} \mu_j - \max_{1 \leq j < i} \mu_j + \max_{i \leq j \leq N} \mu_j - \max_{i < j \leq N} \mu_j\right), \quad 2 \leq i \leq N-1$$

$$\omega_N = \frac{1}{2}\left(\max_{1 \leq j \leq N} \mu_j - \max_{1 \leq j < N} \mu_j + \mu_N\right),$$

and $\mu_j = \text{Pos}\{\boldsymbol{\theta}_j\}$ for $j = 1, 2, \cdots, N$.

Step 1. Sample a sequence $\boldsymbol{\theta}_1, \boldsymbol{\theta}_2, \cdots, \boldsymbol{\theta}_N$ from the universe Θ such that $\text{Pos}\{\boldsymbol{\theta}_i\} \geq \varepsilon$ for $i = 1, 2, \cdots, N$, where ε is a sufficiently small positive number.

Step 2. Estimate $E[T(\mathbf{x}, \xi(\boldsymbol{\theta}_i))]$ by the stochastic simulation, $i = 1, 2, \cdots, N$.

Step 3. Rearrange $\boldsymbol{\theta}_1, \boldsymbol{\theta}_2, \cdots, \boldsymbol{\theta}_N$ such that $E[T(\mathbf{x}, \xi(\boldsymbol{\theta}_i))] \leq E[T(\mathbf{x}, \xi(\boldsymbol{\theta}_{i+1}))]$ for $i = 1, 2, \cdots, N-1$.

Step 4. $E[T(\mathbf{x}, \xi)] = \sum_{i=1}^{N} \omega_i E[T(\mathbf{x}, \xi(\boldsymbol{\theta}_i))]$.

11.4.3.2 Random Fuzzy Simulation for $\overline{T}$

For any given decision $\mathbf{x}$ and any confidence level α and β, in order to find the maximal $\overline{T}$ such that $\text{Ch}\{T(\mathbf{x}, \xi) \geq \overline{T}\}(\alpha) \geq \beta$, we first set $\overline{T} = 0$. Then we randomly generate a crisp point $\boldsymbol{\theta}$ from the universe Θ such that $\text{Pos}\{\boldsymbol{\theta}\} \geq \alpha$. Since $T(\mathbf{x}, \xi(\boldsymbol{\theta}))$ is a random variable, we can use the stochastic simulation (see Liu 1999 and Zhao and Liu 2003) to estimate T' such that $\Pr\{T(\mathbf{x}, \xi(\boldsymbol{\theta})) \geq T'\} \geq \beta$. We set $\overline{T} = T'$ provided that $\overline{T} < T'$. Repeat the above processes for N times.

Step 1. Set $\overline{T} = 0$.
Step 2. Generate a crisp point $\boldsymbol{\theta}$ from Θ such that $\text{Pos}\{\boldsymbol{\theta}\} \geq \alpha$.
Step 3. Find the maximal value T' such that $\Pr\{T(\mathbf{x}, \xi(\boldsymbol{\theta})) \geq T'\} \geq \beta$.
Step 4. Set $\overline{T} = T'$ provided that $\overline{T} < T'$.
Step 5. Repeat Step 2 to Step 4 for N times.
Step 6. Return $\overline{T}$.

11.4.3.3 Random Fuzzy Simulation for System Reliability

For any given decision $\mathbf{x}$ and confidence level α, in order to estimate the system reliability $R = \text{Ch}\{T(\mathbf{x}, \xi) \geq T^0\}(\alpha)$, we first set $R = 0$. Then we generate a crisp point $\boldsymbol{\theta}$ from the universe Θ such that $\text{Pos}\{\boldsymbol{\theta}\} \geq \alpha$. If $\Pr\{T(\mathbf{x}, \xi(\boldsymbol{\theta})) \geq T^0\} > R$, then we set $R = \Pr\{T(\mathbf{x}, \xi(\boldsymbol{\theta})) \geq T^0\}$. Repeat this process for N times; we obtain the estimation of system reliability.

Step 1. Set $R = 0$.
Step 2. Randomly generate a crisp point $\boldsymbol{\theta}$ from the universe Θ such that $\text{Pos}\{\boldsymbol{\theta}\} \geq \alpha$.
Step 3. Compute the probability $\Pr\{T(\mathbf{x}, \xi(\boldsymbol{\theta})) \geq T^0\}$ by the stochastic simulation (see Liu 1999 and Zhao and Liu 2003).
Step 4. Set $R = \Pr\{T(\mathbf{x}, \xi(\boldsymbol{\theta})) \geq T^0\}$ if $R < \Pr\{T(\mathbf{x}, \xi(\boldsymbol{\theta})) \geq T^0\}$.
Step 5. Repeat Step 2 to Step 4 for N times.
Step 6. Return R.

Example 13: Let us consider a system with only one unit. The lifetime of the unit is considered to be a random fuzzy variable defined as $\xi \sim \text{N}(\rho, 10^2)$, with $\rho \sim (20, 60, 100)$. Correspondingly, the system lifetime $T(\mathbf{x}, \xi) = \xi$.

First we use the random fuzzy simulation to estimate the expected system lifetime $E[T(\mathbf{x}, \xi)] = E[\xi]$. The results of random fuzzy simulation are shown in Fig. 8, in which the straight line represents the known expected system lifetime 60.0 and the curve represents the expected system lifetime obtained by different numbers of cycles in simulation. From Fig. 8, we can see that the relative error of the results obtained by performing over 300 cycles is less than 0.5%. We now use random fuzzy simulation to obtain the maximal $\overline{T}$ such that

$$\text{Ch}\{T(\mathbf{x}, \xi) \geq \overline{T}\}(0.9) \geq 0.9, \quad \text{i.e., } \text{Ch}\{\xi \geq \overline{T}\}(0.9) \geq 0.90.$$

The results of random fuzzy simulation are shown in Fig. 9, from which it is easy to know that the relative error of the results obtained by performing over 300 cycles is less than 0.5%.

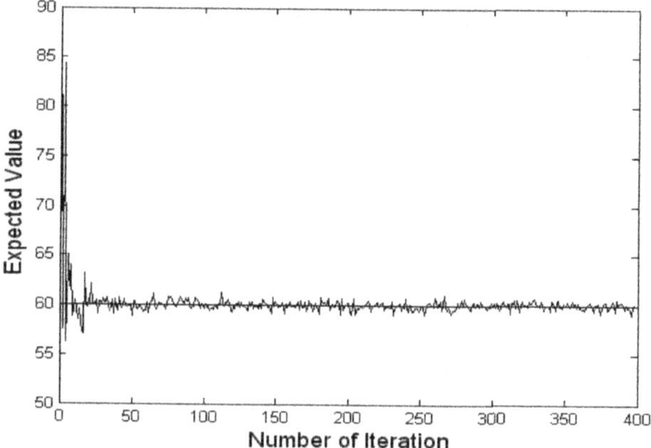

Fig. 8. A random fuzzy simulation process of the expected system lifetime

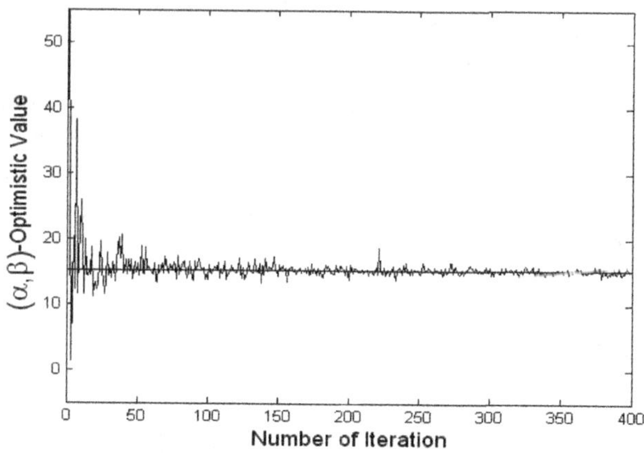

Fig. 9. A random fuzzy simulation process of (α, β)-system lifetime

Finally, we employ the random fuzzy simulation to estimate the value of $\text{Ch}\{T(\mathbf{x}, \xi) \geq \overline{T}\}(0.9) = \text{Ch}\{\xi \geq \overline{T}\}(0.9)$. From the above, it is apparent that its theoretical value is 0.90. The results of random fuzzy simulation are shown in Fig. 10, which implies that the relative error of the results obtained by performing over 400 cycles is less than 1%.

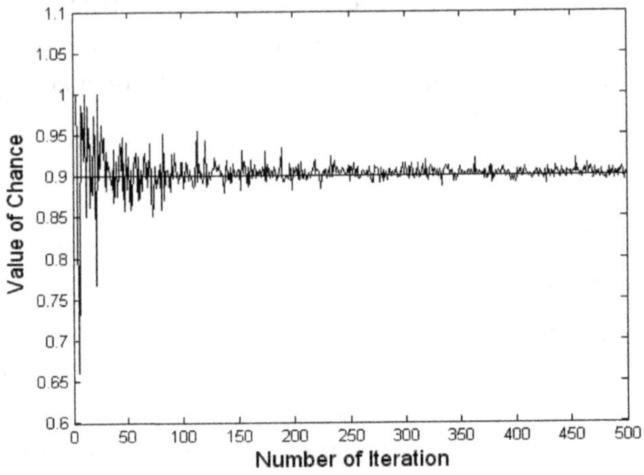

Fig. 10. A random fuzzy simulation process of the system reliability

11.5 Redundancy Optimization Models

In this section, three kinds of models including redundancy expected value models (Redundancy EVMs), redundancy chance-constrained programming (Redundancy CCP) models and redundancy dependent-chance programming (DCP) models are introduced for the different purposes of management (for fuzzy redundancy optimization models, see Zhao and Liu (2005), for fuzzy random redundancy optimization models, see Zhao et al. (2001), for random fuzzy redundancy optimization models, see Zhao and Liu (2004)).

11.5.1 Redundancy EVMs

One of the methods dealing with uncertain lifetimes in redundancy systems is the Redundancy EVM which maximizes the expected system lifetime under the cost constraints. The general form of Redundancy EVM is as follows,

$$\begin{cases} \max\ E[T(\mathbf{x}, \xi)] \\ \text{subject to:} \\ \quad \sum_{i=1}^{n} c_i x_i \le c \\ \mathbf{x} \ge 1, \text{ integer vector,} \end{cases} \tag{23}$$

where c is the maximum capital available, c_i are the costs of the ith type of the redundant elements for components i, $i = 1, 2, \cdots, n$, respectively, and E denotes the expected value operator.

We can formulate the redundancy optimization problem as a redundancy expected value goal programming (Redundancy EVGP) model according to the following priority structure and target level set by the decision-maker.

Priority 1: The expected system lifetime $E[T_1(\mathbf{x}, \xi)]$ of the backbone subsystem should achieve a given level t_1, i.e.,

$$E[T_1(\mathbf{x}, \xi)] + d_1^- - d_1^+ = t_1,$$

where d_1^- will be minimized.

Priority 2: The expected system lifetime $E[T_2(\mathbf{x}, \xi)]$ of the overall system should achieve a given level t_2, i.e.,

$$E[T_2(\mathbf{x}, \xi)] + d_2^- - d_2^+ = t_2,$$

where d_2^- will be minimized.

Priority 3: The total cost $\sum_{i=1}^{n} c_i x_i$ for the redundant elements should not exceed c, i.e.,

$$\sum_{i=1}^{n} c_i x_i + d_3^- - d_3^+ = c,$$

where d_3^+ will be minimized.

Then we have a Redundancy EVGP model,

$$\begin{cases} \text{lexmin } \{d_1^-, d_2^-, d_3^+\} \\ \text{subject to:} \\ \qquad E[T_1(\mathbf{x}, \xi)] + d_1^- - d_1^+ = t_1 \\ \qquad E[T_2(\mathbf{x}, \xi)] + d_2^- - d_2^+ = t_2 \\ \qquad \sum_{i=1}^{n} c_i x_i + d_3^- - d_3^+ = c \\ \qquad \mathbf{x} \geq 1, \text{ integer vector} \\ \qquad d_j^-, d_j^+ \geq 0, \ j = 1, 2, 3, \end{cases} \tag{24}$$

where "lexmin" represents lexicographically minimizing the objective vector. Usually, for a goal programming model, there is following order relationship for the decision vectors: for any 2 decision vectors, if the higher-priority objectives are equal, then, in the current priority level, the one with minimal objective value is better, and if 2 different decision vectors have the same objective values at every level, then there is indifferent between them (Zhao and Liu (2003)).

11.5.2 Redundancy CCP

We follow the idea of CCP proposed by Liu (2001) and Liu and Iwamura (1998) and establish a series of Redundancy CCP model with uncertain lifetimes in which the constraints hold with some predetermined confidence levels.

The general form of Redundancy CCP model is as follows,

$$\begin{cases} \max \overline{T} \\ \text{subject to:} \\ \quad \text{Ch}\{T(\mathbf{x},\xi) \geq \overline{T}\}(\alpha) \geq \beta \\ \quad \sum_{i=1}^{n} c_i x_i \leq c \\ \quad \mathbf{x} \geq 1, \text{ integer vector,} \end{cases} \tag{25}$$

where $\overline{T}$ is the (α, β)-system lifetime, α and β are the predetermined confidence levels. In fuzzy environment, $\text{Ch}\{T(\mathbf{x},\xi) \geq \overline{T}\}(\alpha) \geq \beta$ should be replaced with $\text{Cr}\{T(\mathbf{x},\xi) \geq \overline{T}\} \geq \alpha$, and $\overline{T}$ is the α-system lifetime, α is the predetermined confidence level.

Similar to the Redundancy EVGM, we can also formulate the redundancy optimization problem as a redundancy chance-constrained goal programming (Redundancy CCGP) according to the following priority structure and target level set by the decision maker.

Priority 1: The backbone subsystem lifetime $T_1(\mathbf{x},\xi)$ should achieve a given level t_1 with probability β_1 at credibility α_1, i.e.,

$$\text{Ch}\{t_1 - T_1(\mathbf{x},\xi) \leq d_1^-\}(\alpha_1) \geq \beta_1,$$

where d_1^- will be minimized.

Priority 2: The system lifetime $T_2(\mathbf{x},\xi)$ should achieve a given level t_2 with probability β_2 at credibility α_2, i.e.,

$$\text{Ch}\{t_2 - T_2(\mathbf{x},\xi) \leq d_2^-\}(\alpha_2) \geq \beta_2,$$

where d_2^- will be minimized.

Priority 3: The total cost $\sum_{i=1}^{n} c_i x_i$ should not exceed c, i.e.,

$$\sum_{i=1}^{n} c_i x_i + d_3^- - d_3^+ = c,$$

where d_3^+ will be minimized.

Thus we have

$$\begin{array}{l} \text{lexmin } \{d_1^-, d_2^-, d_3^+\} \\ \text{subject to:} \\ \quad \text{Ch}\{t_1 - T_1(\mathbf{x}, \xi) \le d_1^-\}(\alpha_1) \ge \beta_1 \\ \quad \text{Ch}\{t_2 - T_2(\mathbf{x}, \xi) \le d_2^-\}(\alpha_2) \ge \beta_2 \\ \quad \sum_{i=1}^{n} c_i x_i + d_3^- - d_3^+ = c \\ \quad \mathbf{x} \ge 1, \text{ integer vector} \\ \quad d_j^-, d_j^+ \ge 0, \ j = 1, 2, 3. \end{array} \tag{26}$$

11.5.3 Redundancy DCP

A redundant system is often required to operate during a fixed period $[0, T^0]$, where T^0 is a mission time. The decision maker may want to maximize the chance of satisfying the event $\{T(\mathbf{x}, \xi) \ge T^0\}$. In order to model this type of decision system, Liu (1999, 2000, 2002) provided a theoretical framework of DCP, in which the underlying philosophy was based on selecting the decision with maximal chance of meeting the event. In this section, we try to establish a series of Redundancy DCP with random fuzzy lifetimes.

If decision maker want to maximize the system reliability subject to a cost constraint, then we have the following Redundancy DCP model,

$$\begin{array}{l} \max \text{Ch}\{T(\mathbf{x}, \xi) \ge T^0\}(\alpha) \\ \text{subject to:} \\ \quad \sum_{i=1}^{n} c_i x_i \le c \\ \quad \mathbf{x} \ge 1, \text{ integer vector.} \end{array} \tag{27}$$

In fuzzy environment, $\text{Ch}\{T(\mathbf{x},\xi)\geq T^0\}(\alpha)$ should be replaced with $\text{Cr}\{T(\mathbf{x},\xi)\geq T^0\}$.

We can also formulate the redundancy optimization problem as a redundancy dependent-chance goal programming (Redundancy DCGP) model according to the following priority structure and target level set by the decision maker.

Priority 1: The backbone subsystem reliability that the backbone subsystem lifetime $T_1(\mathbf{x},\xi)$ achieves a given level T_1^0 should be probability β_1 at credibility α_1. Thus, we have

$$\text{Ch}\{T_1(\mathbf{x},\xi)\geq T_1^0\}(\alpha_1)+d_1^- - d_1^+ = \beta_1,$$

where d_1^- will be minimized.

Priority 2: The system reliability that the overall system lifetime $T_2(\mathbf{x},\xi)$ achieves a given level T_2^0 should be probability β_2 at credibility α_2, i.e.,

$$\text{Ch}\{T_2(\mathbf{x},\xi)\geq T_2^0\}(\alpha_1)+d_2^- - d_2^+ = \beta_2,$$

where d_2^- will be minimized.

Priority 3: The total cost for the redundant elements should not exceed c, i.e.,

$$\sum_{i=1}^{n} c_i x_i + d_3^- - d_3^+ = c,$$

where d_3^+ will be minimized.

Then we have

$$\begin{cases} \text{lexmin}\ \{d_1^-, d_2^-, d_3^+\} \\ \text{subject to:} \\ \quad \text{Ch}\{T_1(\mathbf{x},\xi)\geq T_1^0\}(\alpha_1)+d_1^- - d_1^+ = \beta_1 \\ \quad \text{Ch}\{T_2(\mathbf{x},\xi)\geq T_2^0\}(\alpha_2)+d_2^- - d_2^+ = \beta_2 \\ \quad \sum_{i=1}^{n} c_i x_i + d_3^- - d_3^+ = c \\ \quad \mathbf{x}\geq 1, \text{ integer vector} \\ \quad d_j^-, d_j^+ \geq 0,\ j=1,2,3. \end{cases} \tag{28}$$

Remark 5 *In fuzzy environment, we do not consider the standby redundancy optimization problem. Furthermore, the system lifetime and the system reliability should be consistent with the fuzzy variable.*

11.6 Genetic Algorithm Based on Simulation

The genetic algorithm is a stochastic search method for optimization problems based on the mechanics of natural selection and natural genetics. Genetic algorithms have demonstrated considerable success in providing good solutions to many complex optimization problems and received more and more attentions during the past three decades. When the objective functions to be optimized in the optimization problems are multimodal or the search spaces are particularly irregular, algorithms need to be highly robust in order to avoid getting stuck at local optimal solution. The advantage of genetic algorithms is just to obtain the global optimal solution fairly. In addition, genetic algorithms do not require the specific mathematical analysis of optimization problems, which makes genetic algorithms themselves easily coded and used by common users who are not assumed good at mathematics and algorithms. Genetic algorithms (including evolution programs and evolution strategies) have been well discussed and documented by numerous literature, such as Goldberg (1989), Michalewicz (1992) and Fogel (1994), and applied to a wide problems, traveling salesman problems, drawing graphs, scheduling, group technology, facility layout and location as well as pattern recognition. Especially, production plan problems and optimal capacity expansion are also constrained by partial order restrictions and solved by genetic algorithms in Gen and Liu (1995, 1997).

In this section, we will design a genetic algorithm based on simulation for solving redundancy optimization problems. We will discuss representation structure, initialization process, evaluation function, selection process, crossover operation and mutation operation.

11.6.1 Structure Representation

For redundancy optimization problem, we use an integer vector $V=(x_1,x_2,\cdots,x_n)$ as a chromosome to represent a solution $\mathbf{x}$, i.e., the chromosome V and solution $\mathbf{x}$ posses the same form, where $x_i\in\{0,1,2,\cdots\}$, $i=1,2,\cdots,n$. Then we initialize *pop_size* chromosomes. If the constraints include uncertain functions, we will employ the simulation to estimate the values of these functions for each chromosome. For instance, in **Example 14**, a solution $\mathbf{x} = (x_1, x_2, x_3, x_4)$ can be coded as $V = (x_1, x_2, x_3, x_4)$. Then we can use the following subfunction to initialize *pop_size* chromosomes:

```
For i=1 to pop_size do
Mark:
  For j=1 to 4 do
    x[i][j]=rand()%k;
  Endfor
  If 46(x1+1)+55(x2+1)+50(x3+1)+60(x4+1)>400 goto
    mark;
Endfor
Return (x[i][1], x[i][2], x[i][3], x[i][4]).
```

11.6.2 Initialization Process

We define an integer *pop_size* as the number of chromosomes and initialize *pop_size* chromosomes randomly. At first, we set all genes x_i as 0, $i = 1, 2, \cdots, n,$ and form a chromosome V. Then we randomly sample an integer i between 1 and n, and the gene x_i of V will be replaced with $x_i + 1$. We repeat this step until the revised chromosome V is proven to be infeasible by the constraint conditions in which the fuzzy simulation may be used. We take the last feasible chromosome as an initial chromosome. Repeating the above process *pop-size* times, we can make *pop_size* initial feasible chromosomes $V_1, V_2, \cdots, V_{\text{pop\_size}}$.

In **Example 14**, the sub-function of checking the feasibility of the chromosome $V = (x_1, x_2, x_3, x_4)$ can be written as follows,

```
If (x1<0 || x2<0 || x3<0 || x4<0) return 0;
If 46(x1+1)+55(x2+1)+50(x3+1)+60(x4+1)>400 re-
  turn 0;
Return 1;
```

11.6.3 Evaluation Function

Evaluation function, denoted by *eval*(V), is to assign a probability of reproduction to each chromosome V so that its likelihood of being selected is proportional to its fitness relative to the other chromosomes in the population, that is, the chromosomes with higher fitness will have more chance to produce offspring by using *roulette wheel selection*.

Let $V_1, V_2, \cdots, V_{\text{pop\_size}}$ be the *pop-size* chromosomes at the current generation. One well-known method is based on allocation of reproductive trials according to rank rather than actual objective values. At first we calculate the values of the objective functions for all chromosomes by the simulation. According to the objective values, we can rearrange these chromosomes $V_1, V_2, \cdots, V_{\text{pop\_size}}$ from good to bad, i.e., the better the chromosome is, the smaller ordinal number it has. For the sake of convenience, the rearranged chromosomes are still denoted by $V_1, V_2, \cdots, V_{\text{pop\_size}}$. Now let a parameter $a \in (0, 1)$ in the genetic system be given, then we can define the so-called *rank-based evaluation function* as follows:

$$eval(V_i) = a(1-a)^{i-1}, \quad i = 1, 2, \cdots, pop\_size. \tag{29}$$

We mention that $i = 1$ means the best individual, $i = pop\_size$ the worst individual.

11.6.4 Selection Process

The selection process is based on spinning the roulette wheel *pop_size* times, each time we select a single chromosome for a new population in the following way:

Step 1. Calculate the cumulative probability q_i for each chromosome,

$$q_0 = 0, \quad q_i = \sum_{j=1}^{i} eval(V_j), i = 1, 2, \cdots, pop\_size. \tag{30}$$

Step 2. Generate a random real number r in $(0, q_{pop\_size}]$.

Step 3. Select the i-th chromosome V_i for a new population provided that $q_{i-1} < r \le q_i, 1 \le i \le pop\_size$.

Step 4. Repeat the second and third steps for *pop_size* times and obtain *pop_size* copies of chromosomes.

11.6.5 Crossover Operation

We define a parameter P_c of a genetic system as the probability of crossover. This probability gives us the expected number $P_c \cdot pop\_size$ of chromosomes which undergo the crossover operation. In order to determine the parents for crossover operation, let us do the following process repeatedly from i=1 to *pop_size*: generating randomly a real number r from the interval $[0, 1]$, the chromosome V_i is selected as a parent if $r < P_c$.

We denote the selected parents as $V_1', V_2', V_3', \cdots$ and divide them to the following pairs:

$$(V_1', V_2'), (V_3', V_4'), (V_5', V_6'), \cdots$$

Let us illustrate the crossover operator on each pair by (V_1', V_2'). We denote

$$V_1' = (x_1^{(1)}, x_2^{(1)}, \cdots, x_n^{(1)}), \quad V_2' = (x_1^{(2)}, x_2^{(2)}, \cdots, x_n^{(2)}).$$

At first, we randomly generate two integers between 1 and n as the crossover points denoted by n_1 and n_2 such that $n_1 < n_2$. Then we exchange the genes of the chromosomes V_1' and V_2' between n_1 and n_2 and produce two children as follows,

$$V_1'' = (x_1^{(1)}, x_2^{(1)}, \cdots, x_{n_1-1}^{(1)}, x_{n_1}^{(2)}, \cdots, x_{n_2}^{(2)}, x_{n_2+1}^{(1)}, \cdots, x_n^{(1)}),$$

$$V_2'' = (x_1^{(2)}, x_2^{(2)}, \cdots, x_{n_1-1}^{(2)}, x_{n_1}^{(1)}, \cdots, x_{n_2}^{(1)}, x_{n_2+1}^{(2)}, \cdots, x_n^{(2)}).$$

If the child V_1'' is proven to be infeasible by the constraints, then we use the following strategy to repair it and make it feasible. At first, we randomly sample an integer i between 1 and n, and then replace the gene x_i of V_1'' with $x_i - 1$ provided that x_i>1. Repeat this process until the revised chromosome V_1'' is proven to be feasible. If the child V_1'' is proven to be feasible, we also revise it in the following way. We randomly sample an integer i between 1 and n, and the gene x_i of V_1'' will be replaced with $x_i + 1$. We repeat this process until the revised chromosome is proven to be infeasible. We will take the last feasible chromosome as V_1''. A similar revising process will be made on V_2''.

11.6.6 Mutation Operation

We define a parameter P_m of a genetic system as the probability of mutation. This probability gives us the expected number of $P_m \cdot pop\_size$ of chromosomes which undergo the mutation operations.

Similar to the process of selecting parents for crossover operation, we repeat the following steps from i=1 to *pop_size*: generating a random real number r from the interval [0, 1], the chromosome V_i is selected as a parent for mutation if $r < P_m$. For each selected parent, denoted by $V = (x_1, x_2, \cdots, x_n)$, we mutate if by the following way.

We randomly choose two mutation positions n_1 and n_2 between 1 and n such that $n_1 < n_2$, then we set all genes x_j of V as 1 for $j = n_1, n_1 + 1, \cdots, n_2$, and form a new one

$$V' = (x_1, \cdots, x_{n_1 - 1}, 1, \cdots, 1, x_{n_2 + 1}, \cdots, x_n).$$

Then we will modify V' by the following process. We randomly sample an integer i between n_1 and n_2, and the gene x_i of V' will be replaced with $x_i + 1$. We repeat this process until the revised chromosome is proven to be infeasible by the constraints. We will replace the parent V' with the last feasible chromosome.

11.6.7 The Genetic Algorithm Procedure

Following selection, crossover and mutation, the new population is ready for its next evaluation. The genetic algorithm will terminate after a given number of cyclic repetitions of the above steps. We now summarize the genetic algorithm for solving the redundancy optimization problem as follows.

```
  Input parameters: pop_size, Pc, Pm;
  Initialize the chromosomes by Initialization
Process;
 REPEAT
   Update chromosomes by crossover and mutation
operators;
   Compute the evaluation function for all chro-
mosomes;
   Select chromosomes by sampling mechanism;
 UNTIL (termination_condition)
```

It is known that the best chromosome does not necessarily appear in the last generation. So we have to keep the best one from the beginning. If we find a better one in the new population, then replace the old one by it. This chromosome will be reported as the optimal solution after finishing the evolution.

Remark 6 *In the above statements, the simulation may be replaced with fuzzy simulation, fuzzy random simulation and random fuzzy simulation for the different problems.*

11.7 Numerical Experiments

This section provides some numerical examples to illustrate the idea of the modelling and the effectiveness of the proposed algorithm. All examples are performed on a personal computer with the following parameters: the population size is 30, the probability of crossover P_c is 0.3, the probability of mutation P_m is 0.2, and the parameter a in the rank-based evaluation function is 0.05.

Example 14: Let us consider the standby redundancy system shown in Fig. 11. We suppose that there are 4 types of units. The lifetimes of 4 types of units are assumed to be trapezoidal fuzzy variables (100, 108, 112, 120), (158, 164, 168, 173), (165, 172, 177, 185) and (150, 160, 165, 178). The decision vector is $\mathbf{x} = (x_1, x_2, x_3, x_4)$, where x_i is the numbers of the ith type of units selected, $i = 1, 2, 3, 4$, respectively.

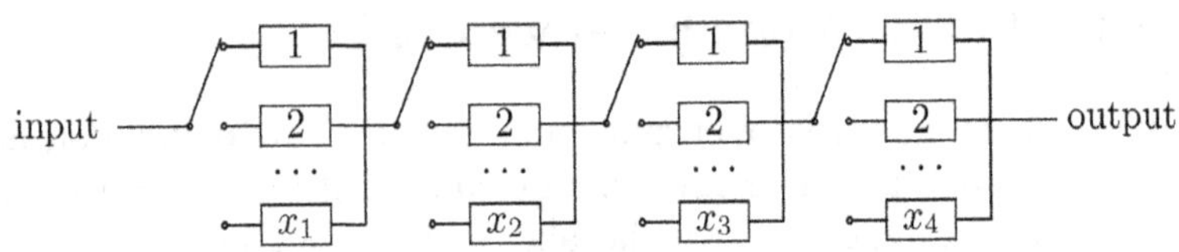

Fig. 11. A 4-stage series standby redundant system

If the prices of the 4 types of units are 46, 55, 50, and 60, and the total capital available is c, then we have a cost constraint,

$$46x_1 + 55x_2 + 50x_3 + 60x_4 \le c.$$

We assume that the system shown in Fig. 11 works if and only if there is a path of working components from the input of the system to the output. Therefore, the system structure function can be expressed as

$$\Psi(\mathbf{y}) = y_1 y_2 y_3 y_4,$$

where y_i are the states of the components i, $i = 1, 2, 3, 4$, respectively.

a). If we wish to maximize the expected system lifetime $E[T(\mathbf{x}, \xi)]$, then we have the following Standby Redundancy EVM,

$$\begin{cases} \max E[T(\mathbf{x}, \xi)] \\ \text{subject to:} \\ \qquad 46x_1 + 55x_2 + 50x_3 + 60x_4 \le c \\ \qquad \mathbf{x}, \text{ nonnegative integer vector.} \end{cases} \tag{31}$$

It is easy to see that the size of the feasible set for the model (31) is mainly determined by the total cost c. The larger the total cost c is, the more large-scale the model (31) is. In order to evaluate the veracity and ability of the genetic algorithm, we consider the two cases: $c = 400$ and $c = 2000$.

First, let us consider the model (31) with $c = 400$. In such a case, we can enumerate and evaluate all possible solutions for the model (31). Accordingly, we can make a comparison between the result obtained by the genetic algorithm and the real solution.

Table 1.1. The feasible solutions for when $c = 400$

Decision **x**	Objective values by simulation	Decision **x**	Objective values by simulation
(0, 0, 0, 0)	110.187838	(0, 0, 0, 1)	110.128405
(0, 0, 0, 2)	109.840205	(0, 0, 0, 3)	109.992156
(0, 0, 1, 0)	109.932146	(0, 0, 1, 1)	109.728742
(0, 0, 1, 2)	109.991412	(0, 0, 2, 0)	110.030478
(0, 0, 2, 1)	110.112720	(0, 0, 3, 0)	110.026445
(0, 1, 0, 0)	110.050711	(0, 1, 0, 1)	110.051400
(0, 1, 0, 2)	109.999129	(0, 1, 1, 0)	110.317962
(0, 1, 1, 1)	110.114414	(0, 1, 2, 0)	109.937235
(0, 2, 0, 0)	109.995286	(0, 2, 0, 1)	110.173055
(0, 2, 1, 0)	110.236544	(0, 3, 0, 0)	109.970437
(1, 0, 0, 0)	162.218055	(1, 0, 0, 1)	165.760442
(1, 0, 0, 2)	165.760512	(1, 0, 1, 0)	162.266755
(1, 0, 1, 1)	165.687772	(1, 0, 2, 0)	162.582525
(1, 1, 0, 0)	163.583805	(1, 1, 0, 1)	174.872788
(1, 1, 1, 0)	163.289253	(1, 2, 0, 0)	163.471459
(2, 0, 0, 0)	162.408408	(2, 0, 0, 1)	165.774582
(2, 0, 1, 0)	162.201530	(2, 1, 0, 0)	163.288448
(3, 0, 0, 0)	162.113450	(3, 0, 1, 0)	162.227875
(4, 0, 0, 0)	162.387951		

A run of genetic algorithm (5000 cycles in simulation and 2500 generations) for this example shows that the optimal solution is

$$\mathbf{x}^* = (2, 2, 1, 2),$$

the expected system lifetime is $E[T(\mathbf{x}^*, \xi)] = 175.604$, and the total cost is $C(\mathbf{x}^*) = 372$.

All of the feasible solutions for the model (25) with $c = 400$ are enumerated in Table 1 in which the optimal solution is just $\mathbf{x}^*$ and its expected system lifetime is 174.873. This indicates that the result obtained by the genetic algorithm is just the optimal solution.

From the above argumentation, the genetic algorithm has exhibited its excellent ability in solving small problems. Now let us take into account the problem with $c = 2000$. In such a case, each decision variable x_i may be range between 0 and 30, $i = 1, 2, 3, 4$. Thus there are about 9.2×10^5 feasible solutions for the model (31).

A run of the genetic algorithm (4000 cycles in simulation, and 3000 generations) shows that the optimal solution is

$$\mathbf{x}^* = (13, 8, 8, 9),$$

the expected system lifetime is $E[T(\mathbf{x}^*, \xi)] = 1330.893$, and the total cost is $C(\mathbf{x}^*) = 1978$.

Remark 7 *The above results imply that the genetic algorithm has ability to solve both small and large problems. However, we must emphasize that the maximal relative error between the results obtained by fuzzy simulation and the real value may be increased with the accretion of the size of the problems. If the values provided by the fuzzy simulation have biases or inaccuracies, then GA may not actually be searching the best solution. This indicates that the ability and precision of the genetic algorithm are reduced with the accretion of the size of problems. One of methods to strengthen the ability and precision of the genetic algorithm is to increase the cycle times in fuzzy simulation. This action will expend much CPU time since a mass of calculation. In addition, the result obtained by genetic algorithm is not surely optimal. Although we cannot guarantee that the result by genetic algorithm is the optimal solution for the model, the result is satisfactory. Numerous papers on genetic algorithms have proved it though numerical examples.*

b). If we want to determine the optimal numbers of the standby units so as to maximize 0.90-system lifetime for system under the cost constraint. Then we obtain the following Standby Redundancy CCP model,

$$\begin{array}{l} \max\ \overline{T} \\ \text{subject to:} \\ \qquad \mathrm{Cr}\{T(\mathbf{x}, \xi) \ge \overline{T}\} \ge 0.90 \\ \qquad 46x_1 + 55x_2 + 50x_3 + 60x_4 \le c \\ \qquad \mathbf{x}, \text{ nonnegative integer vector.} \end{array} \tag{32}$$

0.90-system lifetime $Cr\{T(\mathbf{x}, \xi) \ge \overline{T}\} \ge 0.90$ is estimated by the fuzzy simulation designed in Section 3.1. A run of the genetic algorithm (5000 cycles in fuzzy simulation, 3000 generations) shows that the optimal solution (see Fig. 12) is

$$\mathbf{x}^* = (2, 2, 1, 2)$$

with 0.90-system lifetime $T^* = 166.512$, and $\sum_{i=1}^{4} c_i x_i^* = 372$.

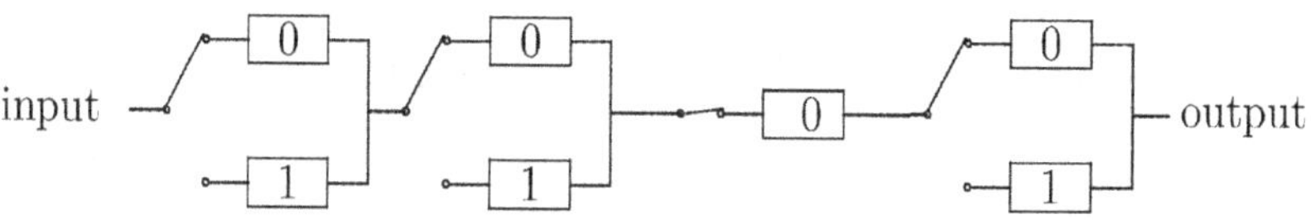

Fig. 12. The optimal solution of standby redundant CCP in Example 14

c). If we want to determine the optimal numbers of the standby units so as to maximize system reliability $Cr\{T(\mathbf{x}, \xi) \geq 170\}$ under the cost constraint. Then we obtain the following Standby Redundancy DCP model,

$$\begin{cases} \max \mathrm{Cr}\{T(\mathbf{x}, \xi) \geq 170 \\ \text{subject to} \\ \quad 46x_1 + 55x_2 + 50x_3 + 60x_4 \leq c \\ \quad \mathbf{x}, \text{ nonnegative integer vector.} \end{cases} \tag{33}$$

The system reliability $Cr\{T(\mathbf{x}, \xi) \geq 170\}$ is estimated by the fuzzy simulation designed in Section 3.1. A run of the genetic algorithm (5000 cycles in fuzzy simulation, 3000 generations) shows that the optimal solution is

$$\mathbf{x}^* = (2, 2, 1, 2)$$

with the system reliability $Cr\{T(\mathbf{x}, \xi) \geq 170\} = 0.72$, and the total cost $\sum_{i=1}^{4} c_i x_i^* = 372$.

Example 15: Consider a 4-stage series standby redundancy system shown in Fig. 11. We suppose that there are 4 types of elements available and their lifetimes are considered to be fuzzy random variables defined as follows,

$$\begin{aligned}
&\xi_1 \sim (\rho_1 - 10, \rho_1 - 5, \rho_1 + 5, \rho_1 + 10), && with\, \rho_1 \sim \mathrm{N}(100, 10^2),\\
&\xi_1 \sim (\rho_2 - 7, \rho_2 - 4, \rho_2 + 6, \rho_2 + 12), && with\, \rho_2 \sim \mathrm{N}(110, 10^2),\\
&\xi_1 \sim (\rho_3 - 12, \rho_3 - 6, \rho_3 + 5, \rho_3 + 13), && with\, \rho_3 \sim \mathrm{N}(115, 10^2),\\
&\xi_1 \sim (\rho_4 - 8, \rho_4 - 3, \rho_4 + 6, \rho_4 + 17), && with\, \rho_4 \sim \mathrm{N}(90, 10^2).
\end{aligned}$$

Correspondingly, their costs are assumed to be 50, 45, 60, and 72, respectively. The decision vector is $\mathbf{x}=(x_1, x_2, x_3, x_4)$, where x_i denote the numbers of the ith type of elements selected, $i=1, 2, 3, 4$, respectively.

If the total capital available is 480, we have the following cost constraint

$$50x_1+45x_2+60x_3+72x_4 \le 480.$$

The system structure function can be expressed as

$$\Psi(\mathbf{y}) = y_1 y_2 y_3 y_4,$$

where y_i are the states of the components i, $i=1, 2, 3, 4$, respectively.

a). If the decision maker wants to maximize the expected system lifetime $E[T(\mathbf{x}, \xi)]$ with the cost constraint, i.e., we have the following Redundancy EVM

$$\begin{cases} \max\ E[T(\mathbf{x}, \xi)] \\ \text{subject to:} \\ \qquad 50x_1+45x_2+60x_3+72x_4 \le 480 \\ \qquad \mathbf{x} \ge 1, \text{ integer vector.} \end{cases} \tag{34}$$

We employ the genetic algorithm to solve the model. A run of the genetic algorithm (3000 cycles in fuzzy random simulation, and 300 generations) shows that the optimal redundancy is

$$\mathbf{x}^* = (2, 2, 2, 2)$$

with the expected system lifetime $E[T(\mathbf{x}^*, \xi)] = 180.124$ and the cost 454.

b). If we want to determine the optimal numbers of the units so as to maximize (0.9, 0.9)-system lifetime under the cost constraint. Then we obtain the following Redundancy CCP model,

$$\begin{cases} \max\ \overline{T} \\ \text{subject to:} \\ \qquad \mathrm{Ch}\{T(\mathbf{x},\xi) \ge \overline{T}\}(\alpha) \ge \beta \\ \qquad 50x_1+45x_2+60x_3+72x_4 \le 480 \\ \qquad \mathbf{x} \ge 1, \text{ integer vector.} \end{cases} \tag{35}$$

The (0.9, 0.9)-system lifetime $\mathrm{Ch}\{T(\mathbf{x},\xi) \ge \overline{T}\}(0.9) \ge 0.9$ is estimated by the fuzzy random simulation designed in Section 4.2. A run of the genetic algorithm (2000 cycles in fuzzy random simulation, 1000 generations) shows that the optimal solution is

$$\mathbf{x}^* = (2, 2, 2, 2)$$

with (0.9. 0.9)-system lifetime $T^* = 141.451$, and the total cost is 454.

c). If we want to determine the optimal numbers of the units so as to maximize system reliability subject to a cost constraint. Then we have the following Redundancy DCP model,

$$\begin{cases} \max \ \mathrm{Ch}\{T(\mathbf{x},\xi) \geq 150\}(0.9) \\ \text{subject to:} \\ \qquad 50x_1 + 45x_2 + 60x_3 + 72x_4 \leq 480 \\ \qquad \mathbf{x} \geq 1, \text{ integer vector.} \end{cases} \tag{36}$$

The system reliability $\mathrm{Ch}\{T(\mathbf{x},\xi) \geq 150\}(0.9)$ can be estimated by the fuzzy random simulation designed in Section 3.3. A run of the genetic algorithm (3000 cycles in fuzzy random simulation, 500 generations) shows that the optimal solution is

$$\mathbf{x}^* = (2, 2, 2, 2)$$

with the system reliability $\mathrm{Ch}\{T(\mathbf{x},\xi) \geq 150\}(0.9) = 0.59$, and the cost 454.

Example 16: Consider a series parallel redundant system consisting of 5 components shown in Fig. 13. For each component, we assume that there is one type of element available.

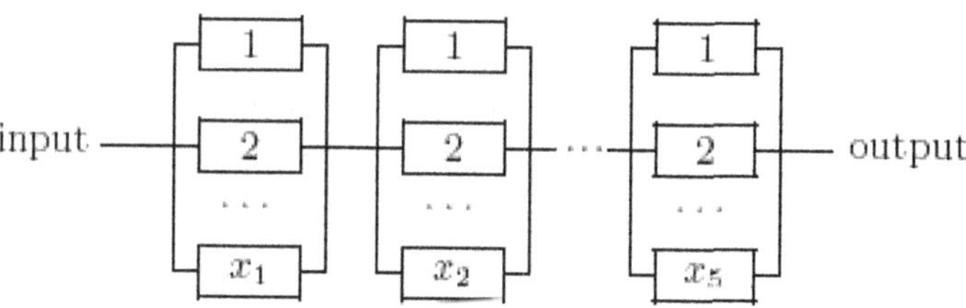

Fig. 13. A series parallel redundant system

The lifetimes of the 5 types of elements are assumed to be random fuzzy variables

$$\xi_1 \sim \mathcal{N}(\rho_1, 10^2), \text{ with } \rho_1 = (120, 132, 142),$$
$$\xi_2 \sim \mathcal{N}(\rho_2, 10^2), \text{ with } \rho_2 = (95, 100, 126),$$
$$\xi_3 \sim \mathcal{N}(\rho_3, 10^2), \text{ with } \rho_3 = (80, 92, 118),$$
$$\xi_4 \sim \mathcal{N}(\rho_4, 10^2), \text{ with } \rho_4 = (74, 98, 116),$$
$$\xi_5 \sim \mathcal{N}(\rho_5, 10^2), \text{ with } \rho_4 = (73, 93, 121),$$

and their costs c_i's are 112, 132, 109, 105 and 123. Thus the total cost of system $C(x) = 112x_1 + 132x_2 + 109x_3 + 105x_4 + 123x_5$. We also suppose that the total capital available should not exceed 950. Then we have the cost constraint $C(x) \le 950$. The system structure function can be expressed as $\Psi(y) = y_1 y_2 y_3 y_4 y_5$, where y_i are the states of the components i, $i = 1, 2, 3, 4, 5$.

a). If the decision maker wants to maximize the expected system lifetime $E[T(\mathbf{x}, \xi)]$ with the cost constraint, i.e., we have the following Redundancy EVM

$$\begin{cases} \max \ E[T(\mathbf{x}, \xi)] \\ \text{subject to:} \\ \quad 112x_1 + 132x_2 + 109x_3 + 105x_4 + 123x_5 \le 950 \\ \quad \mathbf{x} \ge 1, \text{ integer vector.} \end{cases} \tag{37}$$

We employ the genetic algorithm based on random fuzzy simulation to solve the model (37). $E[T(\mathbf{x}, \xi)]$ is estimated by the random fuzzy simulation designed in Section 3.3. A run of the genetic algorithm (3000 cycles in simulation, 500 generations) shows that the optimal solution is

$$\mathbf{x}^* = (1, 1, 2, 2, 2)$$

with the expected system lifetime $E[T(\mathbf{x}^*, \xi)]$=92.314, and the cost 918.

b). If the decision maker wants to maximize $(0.9, 0.9)$-system lifetime under the cost constraint, then we have the following Redundancy CCP model,

$$\begin{cases} \max \ \overline{T} \\ \text{subject to:} \\ \quad \mathrm{Ch}\{T(\mathbf{x}, \xi) \ge \overline{T}\}(0.9) \ge 0.9 \\ \quad 112x_1 + 132x_2 + 109x_3 + 105x_4 + 123x_5 \le 950 \\ \quad \mathbf{x} \ge 1, \text{ integer vector.} \end{cases} \tag{38}$$

A run of the genetic algorithm (3000 cycles in simulation, and 500 generations) shows that the optimal solution is

$$\mathbf{x}^* = (1, 1, 2, 2, 2)$$

with (0.9, 0.9)-system lifetime T^*=82.015, and the cost 918.

c). If the decision maker wants to maximize the system reliability subject to a cost constraint, then we have the following Redundancy DCP model,

$$\begin{cases} \max \mathrm{Ch}\{T(\mathbf{x},\xi)\geq 85\}(0.9) \\ \text{subject to:} \\ \qquad 112x_1+132x_2+109x_3+105x_4+123x_5\leq 950 \\ \qquad \mathbf{x}\geq 1, \text{ integer vector.} \end{cases} \tag{39}$$

A run of the genetic algorithm (3000 cycles in simulation, and 500 generations) shows the optimal solution is

$$\mathbf{x}^* =(1, 1, 2, 2, 2)$$

with the system reliability $\mathrm{Ch}\{T(\mathbf{x},\xi)\geq 85\}(0.9)$ = 0.690, and the total cost 918.

Remark 8 Zhao and Liu (2000, 2003) designed a hybrid intelligent algorithm for solving the above redundancy optimization models. This method used simulation techniques to estimate system performance such as the expected system lifetime, α-system lifetime, or system reliability, and employed the designed simulation techniques to generate a training data and trained a neural network (NN) by the backpropagation algorithm to approximate the system performance, furthermore, embedded the trained NN into a genetic algorithm to form a hybrid intelligent algorithm (the similar algorithm has been designed by Coit and Smith 1996). However, we must emphasize that the relative errors between the results obtained by the simulation techniques (as a test data set) and that by the NN are remarkably increased with the accretion of the size of the problems. One of the reasons is that the NN gets more complex as the problems get larger and the values of uncertain functions provided by the NN will also be less accurate. If the values provided by the NN have biases or inaccuracies, then GA may not actually be searching the best solution. This indicates that the ability and precision of the hybrid intelligent algorithm are reduced with the accretion of the size of problems. In addition, the result obtained by the hybrid intelligent algorithm is not surely optimal.

Acknowledgments

This work was supported by National Natural Science Foundation of-China Grant No. 70571056.

References

Cai KY, Wen CY, Zhang ML (1991) Fuzzy variables as a basis for a theory of fuzzy reliability in the possibility context. Fuzzy Sets and Systems 42:145-172

Campos L, González A (1989) A subjective approach for ranking fuzzy numbers. Fuzzy Sets and Systems 29:145-153

Castellano G, Fanelli AM, Pelillo M (1997) An iterative pruning algorithm for feedforward neural networks. IEEE Transactions on Neural Network 8:519-537

Chanas S, Nowakowski M (1988) Single value simulation of fuzzy variable. Fuzzy Sets and Systems 25:43-57

Chern CS (1992) On the computational complexity of reliability redundancy allocation in the series system. Operations Research Letters 11:309-315

Coit DW (2001) Cold-standby redundancy optimization for nonrepairable systems. IIE Transactions 33:471-478

Coit DW, Smith AE (1996) Reliability optimization of series-parallel systems using a genetic algorithm. IEEE Transactions on Reliability 45:254-260

Coit DW, Smith AE (1998) Redundancy allocation to maximize a lower percentile of the system time-to-failure distribution. IEEE Transactions on Reliability 47:79-87

Dubois D, Prade H (1987) The mean value of a fuzzy number. Fuzzy Sets and Systems 24:279-300

Dubois D, Prade H (1988) Possibility Theory: An Approach to Computerized Processing of Uncertainty. Plenum, New York

Fogel DB (1994) An introduction to simulated evolutionary optimization. IEEE Transactions on Neural Networks 5:3-14

Gao J, Liu B (2001) New primitive chance measures of fuzzy random event. International Journal of Fuzzy Systems 3:527-531

Gen M, Liu B (1995) Evolution program for production plan problem. Engineering Design and Automation 1:199-204

Gen M, Liu B (1997) Evolution program for optimal capacity expansion. Journal of Operations Research Society of Japan 40:1-9

Goldberg DE (1989) Genetic Algorithms in Search. Optimization and Machine Learning, Addison-Wesley

González A (1990) A study of the ranking function approach through mean values. Fuzzy Sets and Systems 35:29-41

Heilpern S (1992) The expected value of a fuzzy number. Fuzzy Sets and Systems 47:81-86

Karwowski W, Mital A (1986) Applications of fuzzy set theory in human factors. Elsevier, Amasterdam

Kaufmann A (1975) Introduction to the theory offuzzy subsets. Academic Press, New York

Kruse R, Meyer KD (1987) Statistics with Vague Data. D. Reidel Publishing Company, Dordrecht

Kuo W, Prasad VR (2000) An annotated overview of system-reliability optimization. IEEE Transactions on Reliability 49:176-187

Kwakernaak H (1978) Fuzzy random variables—I. Information Sciences 15:1-29

Kwakernaak H (1979) Fuzzy random variables—II. Information Sciences 17:253-278

Levitin G, Lisnianski A, Ben-Haim H, Elmakis D (1998) Redundancy optimization for series-parallel multi-state system. IEEE Transactions on Reliability 47:165-172

Liu B (1998) Minimax chance constrained programming models for fuzzy decision systems. Information Sciences 112:25-38

Liu B (1999) Dependent-chance programming with fuzzy decision. IEEE Transactions on Fuzzy Systems 7:354-360

Liu B (1999) Uncertain Programming. John Wiley & Sons, New York

Liu B (1999) Dependent-chance programming with fuzzy decisions. IEEE Transactions on Fuzzy Systems 7:354-360

Liu B (2000) Dependent-chance programming in fuzzy environments. Fuzzy Sets and Systems 109:97-106

Liu B (2001) Fuzzy random chance-constrained programming. IEEE Transactions on Fuzzy Systems 9:713-720

Liu B (2001) Fuzzy random dependent-chance programming. IEEE Transactions on Fuzzy Systems 9:721-726

Liu B (2002) Theory and Practice of Uncertain Programming. Physica-Verlag, Heidelberg

Liu B (2004) Uncertainty Theory: An Introduction to Its Axiomatic Foundations. Springer-Verlag, Berlin

Liu B, Iwamura K (1998) Chance-constrained programming with fuzzy parameters. Fuzzy Sets and Systems 94:227-237

Liu B, Iwamura K (1998) A note on chance-constrained programming with fuzzy coefficients. Fuzzy Sets and Systems 100:229-233

Liu B, Liu Y (2002) Expected value of fuzzy variable and fuzzy expected value model. IEEE Transactions on Fuzzy Systems 10:445-450

Liu Y, Gao J (2005) The independence of fuzzy variables in credibility theory and its applications. Technical Report

Liu Y, Liu B (2002) Random fuzzy programming with chance measures defined by fuzzy integrals, Mathematical and Computer Modelling 36:509-524

Liu Y, Liu B (2003) Expected value operator of random fuzzy variable and random fuzzy expected value models. International Journal of Uncertainty, Fuzziness & Knowledge-Based Systems 11:195-215

Liu Y, Liu B (2003) Fuzzy random variables: a scalar expected value operator. Fuzzy Optimization and Decision Making 2:143-160

Liu Y, Liu B (2005) On minimum-risk problems in fuzzy random decision systems, Computers & Operations Research 32:257-283

Michalewicz Z (1992) Genetic Algorithms + Data Structures = Evolution Programs. Springer-Verlag, New York

Nahmias S (1978) Fuzzy variables. Fuzzy Sets and Systems 1:97-110

Prasad VR, Kuo W, Kim KMO (1999) Optimal allocation of s-identical, multifunctional redundant elements in a series system. IEEE Transactions on Reliability 47:118-126

Puri ML, Ralescu DA (1985) The concept of normality for fuzzy random variables. Ann. probab. 13:1371-1379

Puri ML, Ralescu DA (1986) Fuzzy random variables. Journal of Mathematical Analysis and Applications 114:409-422

Shitaishi N, Furuta H (1983) Reliability analysis based on fuzzy probability. Journal of Engineering Mechanism 109:1445-1459

Venkatech S (1992) Computation and learning in the context of neural network capacity. Neural Networks for Perception 2:173-327

Yager RR (1981) A procedure for ordering fuzzy subsets of the unit interval. Information Sciences 24:143-161

Yager RR (1992) On the specificity of a possibility distribution. Fuzzy Sets and Systems 50:279-292

Yager RR (1993) On the completion of qualitative possibility measures. IEEE Transactions on Fuzzy Systems 1:184-193

Yager RR (2002) On the evaluation of uncertain courses of action. Fuzzy Optimization and Decision Making 1:13-41

Zadeh LA (1978) Fuzzy sets as a basis for a theory of possibility. Fuzzy Sets Systems 1:3-28

Zadeh LA (1979) A theory of approximate reasoning. In: Hayes J, Michie D and Throll DM (eds): Mathematical Frontiers of the Social and Policy Sciences. Westview Press, Boulder Cororado, pp 69-129

Zhao R, Liu B (2003) Stochastic programming models for general redundancy optimization problems. IEEE Transactions on Reliability 52:181-192

Zhao R, Liu B (2004) Redundancy optimization problems with uncertainty of combining randomness and fuzziness. European Journal of Operational Research 157: 716-735

Zhao R, Liu B (2005) Standby Redundancy Optimization Problems with Fuzzy Lifetimes. Computer & Industrial Engineering 49: 318-338

Zhao R, Song K, Zhu J (2001) Redundancy Optimization Problems with Fuzzy Random Lifetimes. IEEE International Conference on Fuzzy Systems: 288-291

Computational Intelligence Methods in Software Reliability Prediction

Liang Tian and Afzel Noore

Lane Department of Computer Science and Electrical Engineering,
West Virginia University, U.S.A.

12.1 Introduction

Computational intelligence methods are evolving collections of methodologies, which adopt tolerance for imprecision, uncertainty, and partial truth to obtain robustness, tractability, and low cost. Fuzzy logic, neural networks, genetic algorithm and evolutionary computation are the most important key methodologies. These advanced knowledge processing methods have recently gained more and more attention and have been successfully applied in wide varieties of application fields, such as aerospace, communication systems, consumer appliances, electric power operation systems, manufacturing automation and robotics, process engineering, transportation systems, and software measurement and estimation.

Fuzzy logic modeling is a white box approach, assuming that there is already human knowledge about a solution. Fuzzy logic modeling can deal with the imprecision of the input and output variables directly by defining them with fuzzy numbers and fuzzy sets, which can be represented in linguistic terms. In this manner, complex process behaviors can be described in general terms without precisely defining the nonlinear phenomena involved. It is an efficient generic decision-making tool by embedding structured human knowledge into useful algorithms, which can approximate multivariate nonlinear function and operate successfully under a lack of precise sensor information.

An artificial neural network can be defined as a data processing system consisting of a large number of simple, highly interconnected processing elements in an architecture inspired by the structure of the cerebral cortex of the brain [39]. These processing elements are usually organized into a sequence of layers with connections between the layers. Multilayer feed-forward neural network is one of the most commonly used networks in various applications. As an example, the three-layer feed-forward neural network architecture is shown in Fig. 1.

L. Tian and A. Noore: *Computational Intelligence Methods in Software Reliability Prediction*, Computational Intelligence in Reliability Engineering (SCI) **39**, 375–398 (2007)
www.springerlink.com

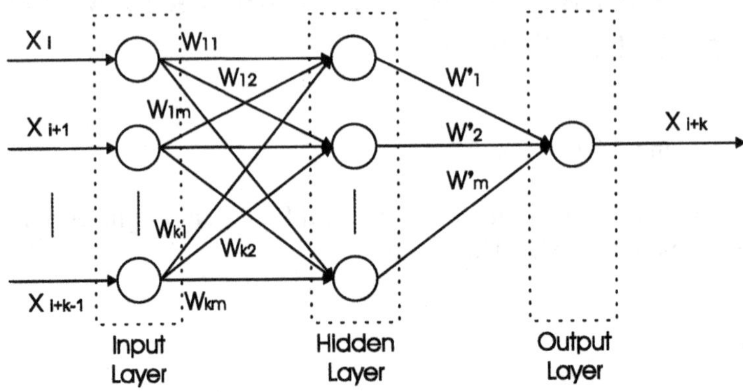

Fig. 1. Three-layer feed-forward neural network

w_{ij} is the weight connecting the ith input neuron and the jth hidden neuron, where $1 \leq i \leq k$, and $1 \leq j \leq m$. w'_j is the weight connecting the jth hidden neuron and the output neuron, where $1 \leq j \leq m$.

By mimicking human learning ability, neural networks learn by example data and they constitute a distributed associative memory. Also, they are fault tolerant, robust to the presence of noisy data, and capable of approximation of multivariate nonlinear functions.

Genetic Algorithms (GA) are global search and optimization techniques modeled from natural genetics, exploring search space by incorporating a set of candidate solutions in parallel. GA maintains a population of candidate solutions where each candidate solution is coded as a binary string called chromosome. A chromosome encodes a parameter set for a group of variables being optimized. A set of chromosomes forms a population, which is ranked by a fitness evaluation function. The fitness evaluation function provides information about the validity and usefulness of each candidate solution. This information guides the search of GA. More specifically, the fitness evaluation results determine the likelihood that a candidate solution is selected to produce candidate solutions in the next generation.

As shown in Fig. 2, the evolution from one generation to the next generation involves three steps. Fitness evaluation in which the current population is evaluated using the fitness evaluation function and ranked based on their fitness values. Next, the genetic algorithm stochastically selects *parents* from the current population with a bias that better chromosomes are more likely to be selected. This process is implemented by using probability that is determined by the fitness value. The reproduction process generates *children* from selected *parents* using genetic operations, such as crossover or resemblance and mutation. This cycle of fitness evaluation, selection, and reproduction terminates when an acceptable solution is found, a convergence criterion is met, or a predetermined limit on the number of iterations is reached.

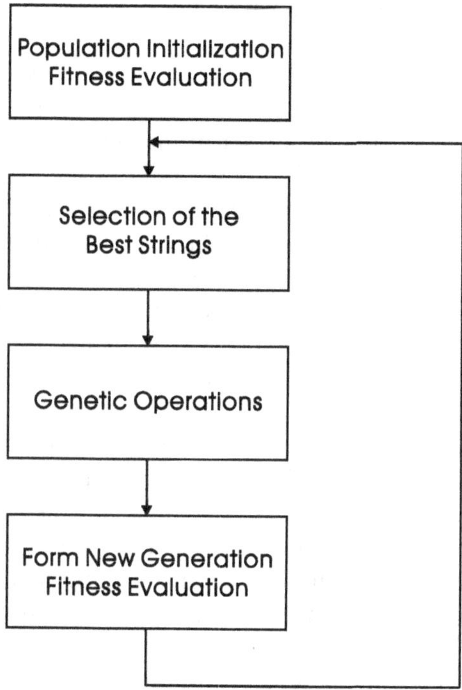

Fig. 2. Illustration of genetic algorithm framework

Today, software plays a very important role in scientific and industrial fields ranging from intelligent washing machine, computer operating systems to NASA safety critical mission operations. They have become larger, more complex and sophisticated in their demand for performance. Modern software systems are designed using the entire engineering processes to solve sophisticated problems. An important issue in developing such a system is to produce high quality software system that satisfies user requirements. Software reliability is generally accepted as the key indicator in software quality since it quantifies software failures. Few software systems are completely free of faults due to design or human errors. Software that contains undetected faults and released to the market incurs much higher failure costs. Thus, before newly developed software is released to the user, it is extensively tested for faults that may have been introduced during development. Debugging and testing will greatly reduce the error content, but at the same time, increase development costs. The problem of developing reliable software at a low cost still remains an open challenge. As a result, software reliability assessment is increasingly important and indispensable with respect to determining optimal time to stop testing, optimizing testing resource allocation, and also ensuring that the reliability requirement is met based on various software reliability measurement metrics.

Software reliability is defined as the probability of a failure free operation of software for a specified period of time in a given environment. A software reliability model is a set of mathematical equations that are used to describe the behavior of software failures with respect to time and predict software reliability performance such as the mean time between failures and the number of residual faults. Many methodologies have been adopted for software reliability modeling, such as random-time approach (time between failures are treated as random variable), stochastic approach (number of faults in a time interval is treated as a stochastic process), Bayesian approach, fuzzy approach, neural network approach and non-parametric approach.

Software reliability models must cover two different types of situations. One is finding faults and fixing them, and the other is referred to as *no fault removal*. *No fault removal* actually means *deferred fault removal*. When the failures are identified, the underlying faults will not be removed until the next release. This situation is simple and usually occurs during validation test and operation phase. Most of software reliability models deal with the process of finding and fixing faults that usually occur during software verification process. Thus, if it is assumed that fault removal process does not introduce new faults, the software reliability will increase with the progress of debugging [37]. A software reliability model describing such fault detection and removal phenomenon is called a software reliability growth model.

Most of the existing analytical software reliability growth models depend on *a priori* assumptions about the nature of software faults and the stochastic behavior of software failure process. As a result, each model has a different predictive performance across various projects. A general model that can provide accurate predictions under multiple circumstances is most desirable [18,19,26]. Cai et al. [8,10,11] first proposed fuzzy software reliability models (FSRMs), based on the assumption that software reliability behavior is fuzzy in nature. They also demonstrated how to develop a simple fuzzy model to characterize software reliability behavior. Mihalache [23], So et al. [31], and Utkin et al. [40] have also made contributions to fuzzy software reliability/quality assessment.

Karunanithi et al. [18,19] were the first to propose a neural network approach for software reliability growth modeling. Adnan et al. [1,2], Aljahdali et al. [3,4], Guo et al. [14], Gupta et al. [15], Ho et al. [16], Park et al. [26], Sitte [30], and Tian et al. [34-36,38] have also made contributions to software reliability growth prediction using neural networks, and have obtained better results compared to existing approaches with respect to predictive performance.

The traditional feed-forward neural network does not have the capability of incorporating dynamic temporal property internally, which may have impact on the network prediction performance. If time can be represented by the effect it has on processing, the network will perform better in terms of responsiveness to temporal sequences. This responsiveness can be obtained by providing feedback of data generated by the network back into the units of the network to be used in future iterations. Recurrent neural network has the inherent capability of developing an internal memory, which may naturally extend beyond the externally provided lag spaces, and hence relaxing the requirements for the determination of external number of inputs in time-related prediction applications.

Those neural network modeling approaches adopt the gradient descent based back-propagation learning scheme to implement the empirical risk minimization (ERM) principle, which only minimizes the error during the training process. The error on the training data set is driven to a very small value for known data, but when out-of-sample data is presented to the network, the error is unpredictably large and yields limited generalization capability. As a novel type of machine learning algorithm, support vector machine (SVM) has gained increasing attention from its original application in pattern recognition to the extended application in function approximation and regression estimation. Based on the structural risk minimization (SRM) principle, the learning scheme of SVM is focused on minimizing an upper bound of the generalization error that includes the sum of the empirical training error and a regularized confidence interval, which eventually results in better generalization performance.

The purpose of this chapter is to introduce how these newly emerging computational intelligence methods can be used in software reliability engineering applications. Unlike traditional software reliability modeling approaches, we introduce three new methods. (1) Dynamic learning model using evolutionary neural networks; (2) Improving generalization capability using recurrent neural network and Bayesian regularization; and (3) Adaptive learning model using support vector machines. The performance of these new models have been tested using the same real-time control application and flight dynamic application data sets as cited in Park et al. [26] and Karunanithi et al. [18]. We choose a common baseline to compare the results with related work cited in the literature. By using the common data sets and performance evaluation metrics, it is possible to quantitatively validate the effectiveness of the new approaches. Experimental results show that these new approaches achieve better prediction accuracy compared to existing approaches.

12.2 Dynamic Evolutionary Neural Network (D–ENN) Learning

It has been shown that a neural network approach is a universal approximator for any non-linear continuous function with an arbitrary accuracy [9,20]. The underlying failure process can be learned and modeled based on only failure history of a software system rather than based on *a priori* assumptions. Consequently, it has become an alternative method in software reliability modeling, evaluation, and prediction.

The published literature uses neural network to model the relationship between software failure time and the sequence number of failures. Some examples are: execution time as input and the corresponding accumulated number of defects disclosed as desired output, and failure sequence number as input and failure time as desired output. Recent studies focus on modeling software reliability based on time-lagged neural network structure. Aljahdali et al. [3] used the recent days' faults observed before the current day as multiple inputs to predict the number of faults initially resident at the beginning of testing process. Cai et al. [9] and Ho et al. [16] used the recent inter-failure time as multiple inputs to predict the next failure time.

The effect of both the number of input neurons and the number of neurons in hidden layers were determined using a selected range of predetermined values. For example, 20, 30, 40, and 50 input neurons were selected in Cai's experiment [9], while 1, 2, 3, and 4 input neurons were selected in Adnan's experiment [2]. The effect on the structure was studied by independently varying the number of input neurons or the number of neurons in hidden layers instead of considering all possible combinations. Genetic algorithm can be used as an optimization search scheme to determine the optimal network architecture.

We model the inter-relationship among software failure time data instead of the relationship between failure sequence number and failure time data. The inputs and outputs of the neural network are all failure time data. During a dynamic failure prediction process, the number of available failure time data increases over time. The fixed neural network structures do not address the effect on the performance of prediction as the number of data increases. As shown in Fig. 3, when a software failure, x_i, occurs, the failure history database is updated and the accumulated failure data $(x_1, x_2, ..., x_i)$ is made available to the evolutionary neural network model. Genetic algorithm is used to globally optimize the neural network architecture after every occurrence of software failure time data. The optimization process determines the number of the delayed input neurons corresponding to the previous failure time data sequence and the number of neurons in the hidden layer. This information is then used to dynamically reconfigure the neural network architecture for predicting the next-step failure, $\hat{x}_{i+1}$. The corresponding globally optimized network architecture will be iteratively and dynamically reconfigured as new failure time data arrives.

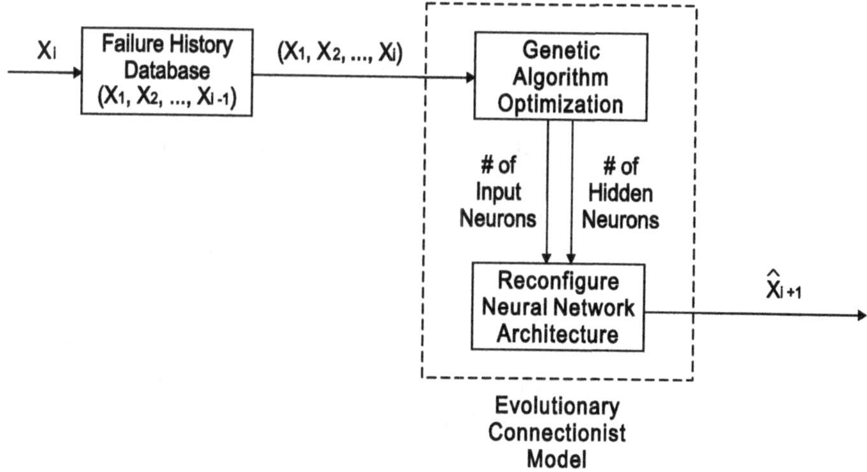

Fig. 3. Dynamic software failure prediction framework

The dynamic learning optimization process for evolutionary neural network model is described by the following procedure:

1. For every occurrence of new data, optimize the neural network structure by finding the optimal number of input neurons and the optimal number of neurons in the hidden layer using the genetic algorithm based on currently available historical data.
2. Apply training data patterns to the optimized neural network structure until the neural network converges, and predict the next-step data.
3. Repeat the above steps as new data arrive.

The fitness evaluation function in genetic algorithm is defined as:

$$fitness = \frac{1}{1 + err} \tag{1}$$

$$err = \sum_{i=1}^{p} \frac{|\hat{x}_i - x_i|^2}{p} \tag{2}$$

where p is the number of exemplars used during the training process. $\hat{x}_i$ and x_i are the predicted output and the actual output respectively during the back-propagation learning process and *err* is the mean square error.

The genetic algorithm optimization process is described in the following procedure:

1. Randomize population.
2. Evaluate the fitness function for each individual in the population.
3. Select the first two individuals with the highest fitness values and copy directly to the next generation without any genetic operation.
4. Select the remaining individuals in the current generation and apply crossover and mutation genetic operations accordingly to reproduce the individuals in the next generation.
5. Repeat from the second step until all individuals in population meet the convergence criteria or the number of generations exceeds the pre-defined maximum values.
6. Decode the converged individuals in the final generation and obtain the optimized neural network structure with optimal number of input neurons, and optimal number of neurons in the hidden layer.

12.3 Recurrent Neural Network with Bayesian Regularization (RNN–BR)

12.3.1 Recurrent Neural Network

One of the major problems for multiple-input single-output purely static feed-forward neural network modeling is that we have to determine the exact number of inputs in advance. Earlier studies have selected this in an ad hoc manner and may not yield a globally optimized solution. This is the reason for using genetic algorithm to optimize the network structure. More importantly, for those applications where time information is involved, feed-forward neural network does not have the capability of incorporating dynamic temporal property internally, which may have impact on the network prediction performance. If time can be represented by the effect it has on processing, the network will perform better in terms of responsiveness to temporal sequences. This responsiveness can be obtained by providing feedback of data generated by the network back into the units of the network to be used in future iterations.

Recurrent neural networks are feedback networks in which the current activation state is a function of previous activation state and the current inputs. This feedback path allows recurrent networks to learn to recognize and generate time-varying patterns. A simple illustration of recurrent network is shown in Fig. 4.

For simplicity and comparison purposes, we first consider the most elementary feed-forward network shown in Fig. 4(a), where the input, hidden, and output layers each has only one neuron. When the input $x(t_0)$ at time t_0 is applied to the input layer, the output $v(t_0)$ of the hidden layer and the output $y(t_0)$ of the output layer are given by:

$$v(t_0) = \Phi(w_{12} \times x(t_0)) \tag{3}$$

$$y(t_0) = \Phi(w_{23} \times v(t_0)) = \Phi(w_{23} \times \Phi(w_{12} \times x(t_0))) \tag{4}$$

where $\Phi(\cdot)$ is the activation function. As shown in Fig. 4(b), recurrent neural network has feedback from the output layer to the hidden layer and feedback from the hidden layer to the input layer through the recurrent neurons labeled *R*. The corresponding feedback weights are w_{32} and w_{21}, respectively. When the input $x(t_1)$ is applied to the input layer, the output $v(t_1)$ of the hidden layer and the output $y(t_1)$ of the output layer are given by:

$$\begin{aligned} v(t_1) &= \Phi(w_{12} \times x(t_1) + w_{21} \times v(t_0)) \\ &= \Phi(w_{12} \times x(t_1) + w_{21} \times \Phi(w_{12} \times x(t_0))) \end{aligned} \tag{5}$$

$$\begin{aligned} y(t_1) &= \Phi(w_{23} \times v(t_1) + w_{32} \times y(t_0)) \\ &= \Phi(w_{23} \times \Phi(w_{12} \times x(t_1) + w_{21} \times \Phi(w_{12} \times x(t_0))) \\ &\quad + w_{32} \times \Phi(w_{23} \times \Phi(w_{12} \times x(t_0)))) \end{aligned} \tag{6}$$

Without loss of generality, we assume that the input layer, the hidden layer, and the output layer each has multiple neurons, and there could be more than one hidden layer.

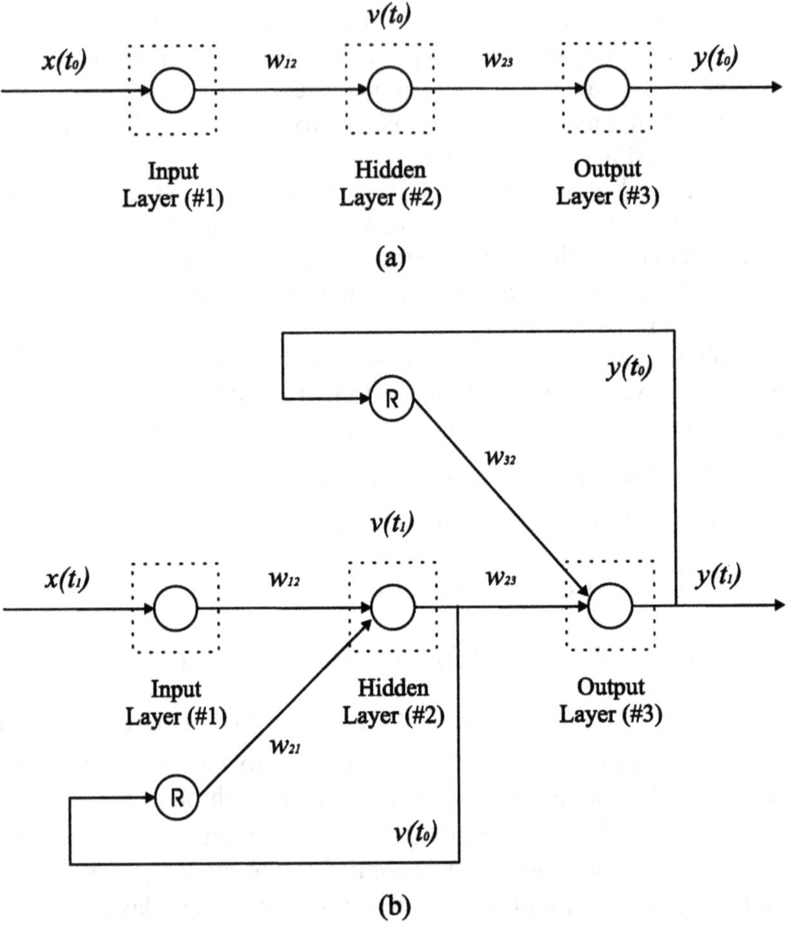

Fig. 4. Static feed-forward neural network in (a) and recurrent neural network in (b) with feedback connections

Each processing element of a recurrent neural network is denoted by the following generalized equations [5]:

$$p_{[l,n]}(t_i)=\sum_{m=1}^{N_{[l]}} w_{[l,m][l,n]}q_{[l,m]}(t_{i-1})+\sum_{m=1}^{N_{[l-1]}} w_{[l-1,m][l,n]}q_{[l-1,m]}(t_i)+b_{[l,n]} \tag{7}$$

$$q_{[l,n]}(t_i)=\Phi_{[l,n]}(p_{[l,n]}(t_i)) \tag{8}$$

where,

$p_{[l,n]}(t_i)$ is the internal state variable of the nth neuron in the lth layer at time t_i

$q_{[l,n]}(t_i)$ is the output of the nth neuron in the lth layer at time t_i

$b_{[l,n]}$ is the bias of the nth neuron in the lth layer

$w_{[l,m][l',n]}$ is the weight associated with the link between the mth neuron of the lth layer to the nth neuron of the l'^{th} layer

$\Phi(\cdot)$ is the activation function

12.3.2 Bayesian Regularization

A desirable neural network model should have small errors not only in the training data set, but also in the validation or testing data set. The ability to adapt to previously known data as well as unknown data requires improving generalization. Regularization constrains the size of the network parameters. When the parameters in a network are kept small, the response of the network will be smooth. With regularization, the performance function is modified by adding a term that consists of the mean of the sum of squares of the neural network weights and biases. The mean squared error with regularization performance, mse_{reg}, is given by:

$$mse_{reg} = \beta \times mse + (1-\beta) \times msw \tag{9}$$

where, β is the performance ratio and represents the relative importance of errors vs. weight and bias values, *mse* is the mean squared error during training, and *msw* is the mean squared weights and biases.

By using this modified performance function, mse_{reg}, the neural network is forced to have smaller weights and biases, which causes the network to respond smoother, represent the true function rather than capture the noise, and is less likely to overfit.

The major problem with regularization is to choose the performance ratio coefficient β. MacKay [22] has done extensive work on the application of Bayes rule for optimizing regularization. Hessian matrix computation is required for regularization optimization. In order to minimize the computational overhead, Foresee and Hagan [12] proposed using a Gauss-Newton approximation to the Hessian matrix, which is readily available while Levenberg-Marquardt algorithm is used as neural network training scheme.

The Bayesian optimization of the regularization coefficient with a Gauss-Newton approximation to the Hessian matrix is described in the following procedure [12]:

1. Initialize β ($\beta = 1$) and the weights.
2. Minimize the performance function mse_{reg} by using Levenberg-Marquardt algorithm.
3. Compute the effective number of parameters using Gauss-Newton approximation to the Hessian matrix available in the Levenberg-Marquardt training algorithm.
4. Derive the new estimate of β.
5. Repeat from Step 2-4 until the convergence is obtained. Thus, the optimized value for β is chosen.

We show how Bayesian regularization can be combined with recurrent training scheme (RNN-BR) for improving the generalization capability.

12.3.3 Modeling Rationale

More specifically, in this failure time modeling, the input-output pattern fed into the network is the software failure temporal sequence. Thus, the recurrent network can learn and recognize the inherent temporal patterns of input-output pair. For one-step-ahead prediction, the input sequence and the desired output sequence should have one step delay during the training process. The desired objective is to force the network to recognize the one-step-ahead temporal pattern. A sample input sequence and the corresponding one-step-ahead desired output sequence is defined as:

$$\textit{Input Sequence: } x(t_0), x(t_1), \cdots, x(t_{i-1}), x(t_i), x(t_{i+1}), \cdots$$
$$\textit{Output Sequence: } x(t_1), x(t_2), \cdots, x(t_i), x(t_{i+1}), x(t_{i+2}), \cdots$$

where $x(t_i)$ is the software failure time in the training data sequence, and t_i is the software failure time sequence index. The activation function in our modeling approach is linear for the output layer, and it is hyperbolic tangent sigmoidal for hidden layer neurons. Once the network is trained based on sufficient training data sequence, the unknown data sequence will be presented to the network to validate the performance.

12.4 Adaptive Support Vector Machine (A–SVM) Learning

12.4.1 SVM Learning in Function Approximation

As a novel type of machine learning algorithm, support vector machine (SVM) has gained increasing attention from its original application in pattern recognition to the extended application in function approximation and regression estimation. Based on the structural risk minimization (SRM) principle, the learning scheme of SVM is focused on minimizing an upper bound of the generalization error that includes the sum of the empirical training error and a regularized confidence interval, which will eventually result in better generalization performance. Moreover, unlike other gradient descent based learning scheme that requires nonlinear optimization with the danger of getting trapped into local minima, the regularized risk function of SVM can be minimized by solving a linearly constrained quadratic programming problem, which can always obtain a unique and global optimal solution. Thus, the possibility of being trapped at local minima can be effectively avoided [33,41,42].

The basic idea of SVM for function approximation is mapping the data x into a high-dimensional feature space by a nonlinear mapping and then performing a linear regression in this feature space. Assume that a total of l pairs of training patterns are given during SVM learning process,

$$(x_1, y_1), (x_2, y_2), \cdots, (x_i, y_i), \cdots, (x_l, y_l)$$

where the inputs are n-dimensional vectors $x_i \in \Re^n$, and the target outputs are continuous values $y_i \in \Re$. The SVM model used for function approximation is:

$$f(x) = w \cdot \phi(x) + b \tag{10}$$

where $\phi(x)$ is the high-dimensional feature space that is nonlinearly mapped from the input space x. Thus, a nonlinear regression in the low-dimensional input space is transferred to a linear regression in a high-dimensional feature space [41]. The coefficients w and b can be estimated by minimizing the following regularized risk function R:

$$R = \frac{1}{2} \| w \|^2 + C \frac{1}{l} \sum_{i=1}^{l} | y_i - f(x_i) |_\varepsilon \tag{11}$$

where

$$|y_i - f(x_i)|_\varepsilon = \begin{cases} 0 & \text{if} |y_i - f(x_i)| \le \varepsilon, \\ |y_i - f(x_i)| - \varepsilon & \text{otherwise.} \end{cases} \tag{12}$$

$\|w\|^2$ is the weights vector norm, which is used to constrain the model structure capacity in order to obtain better generalization performance. The second term is the Vapnik's linear loss function with ε-insensitivity zone as a measure for empirical error. The loss is zero if the difference between the predicted and observed value is less than or equal to ε. For all other cases, the loss is equal to the magnitude of the difference between the predicted value and the radius ε of ε-insensitivity zone. C is the regularization constant, representing the trade-off between the approximation error and the model structure. ε is equivalent to the approximation accuracy requirement for the training data points. Further, two positive slack variables ξ_i and ξ_i^* are introduced. We have

$$|y_i - f(x_i)| - \varepsilon = \begin{cases} \xi_i & \text{for data "above" an } \varepsilon \text{ tube,} \\ \xi_i^* & \text{for data "below" an } \varepsilon \text{ tube.} \end{cases} \tag{13}$$

Thus, minimizing the risk function R in Eq. (11) is equivalent to minimizing the objective function R_{w,ξ,ξ^*}.

$$R_{w,\xi,\xi^*} = \frac{1}{2}\|w\|^2 + C\sum_{i=1}^{l}(\xi_i + \xi_i^*) \tag{14}$$

subject to constraints

$$\begin{cases} y_i - w \cdot \phi(x_i) - b \le \varepsilon + \xi_i & i=1,\ldots,l, \\ w \cdot \phi(x_i) + b - y_i \le \varepsilon + \xi_i^* & i=1,\ldots,l, \\ \xi_i, \xi_i^* \ge 0 & i=1,\ldots,l. \end{cases} \tag{15}$$

This constrained optimization problem is typically solved by transforming into the dual problem, and its solution is given by the following explicit form:

$$f(x) = \sum_{i=1}^{l}(\alpha_i - \alpha_i^*)K(x_i, x) + b \tag{16}$$

12.4.2 Lagrange Multipliers

In Eq. (16), α_i and α_i^* are the Lagrange multipliers with $\alpha_i \times \alpha_i^* = 0$ and $\alpha_i, \alpha_i^* \geq 0$ for any i = 1,... , *l*. They can be obtained by maximizing the following form:

$$-\varepsilon \sum_{i=1}^{l} (\alpha_i + \alpha_i^*) + \sum_{i=1}^{l} y_i(\alpha_i - \alpha_i^*) - \frac{1}{2} \sum_{i=1}^{l} \sum_{j=1}^{l} (\alpha_i - \alpha_i^*)(\alpha_j - \alpha_j^*) K(x_i, x_j \tag{17}$$

subject to constraints

$$\begin{cases} \sum_{i=1}^{l} \alpha_i^* = \sum_{i=1}^{l} \alpha_i \\ 0 \leq \alpha_i, \alpha_i^* \leq C \quad i = 1, \ldots, l. \end{cases} \tag{18}$$

After learning, only some of coefficients $(\alpha_i - \alpha_i^*)$ in Eq. (16) differ from zero, and the corresponding training data points are referred to as support vectors. It is obvious that only the support vectors can fully decide the decision function in Eq. (16).

12.4.3 Kernel Function

In Eq. (16), $K(x_i, x)$ is defined as the kernel function, which is the inner product of two vectors in feature space $\phi(x_i)$ and $\phi(x)$. By introducing the kernel function, we can deal with the feature spaces of arbitrary dimensionality without computing the mapping relationship $\phi(x)$ explicitly. Some commonly used kernel functions are polynomial kernel function and Gaussian kernel function.

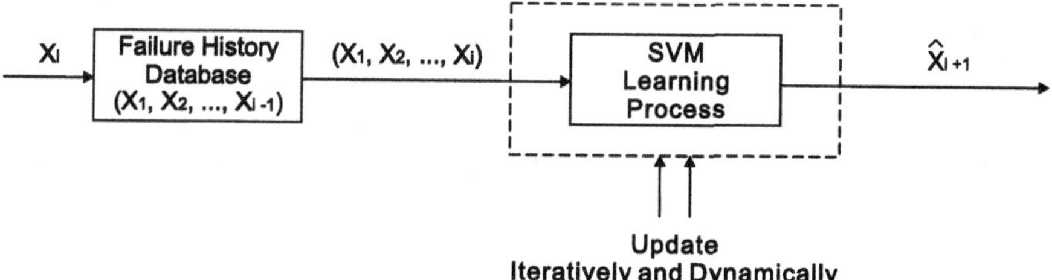

Fig. 5. Dynamic software reliability prediction framework

12.4.4 Formulation of the SVM-Predictor

The software reliability prediction system shown in Fig. 5 consists of a failure history database and an iteratively and dynamically updated SVM learning-predicting process. When a software failure, x_i, occurs, the failure history database is updated and the accumulated failure data $(x_1, x_2, ..., x_i)$ is made available to the SVM learning process. The number of failure data increases over time during a dynamic system. Accordingly, the SVM learning process is iteratively and dynamically updated after every occurrence of new failure time data in order to capture the most current feature hidden inside the software failure sequence. After the SVM learning process is complete based on the currently available history failure data, next-step failure information, $\hat{x}_{i+1}$, will be predicted.

Unlike the existing mapping characteristics, the inter-relationship among software failure time data is investigated. More specifically, the input-output pattern fed into the network is the failure temporal sequence. The SVM learning scheme is applied to the failure time data, forcing the network to learn and recognize the inherent internal temporal property of software failure sequence. For one-step-ahead prediction, the input sequence and the desired output sequence should have one step delay during the learning process. The desired objective is to force the network to recognize the one-step-ahead temporal pattern. Once the network is trained based on all the currently available history failure data using the SVM learning procedure, the one-step-ahead failure time will be predicted. Accordingly, the SVM learning process is iteratively and dynamically updated after every occurrence of new failure time data in order to capture the most current feature hidden in the software failure sequence.

12.5 Validation of New Approaches

Based on the modeling procedure described in D–ENN, RNN–BR, and A–SVM, three sets of experiments with four data sets are conducted. By using the common data sets and performance evaluation metrics, it is possible to quantitatively validate the effectiveness of the new approaches. The data set descriptions and experimental results are summarized in the following sections.

12.5.1 Data Sets Description and Pre-processing

The performance of these models have been tested using the same real-time control application and flight dynamic application data sets as cited in Park et al. [26] and Karunanithi et al. [18]. We choose a common baseline to compare the results with related work cited in the literature. All four data sets used in the experiments are summarized as follows:

Dataset-1: Real-time command and control application consisting of 21,700 assembly instructions and 136 failures.
Dataset-2: Flight dynamic application consisting of 10,000 lines of code and 118 failures.
Dataset-3: Flight dynamic application consisting of 22,500 lines of code and 180 failures.
Dataset-4: Flight dynamic application consisting of 38,500 lines of code and 213 failures.

Dataset-1 is obtained from Musa et al. [25]. Dataset-2, Dataset-3, and Dataset-4 are equivalent to DATA-11, DATA-12, and DATA-13 as cited in Park et al. [26] and Karunanithi et al. [18]. All the inputs and outputs of the network are scaled and normalized within the range of [0.1,0.9] to minimize the impact of absolute values. For this purpose, the actual values are scaled using the following relationship:

$$y=\frac{0.8}{\Delta}x+(0.9-0.8\times\frac{x_{max}}{\Delta}) \tag{19}$$

where, y is the scaled value we feed into our network, x is the actual value before scaling, x_{max} is the maximum value in the samples. x_{min} is the minimum value among all the samples, and Δ is defined as $(x_{max}-x_{min})$. After the training process, we test the prediction performance by scaling back all the network outputs to their actual values using the following equation:

$$x=\frac{y-0.9}{0.8}\times\Delta+x_{max} \tag{20}$$

12.5.2 Experimental Results

Our choice for using specific performance measures for assessing the predictive accuracy was based on similar measures used by other researchers. By using common data sets and performance evaluation metrics, it is possible to quantitatively compare the efficacies of various approaches. In addition, the relative error (*RE*) and/or average relative error (*AE*) are widely used in [9,17-19,24,26,30,32] for assessment of predictive accuracy of cumulative patterns.

Let $\hat{x}_i$ be the predicted value of failure time and x_i be the actual value of failure time. n is the number of data points in the test data set.

Relative Error (*RE*) is given by:

$$RE = \left| \frac{\hat{x}_i - x_i}{x_i} \right| \tag{21}$$

Average Relative Prediction Error (*AE*) is given by:

$$AE = \frac{1}{n} \sum_{i=1}^{n} \left| \frac{\hat{x}_i - x_i}{x_i} \right| \times 100 \tag{22}$$

The larger the value of predictability, or smaller the value of *AE*, the closer are the predicted values to the actual values.

Table 1 summarizes the results of the three approaches when applied to software reliability prediction modeling. Park et al. [26] applied failure sequence number as input and failure time as desired output in feed-forward neural network (FF-NN). Based on the learning pair of execution time and the corresponding accumulated number of defects disclosed, Karunanithi et al. [18] employed both feed-forward neural network (FF-NN) and recurrent neural network (R-NN) structures to model the failure process. These results are also summarized in Table 1.

Table 1. Comparison of Average Relative Prediction Error

Data Set	Comparison of Test Data Sets					
	D-ENN	RNN-BR	A-SVM	FF-NN (Ref. [26])	R-NN (Ref. [18])	FF-NN (Ref. [18])
Dataset-1	2.72	1.83	2.44	2.58	2.05	2.50
Dataset-2	2.65	2.06	1.52	3.32	2.97	5.23
Dataset-3	1.16	0.97	1.24	2.38	3.64	6.26
Dataset-4	1.19	0.98	1.20	1.51	2.28	4.76

For example, using D-ENN approach with Dataset-3, the average relative prediction error (*AE*) is 1.16%; using RNN-BR approach, the average relative prediction error (*AE*) is 0.97%; and using A-SVM approach, the average relative prediction error (*AE*) is 1.24%. These errors are lower than the results obtained by Park et al. [26] (2.38%) using feed-forward neural network, Karunanithi et al. [18] (3.64%) using recurrent neural network, and

Karunanithi et al. [18] (6.26%) using feed-forward neural network. In all four data sets, numerical results show that these modeling approaches have significantly improved prediction performance compared to other existing approaches.

12.6 Discussions and Future Work

In this chapter, we survey the literature on the most commonly used computational intelligence methods, including fuzzy logic, neural networks, genetic algorithms and their applications in software reliability prediction. Secondly, we generalized some of the limitations of the existing modeling approaches. Three emerging computational intelligence modeling approaches are then introduced facing those challenges. Specifically, the new modeling approaches that focus on improving the dynamic learning and generalization capabilities are:

In the dynamic learning model using evolutionary connectionist approach, as the number of available software failure data increases over time, the optimization process adaptively determines the number of input neurons and the number of neurons in the hidden layer. The corresponding globally optimized neural network structure is iteratively and dynamically reconfigured and updated as new data arrive to improve the prediction accuracy.

Regularized recurrent neural network model is used for improving generalization capability. Recurrent neural network has the inherent capability of developing an internal memory, which may naturally extend beyond the externally provided lag spaces. Moreover, by adding a penalty term of sum of connection weights, Bayesian regularization approach is applied to the network training scheme to improve the generalization performance and lower the susceptibility of overfitting.

In the adaptive prediction model using support vector machines, the learning process is focused on minimizing an upper bound of the generalization error that includes the sum of the empirical training error and a regularized confidence interval. This eventually results in better generalization performance. Further, this learning process is iteratively and dynamically updated after every occurrence of new software failure data in order to capture the most current feature hidden inside the data sequence.

All the approaches have been successfully validated on applications related to software reliability prediction. Numerical results show that these modeling approaches have improved prediction performance. Some related discussions and future research directions are summarized in the following sections.

12.6.1 Data Type Transformation

There are two common types of software failure data: time-between-failures data (time-domain data) and failure-count data (interval-domain data). The individual times at which failure occurs are recorded for time-domain data collection. The time can either be actual failure time or time between successive failures. The interval-domain approach is represented by counting the number of failures occurring during a fixed interval period, such as the number of failures per hour.

The D-ENN, RNN-BR, and A-SVM software reliability growth modeling approaches are flexible and can take different types of data as input. Our approaches were originally intended for using time-domain data (actual failure time) as input to make predictions. If it is assumed that the data collected are interval-domain data, it is possible to develop new models by changing the input-output pair of the network. One type of software failure data can be transformed into another type in order to meet the input data requirement for a specific model. Interval-domain data can be obtained by counting the number of failures occurring within a specified time period in time-domain data. However, if it is needed to transform interval-domain data to time-domain data, this conversion can be achieved by randomly or uniformly allocating the failures for the specified time intervals, and then recording the individual times at which failure occurred. Some software reliability tools integrate the capability of data transformation between two data types, such as CASRE (Computer-Aided Software Reliability Estimation) [21].

12.6.2 Modeling Long-Term Behavior

The reason we focused on short-term prediction (one-step-ahead) in this research was to establish a baseline for comparison purposes with other known approaches. We also believe it is more meaningful to make one-step-ahead prediction in certain types of applications in order to make early stage preventive action and avoid catastrophic events.

Meanwhile, it is of great interest for modeling and predicting long-term behavior of software failure process as well. For example, suppose x_i is the number of failures in the first i specified time intervals, and $x_0, x_1, \ldots, x_{i-1}$ are used for predicting x_i. Once the predicted value of x_i, denoted by $\hat{x}_i$, is obtained, it is then used as input to the network to generate the predicted value of x_{i+1}, denoted by $\hat{x}_{i+1}$. Further, $\hat{x}_i$ and $\hat{x}_{i+1}$ are used as input to obtain $\hat{x}_{i+2}$, and so forth. These new modeling approaches are also flexible and can deal with this situation by changing the input-output sequences.

12.6.3 Assessment of Predictive Accuracy

Our choice for using specific performance measures for assessing the predictive accuracy was largely predicated on similar measures used by other researchers. We believe it is reasonable to compare our results with existing work using the same data sets and same performance evaluation metrics. This provides us the opportunity to quantitatively gauge the efficacy of these new approaches.

We also acknowledge that there are other metrics for validating prediction accuracy. Some examples are: sum of squared error (*SSE*) as used in [3,29,45]; mean of the sum of squared error (*MSSE*) as used in [4,13,43]; mean absolute deviation (*MAD*) as used in [16]; *u*-plot and prequential likelihood function as used in [6,7].

12.6.4 Incorporating Environmental Factors

Recent studies show that using software testing time as the only influencing factor may not be appropriate for predicting software reliability [27,28]. Some environmental factors should be integrated. Examples of related environmental factors are program complexity, programmer skills, testing coverage, level of test-team members, and reuse of existing code [27,44]. This is an area of research that can provide comprehensive insight to the problem of software reliability prediction.

Acknowledgements

The authors would like to thank the reviewers for their helpful comments, and Drs. J. Y. Park, N. Karunanithi, Y. K. Malaiya, and D. Whitley for providing data sets DATA-11, DATA-12, DATA-13.

References

[1] Adnan WA, Yaacob MH (1994) An integrated neural-fuzzy system of software reliability prediction. In: Proceedings of the 1st International Conference on Software Testing, Reliability and Quality Assurance. New Delhi, India, pp 154–158

[2] Adnan WA, Yaakob M, Anas R, Tamjis MR (2000) Artificial neural network for software reliability assessment. In: 2000 TENCON Proceedings of Intelligent Systems and Technologies for the New Millennium. Kuala Lumpur, Malaysia, pp 446–451

[3] Aljahdali SH, Sheta A, Rine D (2001) Prediction of software reliability: A comparison between regression and neural network non-parametric models. In: Proceedings of ACS/IEEE International Conference on Computer Systems and Applications. Beirut, Lebanon, pp 470–473

[4] Aljahdali SH, Sheta A, Rine D (2002) Predicting accumulated faults in software testing process using radial basis function network models. In: Proceedings of the ISCA 17th International Conference on Computers and their Applications. San Francisco, CA, pp 26–29

[5] Bhattacharya A, Parlos AG, Atiya AF (2003) Prediction of MPEG-coded video source traffic using recurrent neural networks. IEEE Trans Signal Processing 51:2177–2190

[6] Brocklehurst S, Chan PY, Littlewood B, Snell J (1990) Recalibrating software reliability models. IEEE Trans Software Eng 16:458–470

[7] Brocklehurst S, Littlewood B (1992) New ways to get accurate reliability measures. IEEE Software 9:34–42

[8] Cai KY (1996) Introduction to Fuzzy Reliability, chapter 8. Fuzzy Methods in Software Reliability Modeling. Kluwer International Series in Engineering and Computer Science. Kluwer Academic Publishers

[9] Cai KY, Cai L, Wang WD, Yu ZY, Zhang D (2001) On the neural network approach in software reliability modeling. The Journal of Systems and Software 58:47–62

[10] Cai KY, Wen CY, Zhang ML (1991) A critical review on software reliability modeling. Reliability Engineering and System Safety 32:357–371

[11] Cai KY, Wen CY, Zhang ML (1993) A novel approach to software reliability modeling. Microelectronics and Reliability 33:2265–2267

[12] Foresee FD, Hagan MT (1997) Gauss-Newton approximation to Bayesian learning. In: Proceedings of the 1997 IEEE International Conference on Neural Networks. Houston, TX, pp 1930–1935

[13] Fujiwara T, Yamada S (2002) C0 coverage-measure and testing-domain metrics based on a software reliability growth model. International Journal of Reliability, Quality and Safety Engineering 9:329–340

[14] Guo P, Lyu MR (2004) A pseudoinverse learning algorithm for feedforward neural networks with stacked generalization applications to software reliability growth data. Neurocomputing 56:101–121

[15] Gupta N, Singh MP (2005) Estimation of software reliability with execution time model using the pattern mapping technique of artificial neural network. Computers and Operations Research 32:187–199

[16] Ho SL, Xie M, Goh TN (2003) A study of the connectionist models for software reliability prediction. Computers and Mathematics with Applications 46:1037–1045

[17] Karunanithi N (1993) A neural network approach for software reliability growth modeling in the presence of code churn. In: Proceedings of the 4th International Symposium on Software Reliability Engineering. Denver, CO, pp 310–317

[18] Karunanithi N, Whitley D, Malaiya YK (1992a) Prediction of software reliability using connectionist models. IEEE Trans Software Eng 18:563–574

[19] Karunanithi N, Whitley D, Malaiya YK (1992b) Using neural networks in reliability prediction. IEEE Software 9:53–59

[20] Leung FHF, Lam HK, Ling SH, Tam PKS (2003) Tuning of the structure and parameters of a neural network using an improved genetic algorithm. IEEE Trans Neural Networks 14:79–88

[21] Lyu MR (1996) Handbook of Software Reliability Engineering. McGraw-Hill, New York

[22] MacKay DJC (1992) Bayesian interpolation. Neural Computation 4:415–447

[23] Mihalache A (1992) Software reliability assessment by fuzzy sets. IPB Buletin Stiintific - Electrical Engineering 54:91–95

[24] Musa JD (1998) Software Reliability Engineering, chapter 8 Software Reliability Models. McGraw-Hill Osborne Media, New York

[25] Musa JD, Iannino A, Okumoto K (1987) Software Reliability: Measurement, Prediction, Application. McGraw-Hill Series in Software Engineering and Technology. McGraw-Hill Book Company

[26] Park JY, Lee SU, Park JH (1999) Neural network modeling for software reliability prediction from failure time data. Journal of Electrical Engineering and Information Science 4:533–538

[27] Pham H (2000) Software Reliability, chapter 8 Software Reliability Models with Environmental Factors. Springer

[28] Pham H (2003) Software reliability and cost models: Perspectives, comparison, and practice. European Journal of Operational Research 149:475–489

[29] Pham H, Zhang X (2003) NHPP software reliability and cost models with testing coverage. European Journal of Operational Research 145:443–454

[30] Sitte R (1999) Comparison of software-reliability-growth predictions: Neural networks vs parametric-recalibration. IEEE Trans Reliability 48:285–291

[31] So SS, Cha SD, Kwon YR (2002) Empirical evaluation of a fuzzy logic-based software quality prediction model. Fuzzy Sets and Systems 127:199–208

[32] Stringfellow C, Andrews AA (2002) An empirical method for selecting software reliability growth models. Empirical Software Engineering 7:319–343

[33] Tian L, Noore A (2004a) A novel approach for short-term load forecasting using support vector machines. International Journal of Neural Systems 14:329–335

[34] Tian L, Noore A (2004b) Software reliability prediction using recurrent neural network with Bayesian regularization. International Journal of Neural Systems 14:165–174

[35] Tian L, Noore A (2005a) Dynamic software reliability prediction: An approach based on support vector machines. International Journal of Reliability, Quality and Safety Engineering 12:309–321

[36] Tian L, Noore A (2005b) Evolutionary neural network modeling for software cumulative failure time prediction. Reliability Engineering and System Safety 87:45–51

[37] Tian L, Noore A (2005c) Modeling distributed software defect removal effectiveness in the presence of code churn. Mathematical and Computer Modelling 41:379–389

[38] Tian L, Noore A (2005d) On-line prediction of software reliability using an evolutionary connectionist model. Journal of Systems and Software 77:173–180

[39] Tsoukalas LH, Uhrig RE (1996) Fuzzy and Neural Approaches in Engineering. John Wiley & Sons, New York

[40] Utkin LV, Gurov SV, Shubinsky MI (2002) A fuzzy software reliability model with multiple-error introduction and removal. International Journal of Reliability, Quality and Safety Engineering 9:215–227

[41] Vapnik VN (1999) An overview of statistical learning theory. IEEE Trans Neural Networks 10:988–999

[42] Xing F, Guo P (2005) Support vector regression for software reliability growth modeling and prediction. Lecture Notes in Computer Science 3496:925–930

[43] Yamada S, Fujiwara T (2001) Testing-domain dependent software reliability growth models and their comparisons of goodness-of-fit. International Journal of Reliability, Quality and Safety Engineering 8:205–218

[44] Zhang X, Shin MY, Pham H (2001) Exploratory analysis of environmental factors for enhancing the software reliability assessment. Journal of Systems and Software 57:73–78

[45] Zhang X, Teng X, Pham H (2003) Considering fault removal efficiency in software reliability assessment. IEEE Trans Systems, Man, and Cybernetics-Part A: Systems and Humans 33:114–120